Android 移动平台应用开发高级教程

朱凤山　张建军　编著

清华大学出版社
北京

内 容 简 介

本书由浅入深、循序渐进地介绍了 Android 应用程序开发的主要知识，注重可阅读性和实用性，对于开发过程中经常用到的类、属性、方法和常量都以表格的形式介绍其应用场景和作用。对 Android 开发中概念、方法和应用的介绍主要以 Google 提供的 Android API 文档为参考依据，力求简洁、准确地反映 API 文档中意图。

全书共 12 章，内容如下：Android 开发环境，包括 ADT-Eclipse 和 Android Studio；基本 UI 组件；Activity 和 Intent；项目资源；主要系统组件；二维图像处理；多媒体应用开发；Service 与 BroadcastReceiver 组件；数据存储与 ContentProvider 组件；Android 网络编程；常用传感器与蓝牙通信。最后通过校园 App 应用实例介绍如何设计、开发具备移动端和服务器端的应用程序。

本书可以作为应用型高等院校计算机、软件工程等相关专业的教材，也可以作为自学 Android 移动平台应用程序开发用书，还可以供从事 Android 移动平台应用开发的工程师参考。

本书封面贴有清华大学出版社防伪标签，无标签者不得销售。
版权所有，侵权必究。侵权举报电话：010-62782989 13701121933

图书在版编目（CIP）数据

Android 移动平台应用开发高级教程/朱凤山，张建军编著. —北京：清华大学出版社，2017
（2018.8 重印）
ISBN 978-7-302-46093-0

Ⅰ. ①A⋯ Ⅱ. ①朱⋯ ②张⋯ Ⅲ. ①移动终端－应用程序－程序设计－教材 Ⅳ. ①TN929.53

中国版本图书馆 CIP 数据核字（2017）第 006179 号

责任编辑：刘向威　战晓雷
封面设计：文　静
责任校对：胡伟民
责任印制：刘祎淼

出版发行：清华大学出版社
　　　　网　　址：http://www.tup.com.cn，http://www.wqbook.com
　　　　地　　址：北京清华大学学研大厦 A 座　　邮　　编：100084
　　　　社 总 机：010-62770175　　　　　　　　邮　　购：010-62786544
　　　　投稿与读者服务：010-62776969，c-service@tup.tsinghua.edu.cn
　　　　质量反馈：010-62772015，zhiliang@tup.tsinghua.edu.cn
　　　　课件下载：http://www.tup.com.cn，010-62795954
印 装 者：三河市君旺印务有限公司
经　　销：全国新华书店
开　　本：185mm×260mm　　印　张：23.75　　字　数：595 千字
版　　次：2017 年 4 月第 1 版　　　　　　　　印　次：2018 年 8 月第 3 次印刷
印　　数：3001～4500
定　　价：49.5 元

产品编号：062988-01

前　言

Android 是一款基于 Linux 内核的开源操作系统，主要用于移动设备，如智能手机、智能手表、平板电脑、车载系统、电视等设备。根据 International Data Corporation（国际数据公司）公布的全球智能手机出货量报告，Android 操作系统的智能手机早在 2014 年其市场份额就已经达到 81.5%。该报告称，Android 系统到 2019 年将占据全球 82.6% 的移动系统市场份额。因此，我们有理由相信，在未来一段时间内，Android 系统将牢牢占据智能手机操作系统第一的位置。

这种趋势给熟悉 Java 编程、有意愿从事 Android App 开发的读者提供了新的学习方向和机会。并且，移动互联网行业对专业 Android 应用程序开发人员的需求数量日益增长，对开发人员的技术要求也越来越高。巨大的市场需求提供了更多的创业与就业机会。如果你目前正处于该状态下，本书就是适合你的选择。

本书在编写过程中按照知识的逻辑关系分章，对知识的讲解与介绍力求全面，并给出可以应用于哪种场合的建议。对重点、难点知识，给出演示项目，并按步骤给出实现代码。各位读者在学习过程中一定要循序渐进，切勿急于求成。很多学习者在熟悉了语法知识之后都迫不及待地想一展身手，编写一款自己的软件，这是良好的学习习惯，也是值得肯定的学习态度。但是，如果所选择的项目过于复杂，往往很难实现功能，即使有参考代码和帮助文档，也会陷入"代码海洋"或"文档风暴"中，这样只会收到事倍功半的效果，学习积极性会受到很大的打击。所以，对于初学者，建议选择结构简单、功能单一的项目进行应用实践。

作为 developer.android、CSDN、51CTO、eoeandroid 等技术论坛、社区的忠实用户和学习者，作者在本书的编写过程中也受益匪浅，也建议读者在遇到学习问题时向专业技术论坛或社区求助。

全书共 12 章，内容如下：Android 开发环境，包括 ADT-Eclipse 和 Android Studio；基本 UI 控件；Activity 和 Intent 组件；Android 项目资源；主要系统组件；二维图像处理；多媒体应用开发；Service 与 BroadcastReceiver 组件；数据存储与 ContentProvider 组件；Android 网络编程；常用传感器与蓝牙通信。最后通过校园 App 应用实例介绍如何设计、开发具备移动端和服务器端的应用程序。

本书得到清华大学出版社应用型本科系列教材建设和天津市教育科学"十三五"规划课题的支持，课题名称"慕课在独立学院教育教学中的应用研究与实践"，课题申报号HEYP5021。本书第 1、2、6～12 章由朱凤山编写，第 3～5 章由张建军编写，全书由朱凤山统稿。参与本书编写工作的还有郭园园、张哲。在本书完成之际，特别感谢王慧芳教授、王志军教授、张桂芸教授给予的指导和建议，感谢天津师范大学津沽学院给予的环境支持，感谢新锐 IT 工作室的成员给予的启发和帮助，感谢张新芳、朱思齐的大力支持。

由于作者学术与经验的欠缺，在本书的结构、知识点与难点的选择和解析过程中可能会存在问题与不足，希望广大读者不吝赐教，以便作者对本书加以改进。相关技术问题可以发送邮件到 tj_zhufengshan@163.com。

作　者
2016 年 10 月

目 录

第 1 章 Android 开发环境与项目解析 ································· 1

1.1 Android 介绍 ·· 1
- 1.1.1 Android 发展与智能手机 ·· 1
- 1.1.2 Android 版本说明 ·· 3
- 1.1.3 Android 系统架构 ·· 5

1.2 Android 开发环境 ··· 8
- 1.2.1 使用 Eclipse ··· 8
- 1.2.2 使用 adt-bundle Eclipse ·· 12
- 1.2.3 使用 Android Studio ·· 12

1.3 Android 项目解析 ··· 16
- 1.3.1 创建 Android 项目 ·· 17
- 1.3.2 Android 项目结构 ··· 21

1.4 运行与调试 Android 项目 ·· 25
- 1.4.1 使用 Android 虚拟机 ··· 25
- 1.4.2 使用 Android 真机 ··· 28
- 1.4.3 调试日志的使用 ··· 30

1.5 签名输出 apk 文件 ·· 31
1.6 习题 ·· 32

第 2 章 使用控件创建用户界面 ··· 34

2.1 Android 用户界面设计 ··· 34
- 2.1.1 使用布局文件设计界面 ·· 34
- 2.1.2 使用 Java 代码设计界面 ··· 35

2.2 使用简单控件 ·· 36
- 2.2.1 控件的基本属性 ··· 36
- 2.2.2 TextView ·· 39
- 2.2.3 Button ··· 40
- 2.2.4 ToggleButton 与 Switch ··· 41
- 2.2.5 EditText ·· 43
- 2.2.6 CheckBox ·· 45

	2.2.7	RadioButton 与 RadioGroup	45
	2.2.8	SeekBar	46
	2.2.9	RatingBar	47
	2.2.10	ProgressBar	48

2.3 布局管理器 ... 50
    2.3.1 LinearLayout ... 50
    2.3.2 RelativeLayout ... 51
    2.3.3 FrameLayout ... 54
    2.3.4 GridLayout ... 54
    2.3.5 TableLayout ... 56
    2.3.6 AbsoluteLayout ... 57

2.4 使用图片控件 ... 57
    2.4.1 ImageView ... 57
    2.4.2 ImageButton ... 58

2.5 使用复杂控件 ... 58
    2.5.1 数据适配器 ... 59
    2.5.2 Spinner ... 60
    2.5.3 ListView 与 ListActivity ... 62
    2.5.4 GridView ... 64
    2.5.5 ExpandableListView ... 66
    2.5.6 ScrollView 与 HorizontalScrollView ... 68

2.6 高级控件 ... 68
    2.6.1 TabHost ... 69
    2.6.2 ViewFlipper ... 71
    2.6.3 ImageSwitcher ... 72

2.7 日期和时间控件 ... 75
    2.7.1 DatePicker 和 TimePicker ... 75
    2.7.2 Chronometer ... 77
    2.7.3 AnalogClock 与 TextClock ... 78

2.8 线程机制 ... 79
2.9 习题 ... 82

第3章 Activity 与 Intent ... 84

3.1 Activity 的创建与管理 ... 84
    3.1.1 创建 Activity 与配置信息 ... 84
    3.1.2 Activity 的生命周期 ... 88
    3.1.3 Activity 启动模式 ... 89

3.2 Intent 对象 ... 93
    3.2.1 创建 Intent 对象 ... 94

3.2.2　使用 Intent 启动 Activity ……………………………… 94
　　　3.2.3　使用 Intent 传递数据 …………………………………… 98
　　　3.2.4　Intent 过滤器 ……………………………………………… 98
　　　3.2.5　使用 Intent 启动手机组件 ……………………………… 99
　3.3　Activity 与 Fragment ………………………………………………… 100
　　　3.3.1　Fragment 生命周期 ……………………………………… 100
　　　3.3.2　Fragment 传递数据 ……………………………………… 107
　　　3.3.3　管理 Fragment …………………………………………… 110
　3.4　习题 ……………………………………………………………………… 113

第 4 章　使用项目资源 ……………………………………………………… 114

　4.1　Android 资源类型 ……………………………………………………… 114
　　　4.1.1　资源的创建与引用 ………………………………………… 114
　　　4.1.2　资源的分类 ………………………………………………… 116
　4.2　布局资源 ………………………………………………………………… 116
　4.3　菜单资源 ………………………………………………………………… 117
　　　4.3.1　普通菜单 …………………………………………………… 118
　　　4.3.2　ActionBar 中的菜单 ……………………………………… 121
　4.4　"值"资源 ……………………………………………………………… 123
　　　4.4.1　字符串 ……………………………………………………… 123
　　　4.4.2　颜色资源 …………………………………………………… 123
　　　4.4.3　尺寸资源 …………………………………………………… 124
　4.5　可绘制资源 ……………………………………………………………… 124
　　　4.5.1　Android 中的图片类型 …………………………………… 124
　　　4.5.2　NinePatch 图片格式 ……………………………………… 125
　　　4.5.3　selector 资源 ……………………………………………… 127
　　　4.5.4　shape 资源 ………………………………………………… 130
　4.6　动画资源 ………………………………………………………………… 132
　　　4.6.1　Tween Animation ………………………………………… 132
　　　4.6.2　Frame Animation ………………………………………… 138
　4.7　样式与主题资源 ………………………………………………………… 141
　　　4.7.1　样式资源 …………………………………………………… 141
　　　4.7.2　主题资源 …………………………………………………… 142
　4.8　习题 ……………………………………………………………………… 145

第 5 章　使用系统组件 ……………………………………………………… 146

　5.1　菜单的使用 ……………………………………………………………… 146
　　　5.1.1　创建菜单 …………………………………………………… 146
　　　5.1.2　监听菜单选中 ……………………………………………… 147

5.1.3　子菜单与弹出菜单 …………………………………………… 148
5.2　ActionBar 的使用 ……………………………………………………… 150
　　5.2.1　导航菜单 …………………………………………………… 151
　　5.2.2　导航模式 …………………………………………………… 152
　　5.2.3　Actionbar 与 Fragment ……………………………………… 153
5.3　Toast 与 Notification …………………………………………………… 156
　　5.3.1　创建并显示 Toast …………………………………………… 156
　　5.3.2　自定义 Toast ………………………………………………… 156
　　5.3.3　创建并发出通知 …………………………………………… 157
5.4　对话框的使用 ………………………………………………………… 159
　　5.4.1　普通对话框的创建 ………………………………………… 159
　　5.4.2　选择对话框 ………………………………………………… 160
　　5.4.3　日期与时间对话框 ………………………………………… 161
　　5.4.4　进度条对话框 ……………………………………………… 162
　　5.4.5　自定义对话框 ……………………………………………… 163
5.5　习题 …………………………………………………………………… 164

第 6 章　二维图像的处理 ……………………………………………………… 166

6.1　位图的使用 …………………………………………………………… 166
　　6.1.1　Bitmap 与 BitmapFactory …………………………………… 166
　　6.1.2　位图的缩略图 ……………………………………………… 168
6.2　使用 View 绘制视图 …………………………………………………… 169
　　6.2.1　横竖屏坐标与全屏操作 …………………………………… 169
　　6.2.2　View 类 ……………………………………………………… 171
　　6.2.3　Canvas 类 …………………………………………………… 173
　　6.2.4　Paint 类 ……………………………………………………… 182
　　6.2.5　使用 View 自定义控件 ……………………………………… 184
　　6.2.6　Matrix 变换 ………………………………………………… 185
6.3　使用 SurfaceView 绘制视图 …………………………………………… 187
　　6.3.1　SurfaceHolder 介绍 ………………………………………… 188
　　6.3.2　使用子线程绘制视图 ……………………………………… 189
6.4　线程控制下的动画效果 ……………………………………………… 191
　　6.4.1　属性动画效果 ……………………………………………… 191
　　6.4.2　帧动画效果 ………………………………………………… 193
　　6.4.3　剪切区动画效果 …………………………………………… 195
6.5　习题 …………………………………………………………………… 197

第 7 章　多媒体应用开发 ……………………………………………………… 199

7.1　音频播放 ……………………………………………………………… 199

7.1.1 MediaPlayer 对象的创建 …… 199
7.1.2 MediaPlayer 对象的状态转换 …… 202
7.1.3 SoundPool 的创建和使用 …… 204
7.2 视频播放 …… 206
7.2.1 VideoView 播放本地资源 …… 206
7.2.2 MediaController …… 208
7.2.3 播放网络资源 …… 209
7.3 MediaRecorder …… 209
7.3.1 录制音频 …… 211
7.3.2 同时录制音视频 …… 212
7.4 使用 Camera 拍照 …… 214
7.4.1 启动相机与拍照 …… 215
7.4.2 获取相机返回数据 …… 217
7.4.3 获取原尺寸照片 …… 218
7.4.4 照片缩略图 …… 220
7.5 习题 …… 222

第 8 章 Service 与 BroadcastReceiver …… 223

8.1 创建并配置 Service …… 223
8.1.1 自定义 Service …… 223
8.1.2 Service 的生命周期 …… 225
8.2 Service 的启动模式 …… 226
8.2.1 startService …… 227
8.2.2 bindService …… 230
8.3 远程 Service …… 233
8.4 BroadcastReceiver …… 236
8.4.1 发出广播与接收广播 …… 237
8.4.2 广播的分类与权限 …… 238
8.4.3 注册广播接收器 …… 239
8.4.4 接收系统广播 …… 245
8.5 实现短信拦截 …… 246
8.6 习题 …… 249

第 9 章 数据存储与 ContentProvider …… 251

9.1 以文件形式存储数据 …… 251
9.1.1 读写 XML 文件 …… 251
9.1.2 读写普通文件 …… 254
9.1.3 读写 SD 中的文件 …… 256
9.2 以数据库形式存储数据 …… 258

  9.2.1 SQLiteDatabase 介绍 ……………………………… 258
  9.2.2 执行增删改操作 …………………………………… 260
  9.2.3 Cursor 与查询操作 ………………………………… 262
  9.2.4 SQLiteOpenHelper 的使用 ………………………… 265
 9.3 SQLite 图形化查看工具 …………………………………… 266
 9.4 Content Provider …………………………………………… 268
  9.4.1 使用 ContentProvider ……………………………… 268
  9.4.2 Uri 的组成 ………………………………………… 268
  9.4.3 ContentProvider 基本操作 ………………………… 269
 9.5 管理手机联系人信息 ……………………………………… 274
 9.6 习题 ………………………………………………………… 279

第 10 章　Android 网络编程 ……………………………………… 281

 10.1 基于传输层协议的联网 …………………………………… 281
  10.1.1 传输层协议介绍 …………………………………… 281
  10.1.2 Socket 与 ServerSocket …………………………… 282
  10.1.3 DatagramSocket 与 DatagramPacket ……………… 288
  10.1.4 Android 对联网代码的限制 ……………………… 292
 10.2 基于应用层协议的联网 …………………………………… 293
  10.2.1 URL 介绍 …………………………………………… 293
  10.2.2 GET 请求和 POST 请求 …………………………… 295
  10.2.3 使用 HttpURLConnection 联网 …………………… 296
  10.2.4 使用 HttpClient 联网 ……………………………… 300
 10.3 访问 Web Service …………………………………………… 303
  10.3.1 WSDL 和 SOAP …………………………………… 304
  10.3.2 调用 Web Service ………………………………… 304
 10.4 解析网络传输中的数据 …………………………………… 306
  10.4.1 解析 JSON 格式数据 ……………………………… 306
  10.4.2 解析 XML 格式数据 ……………………………… 310
 10.5 习题 ………………………………………………………… 313

第 11 章　传感器应用与蓝牙通信 ………………………………… 315

 11.1 Android 中的传感器 ……………………………………… 315
  11.1.1 传感器概述 ………………………………………… 315
  11.1.2 测试传感器应用程序 ……………………………… 317
 11.2 加速度传感器 ……………………………………………… 318
 11.3 光线传感器 ………………………………………………… 321
 11.4 距离传感器 ………………………………………………… 323
 11.5 蓝牙通信技术应用 ………………………………………… 324

    11.5.1 近距离通信技术介绍 ……………………………………… 325
    11.5.2 Android 系统中的蓝牙组件 ………………………………… 325
    11.5.3 蓝牙设备间的通信 …………………………………………… 330
  11.6 习题 ………………………………………………………………………… 337

第 12 章 校园 App 项目案例 …………………………………………………… 338

  12.1 校园 App 项目介绍 ……………………………………………………… 338
  12.2 服务器端功能开发 ………………………………………………………… 340
    12.2.1 数据库表 ……………………………………………………… 341
    12.2.2 实体类 ………………………………………………………… 342
    12.2.3 DAO 层 ………………………………………………………… 344
    12.2.4 Action 层 ……………………………………………………… 345
  12.3 Android 客户端开发 ……………………………………………………… 348
    12.3.1 欢迎界面与标题栏样式 ……………………………………… 348
    12.3.2 主界面 Activity ……………………………………………… 349
    12.3.3 自定义 Fragment ……………………………………………… 354
    12.3.4 WebView 加载 HTML5 页面 ………………………………… 363
  12.4 习题 ………………………………………………………………………… 364

参考文献 …………………………………………………………………………………… 365

第 1 章　Android 开发环境与项目解析

本章学习目标
- 了解目前主流移动平台操作系统。
- 掌握搭建 Android 开发环境的方法。
- 掌握 Android 项目的组成部分及作用。
- 掌握安装和调试 APK 文件的方法。

Android 是一款移动设备的操作系统,由 Google 公司主导,被广泛应用于智能手机、平板电脑、智能电视等终端设备。作为目前全球范围内最为流行的智能设备操作系统,Android 在国内的发展如火如荼,众多手机生产厂商,如 HTC、联想、小米、华为、中兴、vivo 等,都借助该操作系统取得了巨大的成功。

1.1　Android 介绍

1.1.1　Android 发展与智能手机

Android 操作系统是由 Android 公司设计开发的,以公司命名,后被 Google 公司收购。Android 主要包括 3 个组成部分,操作系统、中间件、用户界面和应用程序。底层基于 Linux 内核,采用 C 语言开发,提供基本功能;中间层包括调用函数库和运行虚拟机,采用 C++开发;最上层(与 App 开发者最接近的一层)包括各种应用程序,如通话程序、短信程序等,可以自由开发,采用 Java 开发语言。整个系统具有自由开放的特征,不存在阻碍移动产业创新的专有权障碍。2005 年 Google 公司收购注资后,成立了"开放手机联盟",Android 系统的功能得到了进一步完善。Google 公司通过与相关软硬件开发商、设备制造商、运营商等企业结成合作伙伴,在移动产业内建立了开放式环境。图 1.1 是 Android 操作系统的标志性图标。

由于其开源的特性,Android 系统一经推出便受到众多终端制作公司的青睐,纷纷采用 Android 作为手机操作系统。2011 年,该系统市场占有率便跃居全球第一,其市场份额一度超过全球智能手机操作系统的一半以上。中国台湾宏达国际电子

图 1.1　Android 系统图标

(HTC)、摩托罗拉、韩国三星电子、LG电子和中国移动等都是"开放手机联盟"的成员,阵容可谓庞大。随着Android操作系统的普及,该平台的应用软件越来越受欢迎,应用程序的开发速度有待进一步提高,Android应用程序开发者的需求量也越来越大,伴随着巨大的产业空间,国内Android系统开发人才需求不断高涨,Android应用开发及系统开发的工程师已经成为未来几年最为热门的职业之一。

2014年第二季度,分析机构Strategy Analytics公布了智能手机操作系统全球分布情况。其中Android操作系统的全球市场份额已达84.6%,而iOS(苹果手机操作系统)、WP(微软手机操作系统)等系统占比均有所下滑。Android操作系统的爆炸式发展得益于其自身的特点——开源,这使得它能够在全球范围内占据着主导地位。

目前市场上主流智能手机的操作系统主要有iOS、Android和Windows Phone。

(1) iOS是由苹果公司的Mac OS X发展而成的。它结合多种功能于一体,包含网络、桌面级的电子邮件、网页浏览及地图搜索等功能。这个系统原名为iPhone OS,在2010年WWDC大会上宣布改名为iOS。iOS推出的理念是能够使用多点触控屏幕的方式来操控手机。iOS采用Object-C作为开发语言,其内核是C语言的,并基于C语言实现了一些面向对象的特性。图1.2是一款非常经典的苹果手机界面。

(2) Android于2007年发布首款基于Linux平台的开源版本,这标志着移动信息设备的开发平台进入一个崭新的领域。Android采用Java作为开发语言,并提供了专门的SDK,是首个为移动终端打造的真正开放和完整的移动软件。图1.3是一款非常经典的Android手机界面。

(3) Windows Phone(简称WP)是微软公司针对移动手机开发的一款操作系统,其初衷是尽量接近桌面版本的Windows,微软公司按照计算机操作系统的模式来设计这款操作系统,以便能使得它与计算机操作系统更加贴近。WP上应用程序开发使用C♯语言,具有面向对象的特征。图1.4是一款非常经典的WP手机界面。

图1.2　iOS手机界面

图1.3　Android手机界面

图1.4　WP手机界面

1.1.2 Android 版本说明

Android 平台自 2007 年 11 月发布首款商业操作系统(beta 版)开始,不断更新、完善陆续发布了多个版本,这些版本的命名都以甜点为代号,并按照字母表的顺序命名,截至目前,最新版本是 Android 7,代号 Nougat。

(1) Android 1.5,代号 Cupcake,于 2009 年 5 月发布,这是第一个主要版本,用户操作界面得到极大改善,开始吸引开发者的目光,该版本主要完善的功能如下:
- 拍摄和播放视频,并可以上传到 Youtube。
- 支持立体蓝牙耳机。
- 采用 WebKit 技术实现浏览器,支持复制、粘贴和在页面中进行搜索。
- 提高 GPS 性能。
- 提供屏幕虚拟键盘。

(2) Android 1.6,代号 Donut,于 2009 年 9 月发布。搭载该操作系统的 HTC Hero 智能手机获得了意想不到的成功,Android 开始吸引更多人的目光,包括竞争者苹果公司和微软公司。该版本主要完善的功能如下:
- 重新设计 Android Market。
- 增加手势支持。
- 支持 CDMA 网络。
- 增加文本转语音的功能。
- 支持虚拟个人网络(VPN)。
- 支持更多的屏幕分辨率。

(3) Android 2.0/2.1,代号 Éclair,于 2009 年 10 月发布。这是继 1.5 之后的又一个主要版本,主要更新如下:
- 优化硬件速度。
- 用户界面的改良。
- 浏览器支持 HTML5。
- 支持蓝牙 2.1。
- 支持动态桌面设计。
- 改进 Google Maps。

(4) Android 2.2,代号 Froyo,于 2010 年 5 月发布,主要更新如下:
- 支持将软件安装到扩展内存。
- 增加 USB 分享器和 WiFi 热点功能。
- 浏览器集成 Chrome 的 V8 JavaScript 引擎。

(5) Android 2.3,代号 Gingerbread,于 2010 年 12 月发布,主要更新如下:
- 支持更大屏幕尺寸和分辨率。
- 系统级的复制和粘贴。
- 重新设计多点触屏键盘。
- 优化游戏开发支持。
- 支持更多的传感器。

(6) Android 3.x，代号 Honeycomb，于 2011 年 2 月发布，该版本开始支持平板电脑，主要更新如下：
- 优化针对平板电脑的功能。
- 全面支持 Google Maps。
- 3D 加速处理和支持多核心处理器。
- 支持操作杆和游戏控制器。

(7) Android 4.0，代号 Icecream Sandwich，于 2011 年 4 月发布，该版本的主要更新如下：
- 统一手机和平板电脑操作系统，应用可以根据设备选择最佳显示方式。
- 提升硬件的性能以及系统的优化，提升系统运行的流畅度。
- 脸部识别进行锁屏。
- 全新的 3D 驱动，游戏支持能力提升。
- 支持 WiFi 直连功能。

(8) Android 4.1/4.2/4.3，代号 Jelly Bean，于 2012 年 6 月发布，该版本的主要更新内容如下：
- 基于 Android 4.0 版本进行改善。
- 新增脱机语音输入。
- 通知中心显示更多消息。
- 更多的平板优化（主要针对小尺寸平板）。
- 强化 Voice Search（语音搜索），与 S Voice 类近，相当于 Apple Siri。
- 提升反应速度。
- 强化默认键盘。
- 大幅改变用户界面设计。

(9) Android 4.4，代号 KitKat，于 2013 年 9 月发布，该版本的主要更新内容如下：
- 优化存储器使用，在多任务处理时有更佳工作的表现。
- 新的电话通信功能。
- 全新的原生计步器。
- 全新的 NFC 付费集成。
- 全新的非 Java 虚拟机运行环境 ART(Android Runtime)。

(10) Android 5.0，代号 Lollipop，于 2014 年 6 月发布，该版本的主要更新内容如下：
- 采用全新 Material Design 界面。
- 支持 64 位处理器。
- 全面由 Dalvik 转用 ART(Android Runtime)编译，性能可提升 4 倍。
- 改良的通知界面及新增优先模式。
- 强化网络及传输连接性，包括 WiFi、蓝牙及 NFC。
- 改善 Android TV 的支持。

(11) Android 6.0，代号 Marshmallow，于 2015 年 5 月发布，该版本的主要更新内容如下：
- 应用权限管理。

- SD 卡可能和内置存储"合并"。
- Android Pay。
- 原生指纹识别认证。
- 严格的 APK 安装文件验证。
- 支持 MIDI。

(12) Android 7.0，代号 Nougat，于 2016 年 5 月发布，该版本的主要更新内容如下：
- 分屏多任务。
- 全新下拉快捷开关页。
- 通知消息快捷回复。
- 流量保护模式。
- 菜单键快速应用切换。

Google 公司自 2009 年以来不断推出新的"甜品"，逐渐完善了 Android 平台的功能。在选择开发平台时，应该本着使尽量多的人可以使用的原则。在下面的学习开发过程中，主要采用 Android 4.0 这个版本，以便尽可能兼容多数机型。

1.1.3 Android 系统架构

Android 被称作为移动设备打造的首个"真正完整和开放"的移动平台，该平台设计之初便考虑到为应用程序开发者提供二次开发的可能性，具有健壮的应用程序框架，提供丰富的应用程序开发接口。Android 以开放代码为前提，应用程序开发者可以比较自由地获取访问硬件设备的权限，在这个平台上开发应用程序不需要任何许可证和版权费用，这也是能吸引众多开发者参与的因素之一。

1. 免费与开放

Android 是一个完全开放源代码的平台，无论应用程序开发者还是手机制造商，都具有自由的开发空间。这种优势可以吸引众多的开发者和手机制造商加入到该平台的联盟阵营，壮大 Android 平台的影响力，积累人气，丰富应用程序，从而能够赢得更大的消费人群。

应用程序开发者可以任意发布应用程序，既可以编写该平台的免费软件，也可以开发需要授权的软件，获得一定的报酬。系统开发者需要遵循 GPL v2 协议，任何改进必须遵循开放源代码的协议约定。这种开放性在方便开发者的同时，也会导致该平台会出现众多版本不统一的情况，Google 公司有可能会调整相应的开放原则。

2. 开发语言与工具

Android 平台的应用程序可以采用 Java 语言编写，这给众多熟悉 Java 语言的开发者提供了更为广阔的发展空间，在短时间内为 Android 应用程序的开发找到了强有力的支持。Java 语言以开源和开放为特色，有大量的代码模型可以参考。Android 应用程序的集成开发环境是 Eclipse(需要安装 ADT 插件)，该软件具有众多的开源社区和资源。

3. 整合 Google 应用与服务

Android 作为 Google 公司推出的重磅产品，承担了 Google 帝国太多的使命。Android 无缝结合了 Google 公司很多优秀的应用与服务，例如搜索、天气预报、GoogleTalk、地图、Gmail 等。这使得 Android 拥有其他系统无可比拟的优势。用户在使用 Android 在线服务时，可以与计算机上使用的 Google 公司服务进行完全整合，实现 Google 服务的完全同步。

4. 分层的体系结构

Android 平台采用分层的架构，如图 1.4 所示，从上到下分别是应用程序层、应用程序框架层、运行环境与库层、Linux 内核层。其中应用程序层和应用程序框架层是用 Java 语言编写的程序；运行环境与库层是用 C/C++ 编写的，其中虚拟机部分可以运行 Java 语言；Linux 内核层是 Android 的最底层，主要由各种驱动程序组成。随着 Google 公司对 Android 系统的不断革新和完善，这种系统架构也会随着改变。毕竟"只有变化才是永恒"，一成不变的系统只会被淘汰，Nokia 就是一个活生生的例子。

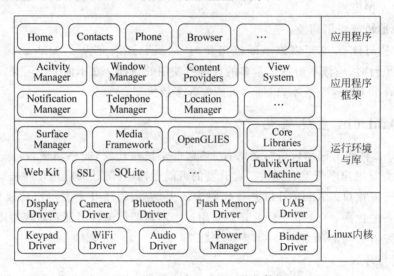

图 1.5　Android 通用系统架构

Android 系统应用程序的开发者所编写的应用程序将位于应用程序层。该层包括 SMS 短消息程序、联系人管理器、浏览器、日历、地图等一系列已经实现的应用程序，所有的应用程序都使用 Java 语言编写。

应用程序框架层为应用程序层的开发者提供 API，是应用程序的支持框架和调用接口。由于上层的应用程序采用 Java 编写，这些框架可以提供 Java 语言直接调用，包含了 UI 程序中所需要的各种控件以及服务，常见的如下：

- View System，包括列表、网格、文本框、按钮等基本控件。
- Content Providers，一个应用程序可以访问另一个应用程序的数据（如联系人数据库），也可以共享自己的数据。
- Resource Manager，资源访问的管理，如字符串、图形和布局文件的访问。
- Notification Manager，应用程序可以在状态栏（手机屏幕的最上方）中显示自定义的提示信息。
- Activity Manager，管理应用程序生命周期并提供常用的导航回退功能。

Android 应用程序在独立的进程中运行，拥有独立的 Dalvik 虚拟机实。Dalvik 可以同时高效地运行多个虚拟系统，Android 5.0 转用 ART 进行编译。程序库包含一些 C/C++

库,可以被 Android 系统中不同的组件使用,通过 Android 应用程序框架为应用程序开发者提供各种服务。

体系结构的最底层是 Linux 内核层,Android 平台的核心系统服务依赖于 Linux 操作系统的内核,如安全性、内存管理、进程管理、网络协议和驱动模型等,这也是硬件和软件栈之间的抽象层。

5. 众多 App Market 与开发社区

Android 平台为开发者提供了开放的开发环境,开发者可以自由开发免费软件、共享软件、试用软件,也可以开发依靠植入广告或直接付费的营利软件。如果想达成这种目标,必须将应用程序推送到用户端才可以,App Market 就是服务开发者和用户的一个平台,它允许开发者发布应用程序。作为一个连接开发者和用户的平台,应用程序商店具有很大的发展空间和诱人的利润,是众多运营商、手机制造商、软件开发商必争之地。短短两年时间,国内的应用程序商店就已达数十个。Android Market 是 Google 官方应用程序商店,面向全球用户,在国内人气比较旺的应用程序商店有机锋市场、安卓市场、安智市场、腾讯应用中心、网易应用、AppChina 应用汇和 360 宝盒等。

很多应用程序商店都有与之匹配的开发社区。如果想快速成为 Android 开发高手,加入 Android 开发者社区是不错的选择,可以获取前沿知识、学习资料、免费资源,同时也可以向行业内高手请教。比较活跃的论坛有 CSDN、51CTO、OSChina、eoe 等,图 1.6 是 eoe 移动开发者社区 Android 板块的首页。选择成熟度较高的国内社区网站可以更快、更便捷地掌握 Android 开发知识。

图 1.6　eoe 论坛移动开发者社区首页

以上介绍可以帮助各位读者了解 Android 平台的发展轨迹,熟悉 Android 平台目前的市场信息,明确学习方法,对 Android 平台的发展前景有一个很好的认识。今后在学习过程中,还需要注意以下几点:

- Android 平台应用程序开发主要采用 Java 语言,所以要熟练掌握 Java 语言的使用。
- 学会学习,目前有很多开源社区提供丰富的资源,可以免费获取,这些都是学习 Android 应用开发的捷径。
- 学习任务任重而道远,需要循序渐进,切不可急于求成。

1.2 Android 开发环境

工欲善其事,必先利其器。开发工作之前的首要任务是搭建 Android 开发环境,目前比较成熟稳定的开发环境是基于 Eclipse 搭建的,另一种开发环境是 Android Studio,这是 Google 官方指定的开发环境。

1.2.1 使用 Eclipse

基于原生态的 Eclipse 搭建 Android 开发环境,过程比较烦琐,更新插件时需要连接 Google 的网站。首先需要下载 Android 的 SDK(Software Development Kit,软件开发工具包),配置环境变量,然后为 Eclipse 安装(说升级更准确)插件 ADT(Android Development Tools),并指定 SDK 的位置,最后创建 Android 虚拟设备 AVD(Android Virtual Device),即虚拟 Android 系统手机。下面给出详细的安装与配置步骤。

1. 下载并安装 JDK

JDK 是开发 Java 类应用程序必备的开发工具包和运行环境,Eclipse 需要以此为基础才可以运行。JDK 可以在 Oracle 官方网站下载,下载前注意查看操作系统是 32 位还是 64 位,网址是 http://www.oracle.com/technetwork/java/javase/downloads/index.html。

2. 下载并解压 Eclipse

安装包可以在 http://www.eclipse.org/downloads/网站下载,由于版本众多,请仔细阅读说明,注意 32 位操作系统和 64 位操作系统是不同的。下载与操作系统匹配的安装包后,直接解压到相应目录即可,Eclipse 是完全绿色的,无须安装。进入解压后的路径,将 eclipse.exe 发送到"桌面快捷方式",方便以后打开。

3. 下载 SDK

SDK 的版本更新得比较快,获取的方式也在改变,连接 Google 网站时会受到一定的限制,如果连接不成功,也可以选择国内的网站下载。Google 提供的 SDK 下载网址是 http://developer.android.com/sdk/index.html,下载页面如图 1.7 所示。关于具体安装注意事项可以阅读该网站的 Installing the SDK,例如系统的需求、软件的需求等。

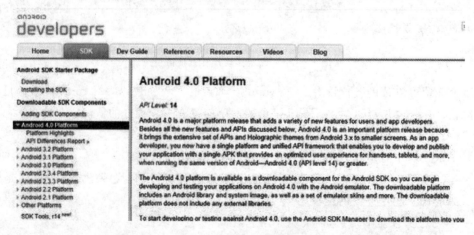

图 1.7 SDK 下载页面

安装结束后，会打开 Android SDK and AVD Manager 窗口，开始更新，选择全部更新即可，更新有时会比较慢，很考验网速，网速不给力的请勿烦躁，更新界面如图 1.8 所示。更新完成后，将安装路径中的 tools 文件夹增加到环境变量 path 中，在本书中是 E:\Android\android-sdk\tools。检验 SDK 是否安装成功可以在"运行"中输入 cmd，打开命令行窗口，输入 android-h，如果识别，表示安装成功。

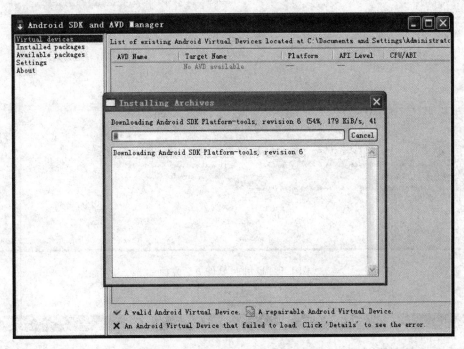

图 1.8 Android SDK and AVD Manager 窗口

4. 安装 ADT

ADT 是借助 Eclipse 开发 Android 应用程序的插件，Eclipse 默认没有该插件，需要手动安装。但是并不是所有 Eclipse 版本都可以安装该插件，需要 Eclipse 3.5 或更高版本。

打开 Eclipse，在 help 菜单中，选择 Install New Software...项，安装界面如图 1.9 所示。在 Work with 输入框中输入 https://dl-ssl.google.com/android/eclipse/，单击 Add 按钮，如果是第一次添加上述地址，会弹出输入名称窗口，随便填写一个即可，例如 adt。

当 Eclipse 查询到更新内容时，会自动列出，从中选择需要安装的插件（建议全选），如图 1.10 所示，单击 Next 按钮进行下载。这将会花费很长时间，相当考验网速。

当漫长的下载工作完成后，会弹出安装窗口，如图 1.11 所示，这表明 Eclipse 所需要的插件已经全部下载完毕，此时单击 Next 按钮进行安装工作。勾选"I accept…"项，开始安装。安装过程如出现提示，请选择 Yes 或 Ok，继续安装就可以。当安装完成后，会提示 Eclipse 需要重启（注意是 Eclipse 软件需要重启，不是计算机需要重启），单击 Restart now 按钮即可。重新启动后的 Eclipse 在工具栏上会新增加一个图标 （提示：opens the Android SDK and AVD manager）。选择 Window 菜单中的 Preferences 项，打开 Preferences 窗口，选择左边的 Android，在 SDK-Location 中输入 SDK 的根目录，如 E:\Android\android-sdk，如图 1.12 所示。

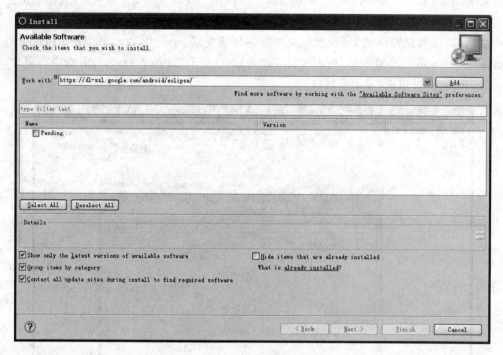

图 1.9 Eclipse 软件更新界面

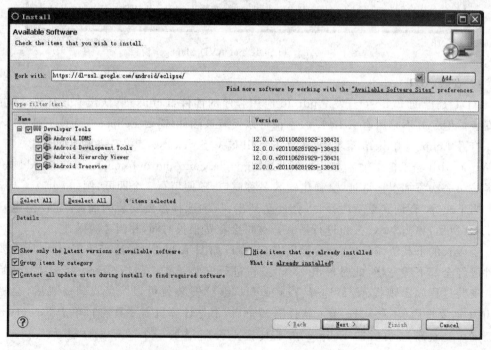

图 1.10 Eclipse 搜索更新内容

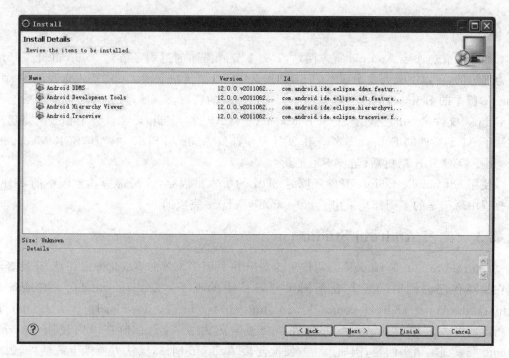

图 1.11　Eclipse 完成下载插件

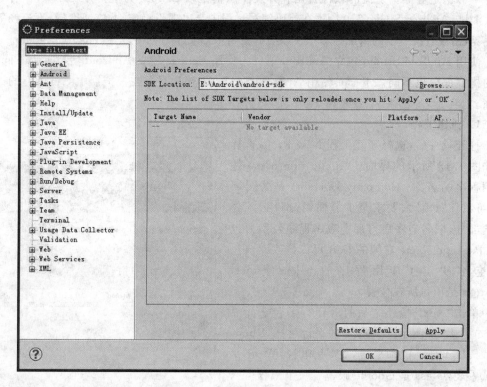

图 1.12　配置 SDK-Location

1.2.2 使用 adt-bundle Eclipse

如果能成功连接 Android 官方网站，1.2.1 节所描述的过程是搭建 Android 开发环境最为理想的，可以下载到各个版本的 SDK 和支持环境。无法联网成功时可以选择使用 adt-bundle 版本的 Eclipse，国内的很多网站都提供相应的资源。该版本的 Eclipse 解压后会得到 Eclipse 文件夹、sdk 文件夹和 SDK Manager.exe 文件。Eclipse 文件夹中存放的是已经安装了 ADT 插件的 Eclipse 软件，sdk 文件夹中存放 Android SDK 和工具，SDK Manager.exe 用于管理 SDK 的更新（建议不用更新）。

使用 adt-bundle Eclipse 开发环境时 SDK 的版本是固定，无法选择其他版本的 SDK。本教程中所编写的案例都是采用了 adt-bundle Eclipse 完成的。

1.2.3 使用 Android Studio

Android Studio 是 Google 公司推出的基于 IntelliJ IDEA 的集成化开发环境，开发者可以在编写程序的同时预览其在不同尺寸屏幕中的样子。Android Studio 下载地址是 https://developer.android.com/sdk/installing/studio.html#download。作为 Google 官方指定的开发环境，Android Studio 势必会逐渐成为主流，但目前在国内使用 Android Studio 开发面临诸多问题，例如无法方便地连接 Android 网站，无法方便地下载插件，版本运行不够稳定等等。主流的软件公司和开发团队仍是以 Eclipse 为主要开发环境，作者经过多次测试，Android Studio 1.0 版本在代码编写过程中会出现不稳定、响应速度慢、不流畅等现象，甚至出现直接退出的情况。相信在以后的更新中 Google 会不断完善 Android Studio。

使用 Android Studio 建立 Android 项目与使用 Eclipse 建立 Android 项目过程不同，项目的构成部分也不同。下面是使用 Android Studio 创建项目并在模拟器中测试运行的详细步骤。

1. 新建工程

打开软件后，选择新建工程 New Project，打开如图 1.13 所示的界面。Application name 是应用程序的名称；Company Domains 是公司域名，可以任意填写，要求至少分两级，即用"."分成两部分，域名逆序后再加上应用程序名作为包名；Package Name 右侧的 Edit 可以修改默认包名，包名决定了应用程序的不同；Project location 是项目所在位置。

单击"Next"按钮之后打开运行平台选择界面，如图 1.14 所示。共有 4 种平台可选，Phone and Tablet 是手机和平板，TV 是电视，Wear 是穿戴手表，Glass 是 Google 眼镜。一个工程可以选择运行在多个平台，每个平台必须各自确定所支持的版本。不同的版本需要下载不同的 SDK，否则项目无法建立。

图 1.13 Studio 新建工程

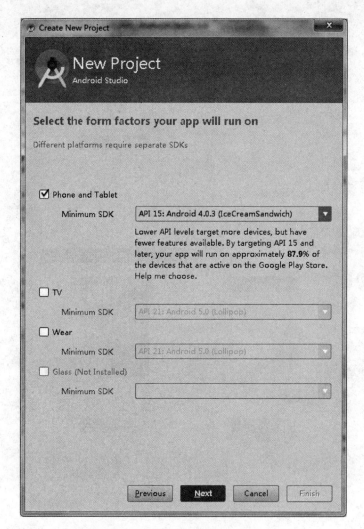

图 1.14　选择项目运行平台

　　单击 Next 按钮，打开 Activity 选择界面，如图 1.15 所示。此处有多种不同布局与不同风格 Activity，如果不需要在项目中生成 Activity，可以选择 Add No Activity，此时 Finish 按钮激活，可以结束项目创建导航。

　　若选择添加了 Activity，则需要对其进行参数配置，如图 1.16 所示。Activity Name 是名称，Layout Name 是 Activity 所采用的布局文件的名称，Tile 是 Activity 的标题，Menu Resource Name 是 Activity 所采用的菜单文件名称，这些参数在项目中可以修改。

2. 项目的构成

　　项目创建导航结束后，打开 Android Studio 软件的开发界面，如图 1.17 所示。在左侧边栏顶部可以切换 Project、Packages 和 Android 三种视图，开发过程中关注 Android 视图即可。

　　在 Android Studio 中一个项目由两部分组成：app 和 Gradle Scripts。app 部分的组成与 Eclipse 中项目的组成大同小异，manifests 文件夹中存放 AndroidManifest.xml 文件，这是应用程序组建的配置文件，Java 中存放应用程序的源代码，res 中是各种资源文件。

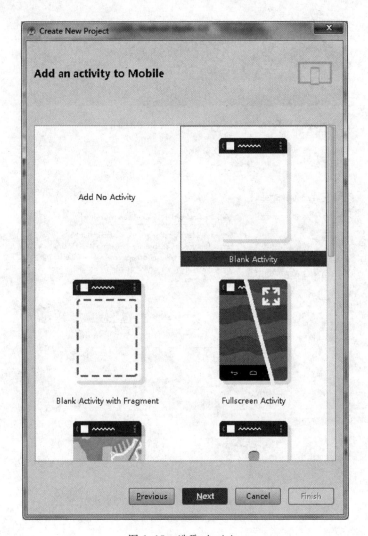

图 1.15 选取 Activity

Gradle Scripts 部分主要是项目相关的配置信息，如支持的最低版本。

3. 创建模拟器并运行项目

单击工具栏中的 AVD Manager 工具 ，打开虚拟机创建窗口，在此可以创建虚拟设备。单击窗口下部的 Create Virtual Device，打开设备选型界面，如图 1.18 所示。选择创建手机设备 Phone，选择机型为 Nexus S。

单击 Next 按钮之后，打开系统选择界面，如图 1.19 所示。该界面会列出已经下载的所有 Android 版本，ABI 标识不同 CUP 类型，选择 armeabi-v7a 可以很好地支持浮点运算。单击 Next 按钮之后，可以在高级配置界面设定其他参数，如存储大小、SD 卡等，若保持默认，单击 Finish 按钮即完成虚拟设备的创建。

虚拟设备创建完成之后就可以测试运行项目了，单击工具栏上的 按钮，就可以将项目安装到虚拟机，并测试运行。运行时要注意虚拟机操作系统版本也应不低于项目所支持的最低版本。

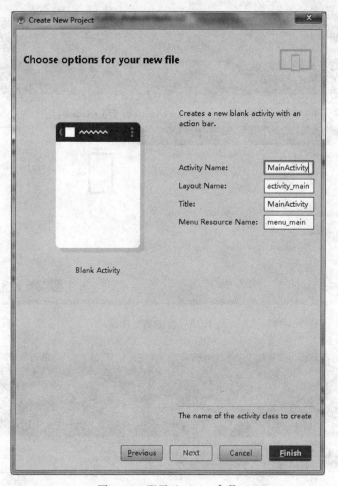

图 1.16 配置 Activity 参数

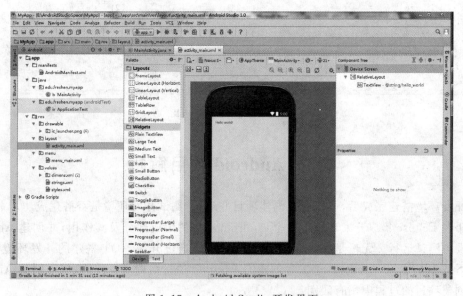

图 1.17 Android Studio 开发界面

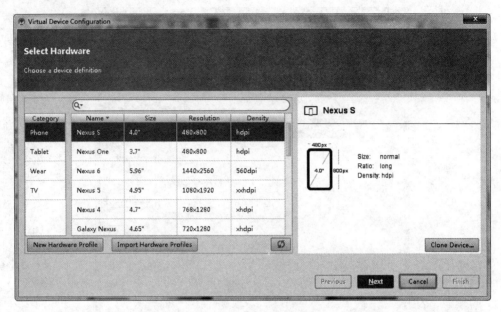

图 1.18 设备类型选择

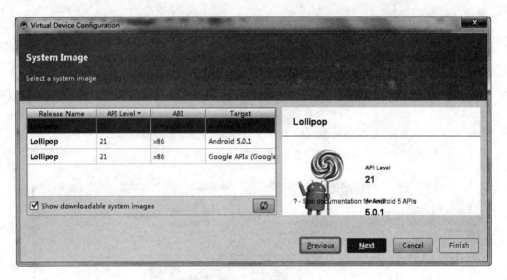

图 1.19 选择系统版本

1.3 Android 项目解析

开发环境搭建完成后,可以通过创建项目并运行,测试开发环境是否搭建成功。本节详细说明在 Eclipse 集成开发环境下一个 Android 项目的组成部分及其作用,并创建 Android 虚拟设备测试运行。无论是在"原生态"的 Eclipse 上通过升级 ADT 搭建的开发环境,还是直接下载 adt-bundle Eclipse,这二者创建的 Android 项目结构是一致的,但不同的 ADT 版本在创建项目的过程中稍有区别。

1.3.1 创建 Android 项目

启动 Eclipse 时,需要配置工作空间(项目所在文件夹)。如果是第一次启动 Eclipse,会出现欢迎界面(Welcome),在此可以学习 Eclipse 的相关知识,关闭后将不再显示。打开 Eclipse 主界面后,依次选择 File 菜单→New→Android Application Project,打开新建项目窗口。如果在 New 子菜单下没有 Android Application Project 选项,可以通过底部的 Other 项打开创建项目的通用窗口,如图 1.20 所示。

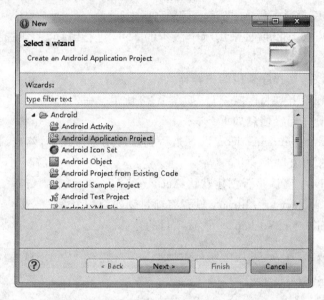

图 1.20 创建项目的通用窗口

单击 Next 按钮打开创建项目窗口,如图 1.21 所示。该窗口的配置信息如下:

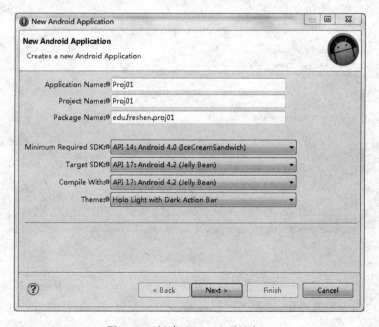

图 1.21 新建 Android 项目窗口

- Application Name 是应用程序的名称,建议以大写字母开头。
- Project Name 是项目的名称,默认与 Application Name 相同,允许修改。
- Package Name 是应用程序的包名,要求至少采用两级包名,这是区分不同 Android 应用程序的依据,不能相同,否则会被视为同一个应用程序。
- Minimum Required SDK 是项目兼容的最低 Android 版本,选择的 SDK 版本越低,兼容的手机数目越多。目前市场主流手机操作系统的版本约有九成都在 4.0 以上,建议选择 API 14,Android 4.0。
- Target SDK 是目标版本,即该项目最高版本。
- Compile With 是指项目采用哪个版本的 SDK 进行编译,一般选择与 Target SDK 一致即可。
- Theme 是该应用程序采用的主题样式。

单击 Next 按钮打开项目配置窗口,如图 1.22 所示。在该窗口下可以配置项目的图标、工作路径等信息,具体配置信息如下:

- Create custom launcher icon 设置是否创建应用程序图标,勾选后下一个界面会出现配置图标的窗口。
- Create activity 确认是否创建默认 Activity。
- Mark this project as a library 设置将项目创建为一个库。
- Create Project in Workspace 设置项目的存储路径。

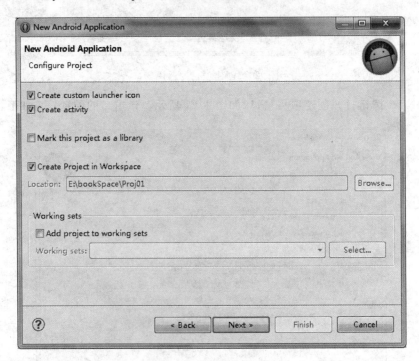

图 1.22 设置项目窗口

单击 Next 按钮打开下一个配置窗口,如果勾选 Create custom launcher icon,则打开如图 1.23 所示的图标配置界面,否则跳过这一步。图标配置窗口的详细信息如下:

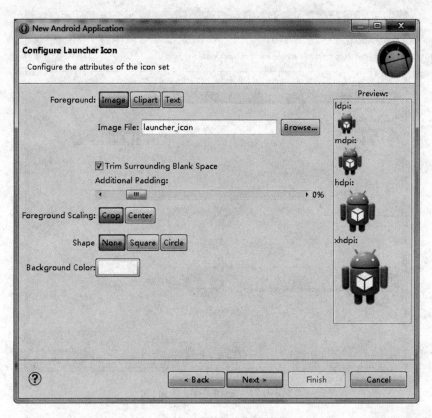

图 1.23 设置应用程序图标窗口

- Foreground 设置图标的前景类型。可以选择位图(Image)、剪贴画(Clipart,这是一种矢量图)和文字(Text)。
- Additional Padding 是图标前景的填充间距。
- Foreground Scaling 设置前景的放置类型,Crop 是剪切,Center 是将前景置中。
- Shape 设置图标的外形,None 会保存前景的形状,Square 是方形的图标,Circle 是圆形图标。
- Background Color 设置图标背景填充颜色。
- Foreground Color 设置图标前景的颜色。当 Foreground 选择了 Clipart 或 Text 时,该设置项才会出现。

Create Activity 窗口可以选择是否要创建默认 Activity 以及 Activity 的类型,如图 1.24 所示。如果不选择创建,则 Finish 按钮激活,可以直接结束项目的创建导航。否则下一步会出现 Activity 属性的配置窗口,如图 1.25 所示。Activity 配置窗口的详细信息如下:

- Activity Name 是所创建 Activity 类的名称,对应源代码的文件名。
- Layout Name 是 Activity 所采用的布局文件名称,对应布局文件的文件名。
- Navigation Type 是 Activity 的导航模式。

所有配置信息都确认后,单击 Finish 按钮结束项目创建导航。如果有配置信息需要修改,可以单击 Back 按钮回到对应的窗口。

图 1.24 创建 Activity 窗口

图 1.25 配置 Activity 属性窗口

1.3.2 Android 项目结构

Android 项目创建后,会默认打开如图 1.26 所示的主界面,这是 Eclipse 中典型的 Java 视图模式。Package Explorer(包视图)是整个项目所具有的结构,由于 Android 应用程序的编译运行机制不同,在项目创建环节就实现了 MVC,项目结构比 Java 应用程序的项目结构要复杂。

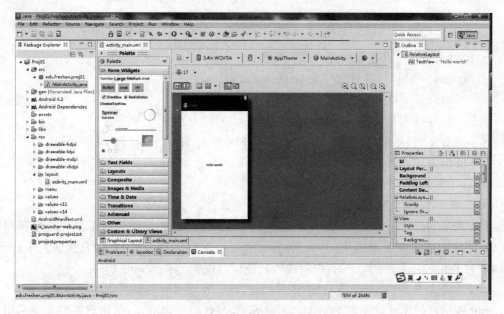

图 1.26 Android 项目默认视图界面

(1) src 是源代码资源,与 Java 项目的 src 类似,存放 Android 应用程序的 Java 代码。其中包(edu.freshen.proj01)和类(MainActivity.java)是在项目创建导航过程中设定,Eclipse 自动生成的。

(2) gen 是 generated Java files 的缩写,表明这是一个由系统自动生成的文件。展开文件夹后有 R.java 和 BuildConfig.java 两个文件,都是自动生成的,禁止修改,分别用于引用后面 res 中的资源和辅助项目检测。当 res 中的文件被修改后,R.java 会自动修改相应的引用值。

(3) Android 4.2 是系统类库,由创建项目是选定的版本决定。

(4) assets 存放外部资源,可以与应用程序打包在一起,与后面 res(也用于存放资源)不同的是,res 中的资源可以与 gen 中的 R 类对应,assets 中的资源不会在 R 类中有引用。

(5) bin 存放编译后的文件。

(6) libs 存放当前项目需要加载的第三方 jar 文件。

(7) res 存放资源文件。对于 Android 项目而言,图片、文本、声音、颜色、样式、动画等内容都可以视为资源,分别存放在 res 中的不同文件夹中。drawable-xxxx 用于存放图片资源,为了兼容不同机型屏幕的分辨率,图片资源文件夹又划分为 4 个:

- drawable-hdpi:高分辨率图片。
- drawable-mdpi:中分辨率图片。

- drawable-ldpi：低分辨率图片。
- drawable-xhdpi：超大分辨率图片。

layout 文件夹用于存放布局文件，其中 activity_main.xml 文件是 MainActivity 默认采用的布局。values 文件夹可以存放多种类型资源，为了方便应用程序的国际化，可以提供 values-en、values-zh 分别实现英语环境和中文环境下的文本显示，应用程序会根据手机操作系统选择的语言自动切换 values 资源包。

- array.xml 可以存放数组。
- strings.xml 可以存放字符串。
- colors.xml 可以存放颜色的字符串值。
- dimens.xml 可以存放尺寸值(dimension value)。
- styles.xml 存放样式(style)对象。

此外，res 中还可以出现以下文件夹：

- anim 文件夹用于存放动画文件，可以被编译进逐帧动画(frame by frame animation)或补间动画(tweened animation)对象。
- xml 文件夹存放任意的 XML 文件。
- raw 文件夹存放直接复制到设备中的任意文件。

(8) AndroidManifest.xml 是每个 Android 项目都必需的文件。用于设置应用程序的 ICON(图标)，声明应用程序中的 Activities(活动)、ContentProviders(内容提供)、Services(服务)和 Intent Receivers(意图/广播接收)，还可以指定 permissions(权限)。

以上 8 个部分是 Android 项目的主要组成，在开发项目时只要关注这些内容就可以了。Android 在项目创建时就实现了将 UI(用户界面)与代码逻辑分离，应用程序的布局采用了类似 Web 开发的 CSS 样式，将布局文件、值的引用与源代码分离，res 文件夹用于存放布局和引用值，存入其中的所谓"资源"，都会生成唯一的 ID 值，对应到 gen 文件夹中的 R 类中。在程序开发过程中，可以直接使用 R 类中的静态属性，引用 res 中任意"资源"，它们与 res 中资源具有一一对应的关系。引用资源的一般形式为"R.资源类型.资源名称"。

  📖 新建 Android 项目中 R.java 文件的源代码如下：

```
/* AUTO - GENERATED FILE. DO NOT MODIFY.
 * This class was automatically generated by the aapt tool from the resource data it found.
 * It should not be modified by hand.
 */
package edu.freshen.proj01;
public final class R {
    public static final class attr {
    }
    public static final class drawable {
        public static final int ic_launcher = 0x7f020000;
    }
    public static final class id {
        public static final int menu_settings = 0x7f070000;
    }
    public static final class layout {
```

```
        public static final int activity_main = 0x7f030000;
    }
    public static final class menu {
        public static final int activity_main = 0x7f060000;
    }
    public static final class string {
        public static final int app_name = 0x7f040000;
        public static final int hello_world = 0x7f040001;
        public static final int menu_settings = 0x7f040002;
    }
    public static final class style {
        public static final int AppBaseTheme = 0x7f050000;
        public static final int AppTheme = 0x7f050001;
    }
}
```

R 类中的所有内部类和成员都用 final 和 static 修饰，这些成员可以直接通过类名引用，无须创建 R 对象。以 layout 内部类为例，该类对应 res 资源文件夹中的 layout 文件夹，该内部类的成员属性 activity_main 对应 layout 文件夹下 activity_main.xml 文件，其值是自动生成的，不能修改。如需在代码中引用 activity_main.xml 文件，则可使用 R.layout.activity_main。

双击打开 activity_main.xml 文件，这是项目创建过程时自动生成的一个布局文件，编辑界面如图 1.27 所示。

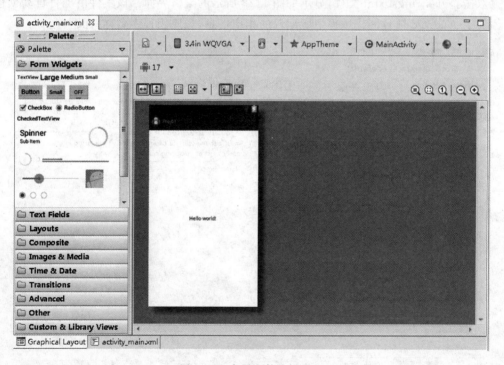

图 1.27　布局文件编辑界面

布局文件的编辑视图有两种：可视化编辑视图和代码视图，通过编辑窗口下方的 Graphical Layout 和 activity_main.xml 标签切换。在可视化编辑视图中，可以实现将 Android 提供的控件拖曳到应用程序界面，实现"所见即所得"的编辑模式。该模式处理简单布局时比较方便，复杂布局应采用代码模式，以求精确。

在可视化编辑视图下，窗口左边是系统控件，分为 Form widgets、Text Fields、Layouts 等不同类别，每种类别下都有多个控件；窗口右边是布局的预览效果，上方的辅助按钮可以浏览当前所编辑的布局在不同尺寸屏幕中的效果。

📖 activity_main.xml 文件的代码如下：

```xml
<RelativeLayout xmlns:android = "http://schemas.android.com/apk/res/android"
    xmlns:tools = "http://schemas.android.com/tools"
    android:layout_width = "match_parent"
    android:layout_height = "match_parent"
    tools:context = ".MainActivity" >
<TextView
        android:layout_width = "wrap_content"
        android:layout_height = "wrap_content"
        android:layout_centerHorizontal = "true"
        android:layout_centerVertical = "true"
        android:text = "@string/hello_world" />
</RelativeLayout>
```

> @string是引用字符串资源的语法

双击打开 strings.xml 文件，显示如图 1.28 所示的编辑界面。该文件由项目自动创建，并添加了 Android 项目创建过程中所配置的相应信息，如应用程序名称、标题等。

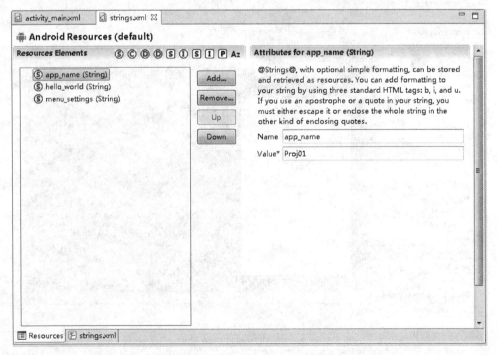

图 1.28　strings.xml 文件编辑窗口

📖 strings.xml 文件的代码如下：

```xml
<?xml version = "1.0" encoding = "utf - 8"?>
<resources>
    <string name = "app_name">Proj01</string>
    <string name = "hello_world">Hello world!</string>
    <string name = "menu_settings">Settings</string>
</resources>
```

如果在代码中引用某种资源，需要借助 R 类，采用"R.资源类型.资源名"的方式。如果在资源文件中引用另外一个资源文件中的资源，需要使用@，采用"@资源类型/资源名"的方式。例如，在布局文件 activity_main.xml 中 TextView 控件的 text 属性（显示文本）引用 strings.xml 文件中的字符串 hello_world 对应的值，采用 android:text = "@string/hello_world"。

1.4 运行与调试 Android 项目

项目创建后不做任何修改，可以直接运行，会显示"Hello World"界面。不涉及录音、录像、传感器等手机硬件的应用程序可以使用 Android 虚拟机测试运行，否则只能借助真机测试。Android 自带的"原生态虚拟机"启动速度比较慢，运行时需要消耗开发机器的资源，建议开发机器的内存应不少于 2GB。除了 ADK 自带的虚拟机，Android 项目的测试运行还可以使用 Genymotion，下载地址为 http://www.genymotion.net/。

1.4.1 使用 Android 虚拟机

创建 Android 虚拟设备（AVD）有多种方式，可以在 Android SDK 目录下直接运行 AVD Manager.exe，也可以在 Eclipse 中选择 Window 菜单→Android Virtual Device Manager 命令，还可以直接单击工具栏上的 ▣（Android Virtual Device Manager）图标，打开 Android Virtual Device（AVD）Manager 窗口，如图 1.29 所示。

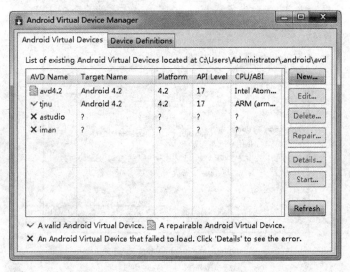

图 1.29　AVD Manager 窗口

在默认的 Android Virtual Devices 标签中单击 New… 按钮，可以打开创建自定义虚拟机的窗口，下方的几个按钮分别实现编辑、删除、修复、查看详情、启动和刷新的功能。选择 Device Definitions 标签，可以创建 ADK 预定义的虚拟设备，如 Nexus 系列模拟器。

在图 11.29 所示的窗口中点击 New 按钮，打开创建虚拟设备窗口，如图 1.30 所示。详细的配置信息如下：

- AVD Name 是虚拟机的名称，只能使用大小写字母、数字、"_"和"."命名。
- Device 是虚拟机的屏幕尺寸。
- Target 是虚拟机的操作系统版本，测试运行的版本应该高于项目所支持的最低 SDK 版本，否则无法运行项目。
- CPU/ABI 是虚拟机采用的 CPU 类型。
- keyboard 和 Skin 是模拟键盘和皮肤，保存默认值即可。
- Front Camera 和 Back Camera 是前后摄像头。
- Memory Options 是运行内存大小，如果开发机器内存够大，可以将该参数调大。
- Internal Storage 是内部存储空间大小。

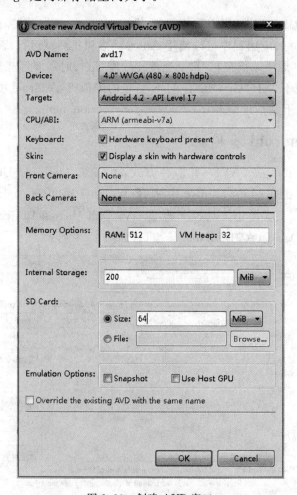

图 1.30 创建 AVD 窗口

- SD Card 是模拟器的 SD 卡空间大小。
- Emulation Options 是模拟器的启动参数,Snapshot 是从快照启动,这样会增加启动速度,但不要修改该模拟器的配置,否则会出错;Use Host GPU 用于指定是否使用图形加速。模拟器启动参数这两项可以保存默认值。

以上所有参数都设定后,单击 OK 按钮完成创建。在 AVD Manager 窗口选择刚才创建的 avd17 虚拟机,单击 Start 启动,弹出 Launcher Options 窗口,选择 Launch 按钮,启动虚拟机。虚拟机的启动过程比较慢,与开发机器的性能有关。在项目调试过程中,应保持虚拟机处于开启状态,无须测试一次启动一次。虚拟机启动后的运行界面如图 1.31 所示,与真机的运行效果基本一致。

图 1.31 虚拟机主界面

在运行的模拟器上可以通过按住左键拖动鼠标实现触屏事件,右侧的模拟按键实现的功能分别是降低音量、增大音量、待机(如果长按可以模拟关机和开机)、显示主页、显示菜单、返回、搜索和上下左右方位。这些模拟按钮只是为了方便调试软件而存在,不同型号的真机不一定具备相同的按键,在项目开发过程中要注意对不同机型的兼容操作。特别提示,如果开发机器处于联网状态下,模拟器也是可以联网的,支持应用软件 App 的安装与卸载,并可以模拟拨打电话和发送短信。虚拟机可以像真机一样进行操作,内置的 Google 公司应用程序,如日历、时钟、浏览器、Google Map 等,都可以使用。

虚拟机启动后,可以通过 Window → Open Perspective→DDMS 打开"设备调试管理系统"视图,在该视图下可以查看虚拟机的运行状态,浏览虚拟机的文件,模拟拨打电话、发送短信和 GPS 定位等操作。测试运行项目时要确保 Eclipse 与虚拟机之间的连接处于活动状态,这个连接由 ADB(Android Debug Bridge)实现,正常情况下 DDMS 视图中会显示设备处于在线状态(Online)。

在项目视图中选择需要运行的项目,执行 Run→Run as→Android Application,或者右击需要运行的项目,在快捷菜单中选择 Run As → Android Application,将项目生成的 apk 文件安装到虚拟机。Proj01 项目的运行效果如图 1.32 所示。

图 1.32　测试项目运行效果图

1.4.2　使用 Android 真机

Android 提供的虚拟机可以解决一般应用程序的测试,但与真机还存在一些不同,在运行速度上与真机无法比拟。若涉及手机状态与传感器操作的应用程序(如检测电池状态、GPS 定位等),虚拟机无法模拟。在项目开发过程中,使用真机测试是不可或缺的,也是发布应用程序之前一定要完成的。

使用手机测试运行程序,需要在手机的"设置"面板中打开"开发人员选项",勾选"USB 调试"(不同机型操作方式稍有区别),然后使用 USB 数据线与开发计算机连接。如果是第一次连接,需要安装手机驱动程序。正确连接后,在"设备管理器"窗口可以看到目前连接手机的信息,如图 1.33 所示。本教程中真机测试使用的是 HTC M8 机型。

图 1.33　设备管理器窗口

在真机上测试运行项目与在模拟器中测试运行项目的步骤类似，选择 Run As→Android Application。如果每次测试运行的设备不同，可以通过执行菜单 Run→Run Configurations 选项打开测试设备配置窗口，如图 1.34 所示。选择 Target 选项卡，选择 Always prompt to pick device，每次运行项目都打开测试设备选择窗口。

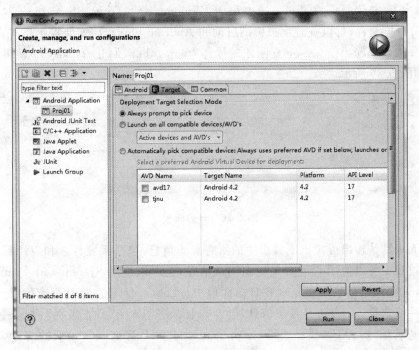

图 1.34　配置测试设备窗口

完成上述配置后，每次运行项目都会提示选择目标设备，如图 1.35 所示。如果存在已经开启的设备，包括已连接的真机和已启动的虚拟机，也会在出现在上面的列表中；下面的列表中是由 AVD 管理器创建但未启动的虚拟机。

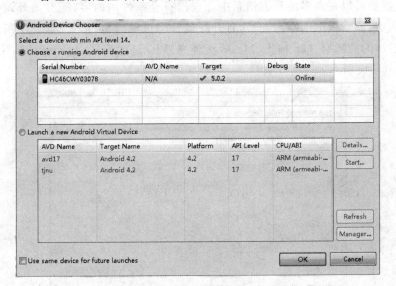

图 1.35　选择测试设备窗口

1.4.3 调试日志的使用

Android 提供了比较完善的程序调试功能,借助 Eclipse 的 ADT 插件,可以轻松地输出程序的提示信息,进行断点调试。不同于 Java 应用程序的 Console(控制台),Android 使用 LogCat 输出日志信息。在 Eclipse 中展开 Window 菜单下的 Show view 选项,选择 Console,打开控制台,可以查看虚拟机的启动和 Android 应用程序安装提示信息,但无法查看应用程序内部的输出信息。重复刚才的操作,单击 other 项,选择 LogCat 就可以打开日志面板,如图 1.36 所示。

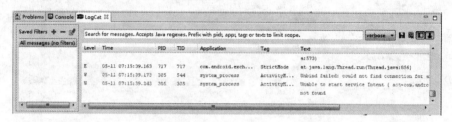

图 1.36 日志面板

日志面板默认输出虚拟机或真机中的所有日志信息,包括系统日志和应用程序中的日志。开发过程中可以使用 Log 对象输出调试信息,也可以使用 System.out.println()输出调试信息,这些调试信息都将呈现在日志面板中。LogCat 面板左边是日志的过滤器,根据不同的条件筛选日志;右边是日志信息,包含 7 个部分:Level(日志级别)、Time(虚拟机时间)、PID(进程编号)、TID(线程编号)、Application(应用程序名)、Tag(日志标题,输出日志时确定)和 Text(日志内容)。

日志信息根据严重等级划分为 5 个级别,分别是 Verbose(全部信息)、Debug(调试信息)、Info(提示信息)、Warn(警告信息)、Error(错误信息),依次使用 Log.v、Log.d、Log.i、Log.w 和 Log.e 输出。使用 System.out 输出的信息属于提示信息。

日志过滤器可以在众多的日志信息中筛选出符合条件的内容,能够提升调试程序的效率。单击日志过滤面板上的绿色加号,打开新建日志过滤器窗口,如图 1.37 所示。过滤器需要命名,过滤的方式有 Tag 标签、Message 消息、PID 进程编号、Application Name 应用程序名称或 Level 级别。

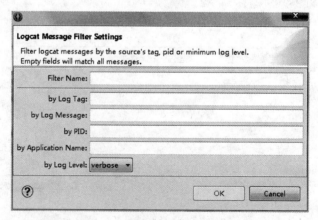

图 1.37 新建日志过滤器窗口

1.5 签名输出 apk 文件

Android 系统应用程序的后缀名是 .apk，即移动平台的安装文件。在 Android 项目的 bin 目录下会产生一个"项目名.apk"文件，该文件由 Eclipse 编译产生，主要用于发布到虚拟机测试。该文件可以被安装到移动平台，但不具有通用性，只有经过数字签名的 apk 文件才能移交给客户或发布到 Android Market。

所谓"签名"是指开发者使用数字证书加密 apk 文件，让开发者与应用程序之间建立信任关系。签名所使用的数字证书不需要权威机构认证，它只用于让应用程序自我认证。Android 系统直接采用 Java 数字证书机制，可以通过命令行使用 keytool 命令签名，也可以通过 Eclipse ADT 插件签名。

使用 ADT 插件签名的操作过程比较简单，先生成一个数字证书，然后使用该数字证书对 apk 文件签名即可。

（1）在需要输出签名文件的项目上右击，在快捷菜单中选择 Android Tools→Export Signed Application Package，打开签名导航过程。如果是第一次签名，需要先生成签名文件，指定签名文件存储路径和密码，如图 1.38 所示。如果存在签名文件，则可以选择 Use existing keystore。

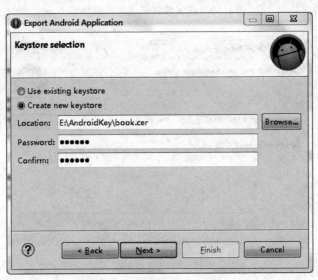

图 1.38　新建签名文件窗口

（2）一个数字证书包含的基本信息有别名、密码、有效期、姓名、单位、组织名称、所在城市、所在州或省、国家代号，如图 1.39 所示。

（3）设置输出 apk 文件的路径，单击 Finish 按钮完成签名，如图 1.40 所示。在指定的路径可以找到相应的证书文件和签名后的 apk 文件。

签名后的 apk 文件可以发布到第三方应用市场，如安卓市场、机锋市场等。发布应用时需要先申请成为开发者，具体按照各平台的流程操作。

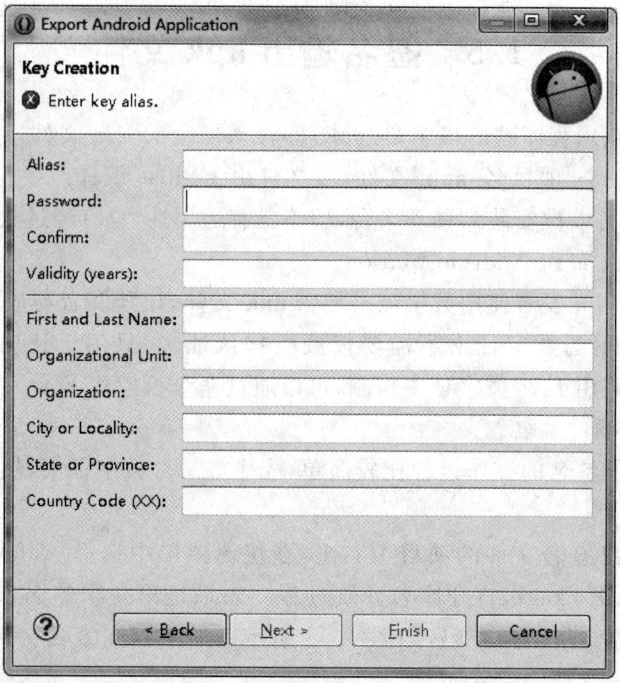

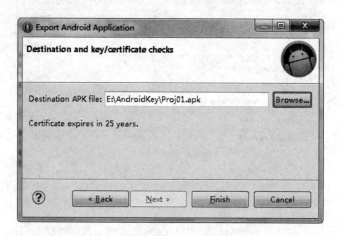

图 1.39 设置数字证书基本信息图

图 1.40 输出签名 apk 文件

1.6 习　　题

1. 选择题

(1) Android 系统从(　　)版本开始全面支持大屏幕设备。
　　A. Android 2.3　　B. Android 3.0　　C. Android 4.0　　D. Android 4.1

(2) 在 Android 项目中用于存放图片、布局等资源的文件夹是(　　)。
　　A. res　　　　　　B. gen　　　　　　C. src　　　　　　D. bin

(3) 以下关于 R.class 文件说法有误的是(　　)。
　　A. 该文件是开发环境生成的,不能对其进行修改
　　B. 该文件主要是为了在代码中引用资源文件夹 res 中的资源
　　C. 引用该文件需要通过 android.R 的方式
　　D. 引用该文件中的成员可以通过"R.资源类型.资源名称"的方式
(4) 调试日志根据紧急程度划分了很多级别,其中错误日志对应的级别是(　　)。
　　A. i　　　　　　B. w　　　　　　C. d　　　　　　D. e
(5) 常用于开发 Android 项目的集成化开发环境不包括(　　)。
　　A. Eclipse with adt　　　　　　B. adt-bundle Eclipse
　　C. Android Studio　　　　　　 D. Dreamweaver

2. 简答题

(1) Android 操作系统的分层体系结构包括哪几层?分别实现什么功能?

(2) Android 应用程序开发在项目创建时就实现了 MVC,请说明 Android 项目如何实现了这一开发模式?

第 2 章 使用控件创建用户界面

本章学习目标
- 了解 Android 系统设计 UI 界面的两种方式：布局文件和代码。
- 掌握 Android 开发中常用控件的属性。
- 掌握数据适配器的使用，可以自定义适配器。
- 掌握 Android 系统对子线程的限制。

用户对一款软件的评价，很大程度上都与该软件的界面是否方便操作、设计是否合理、动画效果是否华丽、色彩搭配是否融洽相关，这种现象在移动应用开发中更加明显。一款功能设计合理、数据结构科学、操作性能优良的软件，如果配上糟糕的界面，也很难赢得很大的用户群体。因此设计制作优美的用户界面非常重要，这往往是能吸引用户的最重要因素之一。

Android ADK 提供了大量的 UI 控件，如文本控件、图片控件、日历控件等，如果系统控件无法满足特殊需求，还可以自定义控件。合理使用系统控件，可以方便快捷地设计出界面美观的应用程序。

2.1 Android 用户界面设计

Android 手机界面设计不同于网页的设计，但是在设计模式上与页面比较类似，支持界面元素与样式和功能的分离，比较好地应用了 MVC 的思想。界面元素就相当于 HTML 中的基本元素，样式相当于 CSS，功能代码相当于 Java Script 脚本。

2.1.1 使用布局文件设计界面

Android 布局文件位于项目结构中的 res/layout 下，是一个 XML 文件，它的命名只能是小写字母、数字和"_"，并且数字不能开头，合法的布局文件会在 R 文件中生成对应的引用。当项目的 R 文件无法自动生成时，很多时候是因为资源文件 res 中命名非法造成的，应仔细检查 res 中的所有资源，是否存在大写字母或其他字符。

布局文件有两种编辑方式，图形化视图(Graphical Layout)和 XML 视图(文件名.xml)。图形化视图可以直接拖放控件，操作简单，但不适合设计比较复杂的界面。XML 视图可以更加灵活地处理界面布局，但需要熟练掌握控件的属性设置。

为了方便引用布局文件中的控件，在添加完控件之后，都应该赋予唯一的标识符，对应的属性是 android:id。这个 id 属性在整个项目中都应该没有重复，否则在引用该控件时会发生错误,建议采用"布局文件名_xxx"的格式命名 id。

Android 布局文件的设计与 Java GUI 开发使用 awt、swing 编程相似,控件应放在布局管理器(或称容器)中,布局管理器之间可以相互嵌套,以实现更加复杂的界面设计。与 Java GUI 开发不同的是,Android 布局文件需要指定给 Activity 或某个视图控件,同一个布局文件可以指定给多个目标。给一个 Activity 指定布局,可以在 Activity 的 onCreate 方法中,使用如下代码:

setContentView(R.layout.布局文件名称)

设置完布局文件之后,可以引用该布局文件中的控件,前提是该控件必须具备 id 属性。布局文件中的控件被成功引用后,可以使用 Java 代码来控制该控件的样式、内容、行为,设置监听器。引用控件的代码如下:

findViewById(R.id.控件 id 属性)

上述方法必须位于设置布局文件 setContentView 方法之后,否则会抛出错误。无论控件还是布局文件,findViewById 方法的返回值都是 View 类型,这是因为控件和布局管理器都是 View 的子类。程序员需要根据某个 id 值对应的具体类型将返回值强制转换为对应控件。如 id 为 tv 的控件是 TextView 控件,在引用时可以采用如下代码:

```
TextView tv = (TextView) findViewById(R.id.tv);
```

2.1.2 使用 Java 代码设计界面

除了使用 XML 布局文件设计界面,Android 也支持类似于 Java GUI 方式的 UI 开发,通过 new 关键字创建控件,再将这些控件放置在相应布局容器中。这种方式创建的控件和容器不会在 R 文件中生成相应的引用。

控件和布局管理器继承自 android.view.View 类,都拥有一个类似的构造方法 View(Context context),参数是 Context 类型。Activity 和 Service 都是 Context 的子类,因此在 Activity 中可以直接传入 this,创建控件和布局管理器。项目 Proj02_1 演示如何使用代码创建布局文件,LinearLayout 是线性布局管理器,ImageView 是图片控件,项目的运行效果如图 2.1 所示。

图 2.1 使用代码创建界面

📖 Proj02_1 项目 MainActivity.java 设置界面

```
…
@Override
protected void onCreate(Bundle savedInstanceState) {
    super.onCreate(savedInstanceState);
    //创建布局管理器
```

```
        LinearLayout ll = new LinearLayout(this);
        //设置布局管理器的相关属性
        ll.setBackgroundColor(Color.GRAY);
        //创建控件
        ImageView iv = new ImageView(this);
        //设置控件的相关属性
        iv.setImageResource(R.drawable.tip);
        //把控件添加到容器中
        ll.addView(iv);
        //把布局设置给 Activity
        setContentView(ll);
    }
    ...
```

从上面的代码可见,使用 Java 代码直接生成布局分为 3 步:

(1) 创建布局管理器,加载构造方法,指明当前 Context(上下文)对象,设置布局管理器的属性。如果界面比较复杂,可以嵌套其他布局管理器。

(2) 创建控件,并设置属性,添加监听器。

(3) 把控件添加给布局管理器,控件在布局管理器中的布局受不同布局管理器的约束。

Android 开发中的两种布局方式各有优缺点。使用 XML 布局文件方便快捷,利于布局代码的重用,但不是很灵活。使用 Java 代码控制布局,可以根据逻辑动态修改,非常灵活,但缺点是代码烦琐,且不利于重用。因此在实际项目开发中,常将两种方式结合使用。对于相对固定的布局,使用布局文件实现;对于需要频繁更改显示内容和方式的布局内容,使用 Java 代码实现。

2.2 使用简单控件

Android 提供了一系列 View 类控件(如按钮(Button)、文本框(TextView)、下拉列表(Spinner)等)和 ViewGroup 布局(如线性布局(LinearLayout)、相对布局(RelativeLayout)等),熟练使用这些控件可以快速开发界面精美的 Android 应用程序。Android 提供的常用控件都位于 android.widget 包中。

2.2.1 控件的基本属性

打开布局文件的图形化视图 Graphic Layout,左侧上方是设备屏幕设定,包括屏幕尺寸设定、横竖屏设定(Portrait 为竖屏,Landscape 为横屏),显示的样式等,右侧是系统控件区域。单击 Palette 右边的下拉三角,可以选择控件的不同预览模式。图 2.2 是控件面板的图标模式和图标及文本模式。

Android 系统控件共分为 10 类,包括了应用程序所需要的大部分控件:

- Form Widgets(表单类控件),包括 TextView(文本标签)、Button(按钮)、RadioButton(单选按钮)、CheckBox(复选框)、SeekBar(拖动条)、Spinner(下拉列表)等。
- Text Fields(文本框控件),根据输入内容的不同分为普通文本框、密码文本框、电话文本框、数字文本框等。
- Layouts(布局管理器控件),包括 LinearLayout(线性布局)、RelativeLayout(相对布局)、FrameLayout(帧布局)、TableLayout(表格布局)、AbsoluteLayout(绝对布局)、GridLayout(网格布局)。
- Composite(组合控件),包括 ListView(列表视图)、GridView(表格视图)、ScrollView(滚动视图)、TabHost(标签容器)等。
- Images & Media(图片和媒体控件),包括 ImageView(图片控件)、ImageButton(图片按钮)、VideoView(视频播放控件)等。
- Time & Date(时间和日期控件),包括 TimePicker(时间选择器)、DatePicker(日期选择器)、AnalogClock(模拟时钟)、DigitalClock(电子时钟)等。
- Transitions(过渡效果控件),包括 ImagesSwitcher(图片切换)、ViewSwitcher(视图切换)等。
- Advanced(高级控件),包括 SurfaceView(动画视图)、ZoomButton(放缩按钮)等。
- Other(其他控件),包括 TextClock(时间文本框)控件。
- Custom & Liberty Views(自定义控件),存放继承 View 类,是由程序员自己实现的控件。

图 2.2　控件面板的不同预览模式

控件的显示效果由属性决定,控件添加到布局文件中后,可以根据需要调整相应的属性。在 Graphic Layout 视图中,右击控件,通过快捷菜单配置控件的各项属性,或者在右侧的 Properties 面板配置属性。在代码视图中配置控件属性时,需要在标签内部指定属性名,然后赋值。Android 中控件的属性比较多,表 2.1 列出了一些常用的属性及其含义和取值信息。

表 2.1 控件基本属性

属 性	说 明	取 值
android：id	控件的 ID,具有唯一性	自定义
android：layout_width	控件宽度	系统值： • fill_parent 填充(充满)父容器 • match_parent 匹配父容器 • wrap_content 包围内容 自定义值：直接指定控件尺寸
android：layout_height	控件高度	
android：text	显示的文本信息	引用 strings.xml 文件中的字符串,或直接指定字符串值
android：editable	是否允许编辑	true、false
android：background	设定背景图片或颜色	引用 drawable 中的图片,或直接给出 RGB 颜色
android：drawableTop android：drawableLeft android：drawableRight android：drawableBottom android：drawableStart android：drawableEnd	在指定位置绘制图片	图片的引用
android：textColor	文字颜色	
android：textSize	文字大小	
android：textStyle	文字风格	normal、bold、italic
android：fontFamily	设置字体	
android：lines	文本框占几行	
android：maxLines	最大行数	
android：singleLine	是否单行	true、false
android：ellipsize	当文本超过控件长度设置时如何处理内容	取值为 none(不做处理)、start(省略开始)、middle(省略中间)、end(省略结尾)、marquee(走马灯显示,需要配合单行模式)
android：gravity	文字的对齐方式	top、bottom、left、right 等
android：password	文本输入框是否是密码	true、false
android：selectAllOnFocus	文本输入框在获得焦点时全选文字	true、false
android：inputType	文本输入框的输入内容	number、date、time 等
android：textAppearance	文字的显示大小	? android：attr/textAppearanceLarge 等系统值
android：padding	内容距控件边缘的填充间距	
android：onClick	控件单击时执行的方法	方法名
android：autoLink	将符合格式的文本转换为超链接,如邮件、电话号码等	web、email、phone、map、all

续表

属 性	说 明	取 值
android:linksClickable	设置 URL、E-mail、电话号码等是否可以单击	true、false
android:layout_gravity	控件在父容器中的对齐方式	top、bottom、left、right 等
android:layout_margin	到其他控件边缘的距离	

2.2.2 TextView

TextView 直接继承 View 类，主要功能是显示文本。TextView 作为常用控件之一，它所具备的属性几乎都是其他控件通用的，其他很多控件都是从 TextView 继承的，如 Button、EditText 等，TextView 的继承关系如图 2.3 所示。

Button 和 EditText 用得比较多，后面会详细介绍。CheckedTextView 是带有复选功能的文本框，可以通过 setChecked、isChecked 方法管理其选择状态。TextClock 可以显示当前时间，是 API 17 开始加入的，推荐 Android 4.2 之后都使用 TextClock，不提倡使用 DigitalClock。

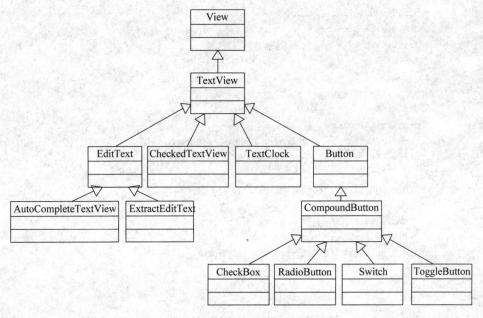

图 2.3 TextView 继承关系类图

📖 Proj02_1 项目 activity_main.xml 文件代码

```
<LinearLayout xmlns:android = "http://schemas.android.com/apk/res/android"
    xmlns:tools = "http://schemas.android.com/tools"
    android:layout_width = "match_parent"
    android:layout_height = "match_parent"
    android:orientation = "vertical"
    tools:context = ".MainActivity" >
```

```xml
<TextView
    android:id = "@+id/tv1"
    android:layout_width = "match_parent"
    android:layout_height = "wrap_content"
    android:autoLink = "email"
    android:textColor = "#CC6666"
    android:textSize = "22sp"
    android:linksClickable = "true"
    android:text = "发送邮件到 test@163.com" />
<TextView
    android:id = "@+id/tv2"
    android:layout_width = "match_parent"
    android:layout_height = "wrap_content"
    android:singleLine = "true"
    android:ellipsize = "end"
    android:text = "当一行内容过多时,结尾会显示省略号!有没有,有没有,有没有" />
<TextView
    android:id = "@+id/tv3"
    android:layout_width = "match_parent"
    android:layout_height = "wrap_content"
    android:singleLine = "true"
    android:drawableStart = "@android:drawable/arrow_up_float"
    android:text = "开始位置有图标了!" />
<CheckedTextView
    android:id = "@+id/tv4"
    android:layout_width = "match_parent"
    android:layout_height = "wrap_content"
    android:singleLine = "true"
    android:checkMark = "@android:drawable/star_on"
    android:text = "可以选中文本"/>
<TextClock
    android:id = "@+id/tv5"
    android:layout_width = "match_parent"
    android:layout_height = "wrap_content"
    android:singleLine = "true"
    android:format24Hour = "yyyy-MM-dd HH:mm"
    />
</LinearLayout>
```

设置字体大小时建议使用单位 sp,设置控件宽高尺寸时建议使用单位 dp 或 dip。对控件的其他属性设置感兴趣的读者可以自己尝试。

2.2.3 Button

Button 控件继承 TextView,主要功能是处理用户的点击操作。在 Android 中按钮的点击事件处理方式有两种,一种是直接给按钮注册监听器(这种方式与 J2SE 中事件处理模型一致),另外一种是设置 android:onClick 属性,指定处理点击事件的方法,方法的参数必须是 View 类型的。

下面布局文件中的两个按钮分别使用两种方式设置监听事件。

📖 Proj02_1 项目 activity_bt.xml 文件代码

```xml
<?xml version = "1.0" encoding = "utf - 8"?>
<LinearLayout xmlns:android = "http://schemas.android.com/apk/res/android"
    android:layout_width = "match_parent"
    android:layout_height = "match_parent"
    android:orientation = "vertical" >
    <Button
        android:id = "@ + id/button1"
        android:layout_width = "wrap_content"
        android:layout_height = "wrap_content"
        android:text = "Java 代码注册监听器" />
    <Button
        android:id = "@ + id/button2"
        style = "?android:attr/buttonStyleSmall"
        android:layout_width = "wrap_content"
        android:layout_height = "wrap_content"
        android:onClick = "pressBt"
        android:text = "onClick 属性设置监听方法" />
</LinearLayout>
```

使用 onClick 属性设置监听方法时，属性的值就是方法名，大小写要一致。该方法参数只能是一个 View 类型的参数，具体写法可以参考下面代码。使用 Java 代码设置监听器与 J2SE 的开发相似，只是监听器名称不同。下面的代码使用的是匿名内部类实现监听器对象，也可以采用其他方法实现监听器对象。

📖 Proj02_1 项目 MainActivity.java 监听按钮点击代码

```java
@Override
protected void onCreate(Bundle savedInstanceState) {
    super.onCreate(savedInstanceState);
    setContentView(R.layout.activity_bt);
    Button bt = (Button) findViewById(R.id.button1);
    //给按钮注册监听器，监听器对象是匿名内部类实现
    bt.setOnClickListener(new OnClickListener(){
        @Override
        public void onClick(View arg0) {
            Log.i("Msg","代码注册的监听器执行!");
        }}
    );
}
//监听 button2 按钮的点击
public void pressBt(View v){
    Log.i("Msg","onClick 属性指定的监听方法!");
}
```

2.2.4 ToggleButton 与 Switch

ToggleButton 是 Button 的子类，主要功能是处理两种状态的点击，又称开关按钮。它具有选中和未选中两种状态，需要给不同状态设置对应显示的文字。常用属性如下：

- android：textOn="开启"，当按钮处于选中状态时显示的文字。
- android：textOff="关闭"，当按钮处于未选中状态时显示的文字。
- android：disabledAlpha="0.1"，当按钮未选中时，按钮的 Alpha 值，取值范围为 0～1。

Switch 也是 Button 的子类，功能与 ToggleButton 类似，显示两种状态，监听器方法相同，都是 android.widget.CompoundButton.OnCheckedChangeListener，二者只是外形不同。Switch 控件是 API 14 中添加的，早期 ADK 中没有此控件。Switch 常用的属性如表 2.2 所示。

表 2.2 Switch 常用属性

属性	说明	取值
android：textOn	开启状态显示的文字	字符串
android：textOff	关闭状态显示的文字	字符串
android：thumb	滑块图片	资源文件 drawable
android：track	滑块下轨道图片	资源文件 drawable
android：checked	是否处于开启状态	true、false

📖 Proj02_1 项目 activity_tgbt.xml 布局文件代码

```xml
<?xml version = "1.0" encoding = "utf-8"?>
<LinearLayout xmlns:android = "http://schemas.android.com/apk/res/android"
    android:layout_width = "match_parent"
    android:layout_height = "match_parent"
    android:orientation = "vertical" >
    <ToggleButton
        android:id = "@+id/toggleButton1"
        android:layout_width = "wrap_content"
        android:layout_height = "wrap_content"
        android:textOn = "打开"
        android:textOff = "关闭"/>
    <Switch
        android:id = "@+id/sw1"
        android:layout_width = "wrap_content"
        android:layout_height = "wrap_content"
        android:thumb = "@drawable/tb"
        android:textOn = "开启"
        android:textOff = "关闭"/>
</LinearLayout>
```

状态按钮运行时的效果如图 2.4 所示。

图 2.4 状态按钮的开关样式

2.2.5 EditText

EditText 控件是 TextView 的子类，主要功能是供用户输入数据。不同类型的文本内容对应不同的文本编辑框，如 Plain Text(普通文本框)、Password(密码输入框)、E-mail(电子邮箱输入框)、Postal Address(通信地址输入框)、Number(数字输入框)等。这些输入框都对应 Android Widget 包中的 EditText 控件，这些类型的划分由属性 android:inputType 决定，其取值类型详见表 2.3。

表 2.3 EditText 控件 inputType 属性取值

android:inputType 取值	说　明	取值类型
android:inputType="number"	输入内容为数字，虚拟键盘切换为数字输入模式	数值
android:inputType="numberDecimal"	输入内容为带小数点的浮点格式	
android:inputType="phone"	输入内容为电话号码，虚拟键盘切换为拨号键盘	
android:inputType="datetime"	输入内容为日期时间	
android:inputType="date"	输入内容为日期键盘	
android:inputType="textCapWords"	输入文本时词首字母大写	字符串
android:inputType="textCapSentences"	输入文本时句首字母大写	
android:inputType="textAutoCorrect"	输入文本时自动更正	
android:inputType="textAutoComplete"	输入文本时自动完成	
android:inputType="textMultiLine"	可以多行输入	
android:inputType="textUri"	输入内容为网址，虚拟键盘会显示 www、com 等内容	
android:inputType="textEmailAddress"	输入内容为电子邮件地址，虚拟键盘会显示@符号	
android:inputType="textPostalAddress"	输入内容为地址	
android:inputType="textPassword"	密码输入框	
android:inputType="textVisiblePassword"	可见密码输入框	

下面的布局文件演示 EditText 控件的几种常用格式。AutoCompleteTextView 是 EditText 的子类，设置适配器后，可以实现自动筛选输入的功能(类似于百度输入框中的自动完成功能)。适配器的应用将在后续章节详细说明。

📖 Proj02_1 项目 activity_et.xml 布局文件代码

```xml
<?xml version = "1.0" encoding = "utf-8"?>
<LinearLayout xmlns:android = "http://schemas.android.com/apk/res/android"
    android:layout_width = "match_parent"
    android:layout_height = "match_parent"
    android:orientation = "vertical" >
    <EditText
        android:id = "@+id/editText1"
        android:layout_width = "match_parent"
        android:layout_height = "wrap_content"
        android:ems = "10" >
        <requestFocus />
```

```
</EditText>
<EditText
    android:id = "@+id/editText2"
    android:layout_width = "match_parent"
    android:layout_height = "wrap_content"
    android:ems = "10"
    android:inputType = "textPassword" />
<EditText
    android:id = "@+id/editText3"
    android:layout_width = "match_parent"
    android:layout_height = "wrap_content"
    android:ems = "10"
    android:inputType = "phone" />
<EditText
    android:id = "@+id/editText4"
    android:layout_width = "match_parent"
    android:layout_height = "wrap_content"
    android:ems = "10"
    android:inputType = "date" />
<AutoCompleteTextView
    android:id = "@+id/autoCompleteTextView1"
    android:layout_width = "match_parent"
    android:layout_height = "wrap_content"
    android:ems = "10"
    android:text = "" />
```

设置 inputType 属性可以实现自动切换输入法,根据输入框内容的不同,输入法自动切换为数字输入或文本输入,如图 2.5 所示,当输入电话号码时,输入法切换为数字键盘。

图 2.5　设置了适配器的输入框效果

2.2.6 CheckBox

CheckBox 是 CompoundButton 类的子类，是 Button 类的间接子类，主要功能是提供复选框。它具有 OnClickListener 监听器和 OnCheckedChangeListener 监听器，通过 isChecked 方法判断该按钮是否处于选中状态。属性 android:checked 设置复选框的默认值，android:text 设置复选框的提示文字。由于复选框的使用比较简单，不再举例演示。

2.2.7 RadioButton 与 RadioGroup

RadioButton 是 CompoundButton 类的子类，是 Button 类的间接子类，主要功能是单选按钮。如果多个单选按钮中只允许选择一个，它们需要放在 RadioGroup 中。RadioGroup 是单选按钮组，继承自 LinearLayout 类，本身不能直接使用，需要与 RadioButton 配合使用，用于将 RadioButton 聚合成一组。

在监听单选按钮时，通常是监听 RadioGroup，而不是 RadioButton，设置的监听方法是 RadioGroup.OnCheckedChangeListener。下面的布局文件演示单选按钮成组的操作。

📖 Proj02_1 项目 activity_rdbt.xml 布局文件代码

```xml
<LinearLayout xmlns:android = "http://schemas.android.com/apk/res/android"
    android:layout_width = "match_parent"
    android:layout_height = "match_parent"
    android:orientation = "horizontal" >
    <TextView
        android:id = "@ + id/textView1"
        android:layout_width = "wrap_content"
        android:layout_height = "wrap_content"
        android:textSize = "20sp"
        android:text = "性别：" />
    <RadioGroup
        android:id = "@ + id/rg01"
        android:layout_width = "wrap_content"
        android:layout_height = "wrap_content"
        android:orientation = "horizontal">
        <RadioButton
            android:id = "@ + id/radioButton1"
            android:layout_width = "wrap_content"
            android:layout_height = "wrap_content"
            android:text = "男" />
        <RadioButton
            android:id = "@ + id/radioButton2"
            android:layout_width = "wrap_content"
            android:layout_height = "wrap_content"
            android:text = "女" />
    </RadioGroup>
</LinearLayout>
```

给 RadioGroup 设置监听器时，为避免歧义（Android 开发中有很多监听器类名一样，但实际上是不同的类），推荐增加所外部类或外部接口，如 RadioGroup.OnCheckedChangeListener。

onCheckedChanged 方法中参数 id 是被选中单选按钮的 id。

📖 Proj02_1 项目 MainActivity.java 添加单选按钮监听器代码

```
@Override
protected void onCreate(Bundle savedInstanceState) {
    super.onCreate(savedInstanceState);
    setContentView(R.layout.activity_rdbt);
    RadioGroup rg = (RadioGroup) findViewById(R.id.rg01);
    rg.setOnCheckedChangeListener(new RadioGroup.OnCheckedChangeListener() {
        @Override
        public void onCheckedChanged(RadioGroup gp, int id) {
            //获取选中单选按钮
            RadioButton rb = (RadioButton) findViewById(id);
            Toast.makeText(MainActivity.this, "当前选中：" + rb.getText(),
                    Toast.LENGTH_LONG).show();
        }
    });
}
```

运行程序，处于单选按钮组中的单选按钮智能地选中一个，每次改变选中状态，都会触发监听器方法，抛出 Toast 提示，如图 2.6 所示。

图 2.6 单选按钮组

2.2.8 SeekBar

SeekBar(拖动条)是 ProgressBar(进度条)的间接子类，ProgressBar 是 View 的子类。拖动条拥有进度条的属性方法，有一个拖动柄，可以拖动，拖动时的监听器是 SeekBar.OnSeekBarChangeListener。属性 android:max 设置最大值，android:progress 设置默认取值,android:thumb 设置拖动柄的图片。

📖 Proj02_1 项目 MainActivity.java,添加拖动条监听器代码

```
@Override
protected void onCreate(Bundle savedInstanceState) {
    super.onCreate(savedInstanceState);
    setContentView(R.layout.activity_seekbar);
    SeekBar sb = (SeekBar) findViewById(R.id.seekBar1);
    final TextView tv = (TextView) findViewById(R.id.seekbarMsg);
    sb.setOnSeekBarChangeListener(new OnSeekBarChangeListener() {
        @Override
        public void onStopTrackingTouch(SeekBar seekBar) {}
        @Override
        public void onStartTrackingTouch(SeekBar seekBar) {}
        @Override
        public void onProgressChanged(SeekBar seekBar, int progress, boolean fmUser) {
            tv.setText("当前取值是: " + progress);
        }
    });
}
```

监听器 OnSeekBarChangeListener 有 3 个方法,分别对应拖动开始、结束和拖动中。比较常用的是 onProgressChanged 方法,progress 参数是拖动条被修改的取值。

2.2.9 RatingBar

RatingBar 的继承关系与 SeekBar 一样,它是 SeekBar 和 ProgressBar 的一种扩展,默认使用星形图片来显示等级评定,触摸、拖动或使用键来设置评分。

RatingBar 有 3 种样式,使用 RatingBar 的默认样式时,用户可以触摸、拖动修改取值。另外两种样式是 ratingBarStyleSmall 和 ratingBarStyleIndicator,它们只适合指示,不能修改取值。

- style="?android:attr/ratingBarStyleSmall",指定为小尺寸 RatingBar。
- style="?android:attr/ratingBarStyleIndicator",指定为大尺寸 RatingBar。

RatingBar 控件的属性见表 2.4。

表 2.4 RatingBar 控制的属性

属性	说明	取值
android:isIndicator	是否只作为指示器使用,指示器不能修改取值	true、false
android:numStars	星的个数	
android:rating	默认取值	
android:stepSize	允许修改的最小单位,取值 0.5 可设置半颗星	

RatingBar 的监听器是 RatingBar.OnRatingBarChangeListener,它只有一个方法: onRatingChanged(RatingBar ratingBar, float rating, boolean fromUser),参数 rating 表示修改时的值。下面的代码是 3 种不同样式的 RatingBar 的使用。

📖 Proj02_1 项目 activity_ratingbar.xml,3 种样式的 RatingBar

```xml
<LinearLayout xmlns:android="http://schemas.android.com/apk/res/android"
    android:layout_width="match_parent"
    android:layout_height="match_parent"
    android:orientation="vertical" >
    <RatingBar
        android:id="@+id/ratingBar1"
        android:layout_width="wrap_content"
        android:layout_height="wrap_content"
        android:numStars="5"
        android:rating="2.5"
        android:stepSize="0.5" />
    <RatingBar
        android:id="@+id/ratingBar2"
        style="?android:attr/ratingBarStyleSmall"
        android:layout_width="wrap_content"
        android:layout_height="wrap_content"
        android:numStars="10"
        android:rating="5" />
    <RatingBar
        android:id="@+id/ratingBar3"
        style="?android:attr/ratingBarStyleIndicator"
        android:layout_width="wrap_content"
        android:layout_height="wrap_content"
        android:numStars="6"
        android:rating="3.6"
        android:stepSize="1.2"/>
</LinearLayout>
```

运行项目,ratingBar2 和 ratingBar3 是无法修改取值的,只能作为取值的指示器使用。stepSize 的取值决定了最小改动单元,ratingBar1 的 stepSize 是 0.5,所以可以取半颗星值;ratingBar3 的 stepSize 是 1.2,所以 3.6 的显示是 3 颗星和大半颗星,如图 2.7 所示。作为指示器使用时,rating 的取值应是 stepSize 的整数倍。

图 2.7 RatingBar 的样式

2.2.10 ProgressBar

ProgressBar 继承自 View,表示进度条,通常用于向用户显示一个耗时操作的进度,降低用户的焦虑感,改善用户体验。Android 系统中的进度条通过 style 属性划分为不同的样

式,该属性的取值如下:
- Widget.ProgressBar.Horizontal,水平进度条样式。
- Widget.ProgressBar.Small,环形进度条,小尺寸。
- Widget.ProgressBar.Large,环形进度条,大尺寸。
- Widget.ProgressBar.Inverse,环形进度条,普通尺寸。

ProgressBar 的常用属性如表 2.5 所示。

表 2.5 ProgressBar 常用属性

属 性	说 明	取 值
android:max	进度条的最大值	数值
android:progress	当前主进度值	数值
android:secondaryProgress	当前次进度值	数值
android:progressDrawable	进度条背景	drawable 对象

下面的布局文件演示了如何设置 4 种样式的 ProgressBar。

📖 Proj02_1 项目 activity_progressbar.xml,ProgressBar 控件

```xml
<LinearLayout xmlns:android = "http://schemas.android.com/apk/res/android"
    android:layout_width = "match_parent"
    android:layout_height = "match_parent"
    android:orientation = "vertical" >
    <ProgressBar
        android:id = "@ + id/progressBar1"
        style = "?android:attr/progressBarStyleLarge"
        android:layout_width = "wrap_content"
        android:layout_height = "wrap_content" />
    <ProgressBar
        android:id = "@ + id/progressBar2"
        android:layout_width = "wrap_content"
        android:layout_height = "wrap_content" />
    <ProgressBar
        android:id = "@ + id/progressBar3"
        style = "?android:attr/progressBarStyleSmall"
        android:layout_width = "wrap_content"
        android:layout_height = "wrap_content" />
    <ProgressBar
        android:id = "@ + id/progressBar4"
        style = "?android:attr/progressBarStyleHorizontal"
        android:layout_width = "match_parent"
        android:max = "100"
        android:progress = "70"
        android:secondaryProgress = "50"
        android:layout_height = "wrap_content" />
</LinearLayout>
```

设置进度值和读取进度值操作仅对水平进度条起作用,环形进度条没有进度值。上述

布局文件的效果如图 2.8 所示。在 2.8 节中将会介绍如何使用线程修改进度条取值。

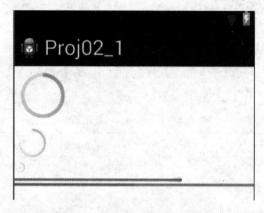

图 2.8　ProgressBar 运行效果

2.3　布局管理器

Android 布局管理器都继承自 ViewGroup 类，间接继承 View 类，用于存放控件或嵌套其他布局。Android 提供了 6 种布局管理器，分别是线性布局（LinearLayout）、相对布局（RelativeLayout）、表格布局（TableLayout）、帧布局（FrameLayout）、绝对布局（AbsoluteLayout）和网格布局（GridLayout）。

2.3.1　LinearLayout

线性布局是开发者最常用的布局管理器之一。放入其中的控件按照垂直或水平方向来排列。纵向布局时，整个容器是一列，其中的元素从上到下分布，一行只能放置一个元素（因为布局管理器中既可以放置控件，又可以嵌套布局，因此本书将放入布局管理器的内容都称为元素）。横向布局时，整个容器是一行，其中的元素默认从左到右分布，一列上只能放置一个元素。内容排列到窗体边缘后，后面的元素将会被掩盖，不会显示出来。

通过属性 android：orientation 设定线性的方向，取值 vertical 为垂直线性，取值 horizontal 为水平线性。线性布局的 android：gravity 属性可以设定其中元素的对齐方式。与 LinearLayout 相关的常用属性如表 2.6 所示。

表 2.6　LinearLayout 常用属性

属　性	说　明	取　值
android：gravity	设置元素的对齐方式	top、bottom、left、right、center_vertical、center_Horizontal 等
android：orientation	设置线性方向	vertical、horizontal
android：layout_gravity	元素属性，对齐方式	
android：layout_weight	元素，表示所占位置的权重	

下面的布局代码演示 LinearLayout 的使用，外层 LinearLayout 采用竖直布局，内层 LinearLayout 采用水平布局。

📖 Proj02_1 项目 activity_linear.xml，线性布局

```xml
<LinearLayout xmlns:android = "http://schemas.android.com/apk/res/android"
    android:layout_width = "match_parent"
    android:layout_height = "match_parent"
    android:orientation = "vertical" >
    <Button
        android:layout_width = "wrap_content"
        android:layout_height = "wrap_content"
        android:layout_gravity = "right"
        android:text = "Button1" />
    <LinearLayout
        android:layout_width = "match_parent"
        android:layout_height = "wrap_content"
        android:orientation = "horizontal">
        <Button
            android:layout_width = "0dp"
            android:layout_height = "wrap_content"
            android:layout_weight = "1"
            android:text = "Button2" />
        <Button
            android:layout_width = "0dp"
            android:layout_height = "wrap_content"
            android:layout_weight = "2"
            android:text = "Button3" />
    </LinearLayout>
</LinearLayout>
```

当一个元素位于线性布局中时，能够获得对应的额外属性，如 Button1 的属性取值 android：layout_gravity＝"right"，实现右对齐的效果。内层 LinearLayout 中，两个按钮设定的宽度为 0dp，使用权重分配整个水平空间，运行效果如图 2.9 所示。

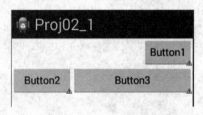

图 2.9 线性布局效果

2.3.2 RelativeLayout

相对布局管理器的布局效果最为丰富。元素可以使用它们彼此之间的位置关系，确定摆放位置。相对布局中指明一个元素的布局位置要以其他元素为参考，元素之间相互关联，如果有一个元素改变其位置，以它为参考的元素位置都会发生改变。每一个元素都要定义 id 值，以便于其他元素引用。使用相对布局管理器，其中的元素会增加很多额外属性，用于定义它们的位置，这些常用属性如表 2.7 所示。

表 2.7　RelativeLayout 常用属性

属　性	说　明	取　值
android：layout_below	在某元素的下方	
android：layout_above	在某元素的上方	
android：layout_toLeftOf	在某元素的左边	
android：layout_toRightOf	在某元素的右边	
android：layout_alignTop	本元素的上边缘和某元素的上边缘对齐	属性值为其他控件的 id 引用，形如@id/id-name
android：layout_alignLeft	本元素的左边缘和某元素的左边缘对齐	
android：layout_alignBottom	本元素的下边缘和某元素的下边缘对齐	
android：layout_alignRight	本元素的右边缘和某元素的右边缘对齐	
android：layout_centerHrizontal	水平居中	
android：layout_centerVertical	垂直居中	
android：layout_centerInparent	相对于父元素完全居中	
android：layout_alignParentBottom	贴紧父元素的下边缘	取值为 true 或 false
android：layout_alignParentLeft	贴紧父元素的左边缘	
android：layout_alignParentRight	贴紧父元素的右边缘	
android：layout_alignParentTop	贴紧父元素的上边缘	
android：layout_marginBottom	底边缘的距离	
android：layout_marginLeft	左边缘的距离	取值为具体的像素值，如 5px
android：layout_marginRight	右边缘的距离	
android：layout_marginTop	上边缘的距离	

这些属性可以划分为 4 类：位置属性、对齐属性、与父容器位置关系属性和边缘距离属性。下面的布局内容是使用相对布局管理器实现登录界面。

📖 Proj02_1 项目 activity_relative.xml，相对布局

```
< RelativeLayout xmlns:android = "http://schemas.android.com/apk/res/android"
    android:layout_width = "match_parent"
    android:layout_height = "match_parent" >
< TextView
    android:id = "@ + id/rl_tv01"
    android:layout_width = "wrap_content"
    android:layout_height = "wrap_content"
    android:text = "账号："
    android:textSize = "20sp"/>
< EditText
    android:id = "@ + id/rl_et01"
    android:layout_width = "match_parent"
    android:layout_height = "wrap_content"
    android:layout_toRightOf = "@id/rl_tv01"/>
```

```xml
<TextView
    android:id = "@+id/rl_tv02"
    android:layout_width = "wrap_content"
    android:layout_height = "wrap_content"
    android:layout_below = "@id/rl_et01"
    android:textSize = "20sp"
    android:text = "密码："/>
<EditText
    android:id = "@+id/rl_et02"
    android:layout_width = "match_parent"
    android:layout_height = "wrap_content"
    android:layout_toRightOf = "@id/rl_tv02"
    android:layout_below = "@id/rl_et01"/>
<Button
    android:id = "@+id/rl_bt01"
    android:layout_width = "220dp"
    android:layout_height = "wrap_content"
    android:text = "登录"
    android:layout_centerHorizontal = "true"
    android:layout_below = "@id/rl_et02"
    android:layout_marginTop = "5dp"/>
<CheckBox
    android:id = "@+id/rl_ck01"
    android:layout_width = "wrap_content"
    android:layout_height = "wrap_content"
    android:text = "记住登录"
    android:textSize = "16sp"
    android:checked = "true"
    android:layout_below = "@id/rl_bt01"
    android:layout_alignLeft = "@id/rl_bt01"/>
<TextView
    android:id = "@+id/rl_tv03"
    android:layout_width = "wrap_content"
    android:layout_height = "wrap_content"
    android:text = "快捷注册"
    android:textSize = "16sp"
    android:layout_toRightOf = "@id/rl_ck01"
    android:layout_alignBaseline = "@id/rl_ck01"
    android:layout_marginLeft = "56dp"/>
</RelativeLayout>
```

通过设置位置属性、对齐属性就可以确定布局。相对布局管理器的实现方式非常丰富，同一种布局有多种参考方案可选。例如上述布局中的密码框 rl_et02，定义其布局的一种方式是设置其位置在账号框 rl_et01 下方，并在文本框 rl_et02 的左侧；另一种定义方式是设置其位置在账号框 rl_et01 下方，并左侧与 rl_et01 对齐。相对布局管理器的参考位置比较灵活，每个位置都可以有多种参考标准。上述布局代码的运行效果如图 2.10 所示。

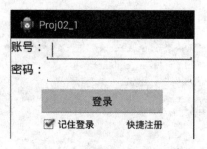

图 2.10 相对布局效果

2.3.3 FrameLayout

帧布局管理器的布局单一,所有放入其中的元素都以左上角为参考,对齐父容器的左边和顶部,后放入的元素将覆盖在上面。帧布局最大的特点就是元素会叠加显示,每添加一个元素,Android 系统都会将其作为一个独立的坐标页面,通常称为显示的一帧。帧布局的各个控件在 Android 系统中都对应一个独立的坐标系统。

2.3.4 GridLayout

网格布局管理器是 Android 4.0 新添加的布局管理器,它使用行列管理放入其中的元素,每个元素都放入指定单元格,同时支持跨行、跨列,与 HTML 中的 Table 标签类似。GridLayout 常用属性如表 2.8 所示。

表 2.8 GridLayout 常用属性

属 性	说 明	取 值
android:columnCount	网格的列数	数值
android:rowCount	网格的行数	数值
android:layout_column	元素属性,所在列(从 0 开始)	
android:layout_columnSpan	元素属性,跨几列	
android:layout_gravity	元素属性,在单元格中的对齐方式	top、bottom、left、right、fill 等
android:layout_row	元素属性,所在行(从 0 开始)	
android:layout_rowSpan	元素属性,跨几行	

下面的布局文件定义了一个 5 行 4 列的网格布局,并指定第一行(下标为 0)元素是 EditText,跨 4 列。

📖 Proj02_1 项目 activity_gridlayout.xml.xml,相对布局

```
<GridLayout xmlns:android="http://schemas.android.com/apk/res/android"
    android:layout_width="match_parent"
    android:layout_height="match_parent"
    android:rowCount="5"
    android:columnCount="4"
    android:id="@+id/gridlayout">
    <EditText
        android:id="@+id/editText1"
```

```
        android:layout_width = "match_parent"
        android:layout_height = "wrap_content"
        android:layout_column = "0"
        android:layout_row = "0"
        android:layout_columnSpan = "4"
        android:ems = "10" >
        <requestFocus />
    </EditText>
</GridLayout>
```

下面的代码使用 Java 代码动态创建控件,添加到网格布局。网格布局在添加元素时使用 GridLayout.LayoutParams 指定布局参数,GridLayout.spec(i/4+1) 和 GridLayout.spec(i%4) 可以计算元素所在行列值,行号加 1 是因为第一行有 EditText 控件。

📖 Proj02_1 项目 MainActivity.java,使用代码创建控件

```
@Override
protected void onCreate(Bundle savedInstanceState) {
    super.onCreate(savedInstanceState);
    setContentView(R.layout.activity_gridlayout);
    GridLayout gl = (GridLayout) findViewById(R.id.gridlayout);
    //动态创建控件
    char flags[] = {'7','8','9','+','4','5','6','-','1','2','3','*','0','.','=','/'};
    for (int i = 0; i < flags.length; i++) {
        Button bt = new Button(this);
        bt.setText(String.valueOf(flags[i]));
        //设置参数,并添加到布局管理器
        GridLayout.LayoutParams params = new GridLayout.LayoutParams(
                GridLayout.spec(i/4 + 1), GridLayout.spec(i % 4));      //计算行列值
        gl.addView(bt, params);
    }
}
```

运行代码可以看到如图 2.11 所示的效果。

图 2.11 网格布局效果

2.3.5 TableLayout

表格布局是以行为单位的布局方式，它不需要指明有多少行、多少列，通过添加 TableRow 或其他元素动态确定行数和列数。每次向 TableLayout 添加一个 TableRow，就是一个表格行，TableRow 是容器，其他元素可以添加进来。除了使用 TableRow 增加一行，TableLayout 还允许直接添加元素，该元素会占用一行。TableLayout 的常用属性如表 2.9 所示。

表 2.9 TableLayout 属性

属性	说明	取值
android:shrinkColumns	允许收缩的列，多个列号用逗号隔开	列号
android:stretchColumns	允许拉伸的列，多个列号用逗号隔开	列号
android:collapseColumns	允许被隐藏的列	列号
android:layout_column	元素属性，所处列	
android:layout_span	元素属性，跨越列数	

下面的布局代码使用 TableLayout，共有 3 行，指定了收缩列和拉伸列，下标从 0 开始。第一行使用 TableRow，内有 3 个元素；第二行是独立元素，占用一行；第三行使用 TableRow，其中第二个元素跨两列。

📖 Proj02_1 项目 activity_tablelayout.xml，表格布局

```
<TableLayout xmlns:android="http://schemas.android.com/apk/res/android"
    android:layout_width="match_parent"
    android:layout_height="match_parent"
    android:stretchColumns="1"
    android:shrinkColumns="0" >
    <TableRow
        android:id="@+id/tableRow1"
        android:layout_width="wrap_content"
        android:layout_height="wrap_content" >
        <Button
            android:layout_width="wrap_content"
            android:layout_height="wrap_content"
            android:text="Button"/>
        <Button
            android:layout_width="wrap_content"
            android:layout_height="wrap_content"
            android:text="Button"/>
        <Button
            android:layout_width="wrap_content"
            android:layout_height="wrap_content"
            android:text="Button"/>
    </TableRow>
    <Button
        android:layout_width="wrap_content"
        android:layout_height="wrap_content"
```

```
                android:text = "独占一行的单一控件"/>
        <TableRow
                android:id = "@ + id/tableRow2"
                android:layout_width = "wrap_content"
                android:layout_height = "wrap_content" >
            <Button
                android:layout_width = "wrap_content"
                android:layout_height = "wrap_content"
                android:text = "内容少"/>
            <Button
                android:layout_width = "wrap_content"
                android:layout_height = "wrap_content"
                android:layout_span = "2"
                android:text = "占两列的按钮"/>
        </TableRow>
</TableLayout >
```

上述布局文件的效果如图 2.12 所示。

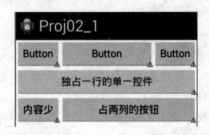

图 2.12　表格布局效果

2.3.6　AbsoluteLayout

绝对布局管理器使用屏幕坐标系规定元素位置,从 API 3 开始不再鼓励使用,建议以 FrameLayout 和 RelativeLayout 代替。以屏幕左上角为坐标原点(0,0),水平向右为 X 轴正方向,垂直向下为 Y 轴正方向。绝对布局管理器中的元素会获得 android:layout_x 属性和 android:layout_y 属性,用于定义元素左上角顶点坐标。

绝对布局管理不会根据屏幕尺寸调整布局效果,无法自适应屏幕尺寸,使用场景比较少,在此不作过多介绍。

2.4　使用图片控件

上面介绍的很多 Android 控件都可以通过指定 android:background 属性设置背景图片,但主要用于显示图片的控件是 ImageView。ImageView 继承 View 类,它的子类有 ImageButton、QuickContactBadge。

2.4.1　ImageView

ImageView 不但可以显示图片,还可以显示任何 Drawable 对象。ImageView 设置图片资源的属性是 android:src。其常用属性如表 2.10 所示。

表 2.10 ImageView 常用属性

属 性	说 明	取 值
android:adjustViewBounds	是否保持宽高比，与 maxWidth、MaxHeight 一起使用	true、false
android:maxHeight android:maxWidth	最大宽高，与 setAdjustViewBounds 一起使用	数值
android:scaleType	图片的缩放模式	matrix、fitXY、fitStart、fitCenter、fitEnd、center、centerCrop、centerInside
android:src	设置 View 的 drawable	Drawable 对象

ImageView 在设置显示内容时可以采用下面 4 种方法：

void setImageBitmap(Bitmap bm)

void setImageDrawable(Drawable drawable)

void setImageResource(int resId)

void setImageURI(Uri uri)

2.4.2 ImageButton

图片按钮继承 ImageView，其本质是 ImageView 的特例，但具备按钮按下、抬起的状态，不过它不能像 Button 那样设置 android:text 属性。ImageButton 常与 selector 资源一起使用，使其根据按下、抬起状态切换不同的图片。如何创建 selector 资源文件会在 4.5.3 节介绍。

2.5 使用复杂控件

复杂控件不同于单一控件的使用，它需要数据的填充和样式的设定，遵循 MVC 的思想，把数据模型和呈现分离。本节介绍的复杂控件都是从 AdapterView 类继承来的，继承关系如图 2.13 所示。

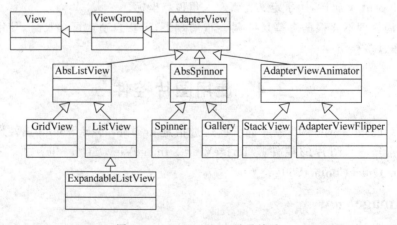

图 2.13 AdapterView 继承关系

2.5.1 数据适配器

数据适配器顾名思义就是把不同格式的数据(数组或集合等)按统一格式填充给控件,使得这些控件在呈现数据时不用关心数据的差异性。经常用到的适配器类有 ArrayAdapter、SimpleAdapter、SimpleCursorAdapter 和 BaseAdapter,它们的继承关系如图 2.14 所示。

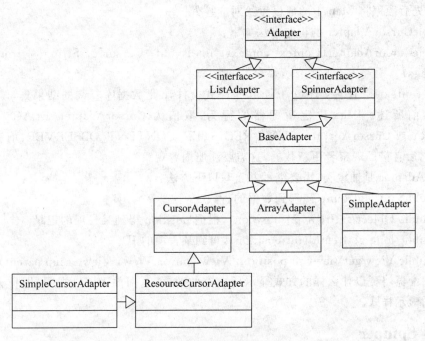

图 2.14 Adapter 继承关系

不同的数据适配器在填充数据和设置样式时不同。ArrayAdapter 填充数组或 List 集合数据,常用于显示文本列表。SimpleAdapter 填充 List<? extends Map<String,?>>格式的数据,可以显示更加丰富的列表。SimpleCursorAdapter 填充 Cursor(数据库查询结果集)数据。BaseAdapter 是抽象类,需要被继承,自定义填充数据方式和显示方式。

ArrayAdapter 常用构造方法如下:

ArrayAdapter(Context context, int resource, int textViewResourceId)

ArrayAdapter(Context context, int resource, T[] objects)

ArrayAdapter(Context context, int resource, int textViewResourceId, T[] objects)

ArrayAdapter(Context context, int resource, List<T> objects)

ArrayAdapter(Context context, int resource, int textViewResourceId, List<T> objects)

参数 context 指上下文环境。resource 是适配器的样式文件,可以使用"R. layout. 布局文件"引用项目中的布局文件,使用"android. R. layout. 布局文件"引用 Android 系统中的布局文件。textViewResourceId 布局文件中的某个控件 ID,适配器中的数据将被设置到该控件。objects 是待填充的数据,可以是数组或 List 集合。数据也可以通过方法 addAll 添加到适配器中。

SimpleAdapter 的构造方法如下：

SimpleAdapter(Context context, List <? extends Map < String, ? >> data, int resource, String[] from, int[] to)

参数 context 是上下文环境。data 是填充数据，每一项数据对应一个 Map，封装多个信息，如一项数据中可能包含文本信息、图片信息等。resource 是样式文件。from 和 to 两个参数共同决定数据的映射关系，from 是 Map 中键的数组，to 是样式文件中控件 ID 的数组，二者按照先后顺序将 Map 中的值设置给对应控件。

SimpleCursorAdapter 的构造方法如下：

SimpleCursorAdapter(Context context, int layout, Cursor c, String[] from, int[] to, int flags)

参数 context 是上下文环境，layout 是样式文件，c 是数据库查询的结果集，from 和 to 的作用如前所述。flags 决定适配器的行为，取值为 CursorAdapter.FLAG_AUTO_REQUERY 或 CursorAdapter.FLAG_REGISTER_CONTENT_OBSERVER，前者会重新请求数据，后者借助观察者更新数据，这需要注册内容观察者。

BaseAdapter 是抽象类，需要继承并重写以下方法：

- public int getCount()，得到数据的行数。
- public Object getItem(int position)，根据 position 得到某一项的记录。
- public long getItemId(int position)，得到某一项的 ID。
- public View getView(int position, View convertView, ViewGroup parent)，定义了适配器将要以什么样的方式显示所填充的数据，返回的 View（视图）即作为每一项的显示样式。

2.5.2 Spinner

Spinner 本质上是一个下拉列表，继承 AbsSpinner，间接继承 AdapterView。用户可以选择列表中的选项，选项数据使用适配器填充。Spinner 常用的属性如表 2.11 所示。

表 2.11 Spinner 常用属性

属性	说明	取值
android:entries	从 AbsSpinner 继承的属性，指定填充数据	资源文件中的数组
android:dropDownHorizontalOffset	水平偏移	数值
android:dropDownVerticalOffset	垂直偏移	数值
android:dropDownWidth	下拉列表框的宽度	数值
android:gravity	列表项的对齐	top、left 等
android:popupBackground	背景颜色	如#99CCDD
android:prompt	对话框模式时的提示信息	文本
android:spinnerMode	下拉列表框的模式，对话框或列表	dialog、dropdown

下面的布局文件包含两个 Spinner。spinner1 设置 android:entries 属性，引用 strings.xml 资源文件中的数组。当 spinnerMode 为 dialog 时，可以设置 prompt 提示信息。spinner2 设置下拉列表框模式，这是默认模式。

📖 Proj02_1 项目 activity_spinner.xml，布局文件

```xml
<LinearLayout xmlns:android = "http://schemas.android.com/apk/res/android"
    android:layout_width = "match_parent"
    android:layout_height = "match_parent"
    android:orientation = "vertical" >
    <Spinner
        android:id = "@+id/spinner1"
        android:layout_width = "match_parent"
        android:layout_height = "wrap_content"
        android:spinnerMode = "dialog"
        android:prompt = "@string/spinnerTitle"
        android:entries = "@array/books"/>
    <Spinner
        android:id = "@+id/spinner2"
        android:layout_width = "match_parent"
        android:layout_height = "wrap_content"
        android:spinnerMode = "dropdown"
        android:dropDownVerticalOffset = "5dp"/>
    <TextView
        android:id = "@+id/spinner_msg"
        android:layout_width = "match_parent"
        android:layout_height = "wrap_content"/>
</LinearLayout>
```

spinner2 的数据采用适配器进行填充。文本数据使用 ArrayAdapter 即可，数据可以存放在数组中或 List 集合中。构造适配器时使用 android.R.layout.simple_list_item_1 布局样式，这是 Android 系统中的布局。监听 Spinner 的选择操作需要使用 OnItemSelectedListener 监听器，Spinner 不支持 OnItemClickListener 监听器。

📖 Proj02_1 项目 MainActivity.java，设置适配器代码

```java
@Override
protected void onCreate(Bundle savedInstanceState) {
    super.onCreate(savedInstanceState);
    setContentView(R.layout.activity_spinner);
    Spinner sp1 = (Spinner) findViewById(R.id.spinner1);
    Spinner sp2 = (Spinner) findViewById(R.id.spinner2);
    final TextView tv = (TextView) findViewById(R.id.spinner_msg);
    final String citys[] = new String[]{"北京","上海","广州"};
    //创建适配器，并设置给 spinner
    ArrayAdapter aa = new ArrayAdapter(this,android.R.layout.simple_list_item_1,citys);
    sp2.setAdapter(aa);
    //添加监听器
    sp2.setOnItemSelectedListener(new OnItemSelectedListener(){
        @Override
        public void onItemSelected(AdapterView<?> parent, View view, int position, long id) {
            tv.setText("选择：" + citys[position]);
        }
        @Override
        public void onNothingSelected(AdapterView<?> parent) {
        }
    });
}
```

运行项目,效果如图 2.15 所示。

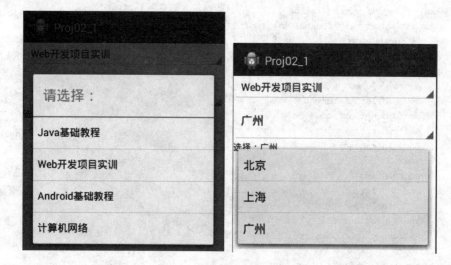

图 2.15　Spinner 运行效果

2.5.3　ListView 与 ListActivity

ListView 是比较常用的一种控件,以垂直列表的方式显示数据。它直接继承 AbsListView,间接继承 AdapterView。使用 ListView 可以直接应用控件,也可以继承 ListActivity(Activity 的子类),借助 setListAdapter(ListAdapter adapter)方法设置适配器。ListView 的常用属性如表 2.12 所示。

表 2.12　ListView 常用属性

属　性	说　明	取　值
android:divider	列表项之间的分隔条	drawable 或颜色
android:dividerHeight	分隔条高度	数值
android:entries	指定填充的数据	引用资源中的数组
android:footerDividersEnabled	底部之前是否绘制分隔条	true、false
android:headerDividersEnabled	头部之前是否绘制分隔条	true、false
android:choiceMode	继承 AbsListView,选择模式	singleChoice:单选 multipleChoice:多选
android:transcriptMode	继承 AbsListView,滚动模式	disabled:关闭滚动 normal:末项可见时滚动 alwaysScroll:自动滚动

1. 使用 ListView 显示单项文本数据

文本数据使用 ArrayAdapter 填充即可,列表项的样式可以自定义布局,也可以使用 Android 系统中的布局。以下 4 种布局样式都是常用的 Android 系统布局,单选、多选和勾选模式的效果如图 2.16 所示。多选和勾选模式必须设定 android:choiceMode 属性,取值为 multipleChoice。

android.R.layout.simple_list_item_1

android.R.layout.simple_list_item_single_choice

android.R.layout.simple_list_item_multiple_choice
android.R.layout.simple_list_item_checked

ListView 的列表项监听器是 AdapterView.OnItemClickListener，当多选模式开启时，可以使用监听器 AbsListView.MultiChoiceModeListener。

图 2.16　ListView 单选、多选和勾选效果

2. 使用 ListView 显示多项图文数据

当列表的每一项都有多个数据填充，且填充的控件各不相同时，需要用到 SimpleAdapter 适配器。该适配器只有一个构造方法，用法比较简单。首先，将多个数据封装到一个 Map 对象中，然后再填充到 List 集合。其次，自定义列表项的布局文件。最后，将 Map 中键对应的值映射到布局文件中的控件上。

📖 Proj02_1 项目 listview_item.xml，自定义列表项布局

```xml
<RelativeLayout xmlns:android = "http://schemas.android.com/apk/res/android"
    android:layout_width = "match_parent"
    android:layout_height = "match_parent">
    <ImageView
        android:id = "@+id/listview_item_img"
        android:layout_width = "60dp"
        android:layout_height = "60dp" />
    <TextView
        android:id = "@+id/listview_item_name"
        android:layout_width = "wrap_content"
        android:layout_height = "wrap_content"
        android:textSize = "16sp"
        android:layout_toRightOf = "@id/listview_item_img"
        android:layout_marginLeft = "5dp" />
    <TextView
        android:id = "@+id/listview_item_cdate"
        android:layout_width = "wrap_content"
        android:layout_height = "wrap_content"
```

```
            android:textSize = "12sp"
            android:layout_below = "@id/listview_item_name"
            android:layout_alignLeft = "@id/listview_item_name"
            android:layout_marginTop = "5dp" />
</RelativeLayout>
```

上面的布局文件决定了列表项的样式,该布局文件中有 3 个控件:ImageView 和两个 TextView,即每一项都需要 3 个数据。下面的代码使用循环创建 Map 对象,每个 Map 中封装 3 个数据,分别使用 img、name、cdate 作为键。在对 img 赋值时采用了 R.drawable.role01+i 的方式,原因是图片资源的命名都是连续的,如 role01.png、role20.png 等,因此资源在 R 中的 id 值也是连续的,可以通过运算获取(注意:id 的取值并不是固定的)。

📖 Proj02_1 项目 MainActivity.java,使用 SimpleAdapter 适配器

```
@Override
protected void onCreate(Bundle savedInstanceState) {
    super.onCreate(savedInstanceState);
    setContentView(R.layout.activity_listview);
    ListView lv = (ListView) findViewById(R.id.listView1);
    //获取资源文件中的数组
    final String roles[] = getResources().getStringArray(R.array.roles);
    //构造数据集合
    List<Map<String,Object>> data = new ArrayList<Map<String,Object>>();
    for (int i = 0; i < roles.length; i++) {
        Map<String,Object> item = new HashMap<String,Object>();
        item.put("img", R.drawable.role01 + i);
        item.put("name", roles[i]);
        item.put("cdate", "创建日期: 2016 - " + (i + 1));
        data.add(item);
    }
    //创建适配器
    SimpleAdapter sa = new SimpleAdapter(this,data,R.layout.listview_item,
            new String[]{"img","name","cdate"},
            new int[]{R.id.listview_item_img,R.id.listview_item_name,
                    R.id.listview_item_cdate});
    lv.setAdapter(sa);
}
```

Map 中的值与布局文件中控件的映射是按照数组元素的先后顺序进行的。img 的值设置给 R.id.listview_item_img,name 的值设置给 R.id.listview_item_name,依次完成控件的赋值。这种赋值是适配器调用 setText、setImageResource 方法自动实现的。如果填充的数据需要通过其他方式设置给控件,如 ProgressBar 需要使用 setProgress,需要继承 BaseAdapter,自定义数据的填充。上述代码的运行效果如图 2.17 所示。

2.5.4 GridView

GridView 与 ListView 并列,作为 AbsListView 的直接子类。与 ListView 垂直显示数

图 2.17 SimpleAdapter 填充效果

据项不同的是,GridView 按照二维表格显示数据项。除了数据呈现方式不同,二者的用法基本一致。GridView 控件的常用属性如表 2.13 所示。

表 2.13 GridView 常用属性

属　　性	说　　明	取　　值
android:columnWidth	列的宽度	数值
android:gravity	元素对齐方式	top、left 等
android:horizontalSpacing	元素水平放方向间距	数值
android:numColumns	列数,与 ListView 一样,行数由数据多少决定	数值
android:stretchMode	拉伸模式	none:不拉伸 spacingWidth:拉伸元素间距 columnWidth:拉伸单元格 spacingWidthUniform:全部拉伸
android:verticalSpacing	元素垂直方式间距	数值

下面的代码将 2.5.3 节中的数据按照 GridView 呈现,布局文件 R.layout.activity_gridview 包括 GridView 控件和 ImageView 控件,当单击某一项时,将图片显示在 ImageView 中。创建适配器时使用布局文件 R.layout.gridview_item 作为每一项的样式。

📖 Proj02_1 项目 MainActivity.java,GridView 控件的使用

```java
@Override
    protected void onCreate(Bundle savedInstanceState) {
        super.onCreate(savedInstanceState);
        setContentView(R.layout.activity_gridview);      //设置布局文件
        GridView gv = (GridView) findViewById(R.id.gridView1);
        final ImageView iv = (ImageView) findViewById(R.id.gridview_img);
        //获取资源文件中的数组
        final String roles[] = getResources().getStringArray(R.array.roles);
```

```
        //构造数据集合
        List<Map<String,Object>> data = new ArrayList<Map<String,Object>>();
        for (int i = 0; i < roles.length; i++) {
            Map<String,Object> item = new HashMap<String,Object>();
            item.put("img", R.drawable.role01 + i);
            item.put("name", roles[i]);
            data.add(item);
        }
        //创建适配器
        SimpleAdapter sa = new SimpleAdapter(this,data,R.layout.gridview_item,
                new String[]{"img","name"},
                new int[]{R.id.gridview_item_img,R.id.gridview_item_name});
        gv.setAdapter(sa);
        gv.setOnItemClickListener(new OnItemClickListener(){
            @Override
            public void onItemClick(AdapterView<?> parent, View view, int position, long id) {
                iv.setImageResource(R.drawable.role01 + position);
            }
        });
    }
```

GridView 添加监听器时与 ListView 一样，使用 OnItemClickListener。代码运行效果如图 2.18 所示。

图 2.18 GridView 运行效果

2.5.5 ExpandableListView

ExpandableListView 继承 ListView，可以视为带有两级功能的 ListView，即每个列表

项又是一个 ListView。第一级列表称为父列表或组列表，第二级列表称为子列表或成员列表。ExpandableListView 采用的适配器是 ExpandableListAdapter 接口的两个实现类，分别是 BaseExpandableListAdapter 和 SimpleExpandableListAdapter，前者是抽象类，需要继承并实现抽象方法，后者可以直接使用。

SimpleExpandableListAdapter 构造方法的参数非常多，它需要同时适配父列表和子列表。这些参数的含义可以对照 SimpleAdapter 构造方法中的参数，同时注意区分两级。

SimpleExpandableListAdapter(Context context, List <? extends Map < String, ? >> groupData, int groupLayout, String[] groupFrom, int[] groupTo, List <? extends List <? extends Map < String, ? >>> childData, int childLayout, String[] childFrom, int[] childTo)

SimpleExpandableListAdapter (Context context, List <? extends Map < String, ? >> groupData, int expandedGroupLayout, int collapsedGroupLayout, String[] groupFrom, int[] groupTo, List <? extends List <? extends Map < String, ? >>> childData, int childLayout, String[] childFrom, int[] childTo)

SimpleExpandableListAdapter 构造方法的参数比较多，主要是对父级列表和子级列表的数据适配，如同需要适配两个 ListView。下面的代码演示如何填充数据。

📖 Proj02_1 项目 MainActivity.java，ExpandableListView 控件的使用

```java
@Override
protected void onCreate(Bundle savedInstanceState) {
    super.onCreate(savedInstanceState);
    setContentView(R.layout.activity_expandlv);
    ExpandableListView elv = (ExpandableListView) findViewById(R.id.elv01);
    //准备填充数据
    String []provinces = new String[]{"北京","天津","河北","山东"};
    String [][]citys = new String[][]{{"北京"},{"天津"},{"石家庄","保定","廊坊"},
            {"济南","青岛","烟台","临沂","日照"}};
    //父级列表数据
    List < Map < String, Object >> groupData = new ArrayList < Map < String, Object >>();
    for (int i = 0; i < provinces.length; i++) {
        Map < String, Object > group = new HashMap < String, Object >();
        group.put("title", provinces[i]);
        groupData.add(group);
    }
    //子列表数据
    List < List < Map < String, Object >>> childData = new ArrayList < List < Map < String, Object >>>();
    for (int i = 0; i < citys.length; i++) {
        List < Map < String, Object >> childList = new ArrayList < Map < String, Object >>();
        for (int j = 0; j < citys[i].length; j++) {
            Map < String, Object > child = new HashMap < String, Object >();
            child.put("city", citys[i][j]);
            childList.add(child);
        }
```

```
            childData.add(childList);
        }
        //创建适配器
        SimpleExpandableListAdapter sela = new SimpleExpandableListAdapter(this,
                groupData,R.layout.expandlv_groupitem,
new String[]{"title"},new int[]{R.id.expandlv_groupitem_title},
                childData,R.layout.expandlv_childitem,
new String[]{"city"},new int[]{R.id.expandlv_childitem_city});
        //设置适配器
        elv.setAdapter(sela);
}
```

R.layout.activity_expandlv 是 Activity 布局文件,包含一个 ExpandableListView 控件。一维数组 provinces 是父级列表采用的数据,二维数组 citys 是子级列表采用的数据。父级列表的布局文件是 R.layout.expandlv_groupitem,子级列表的布局文件是 R.layout.expandlv_childitem,都只包含一个 TextView 控件,项目运行效果如图 2.19 所示。

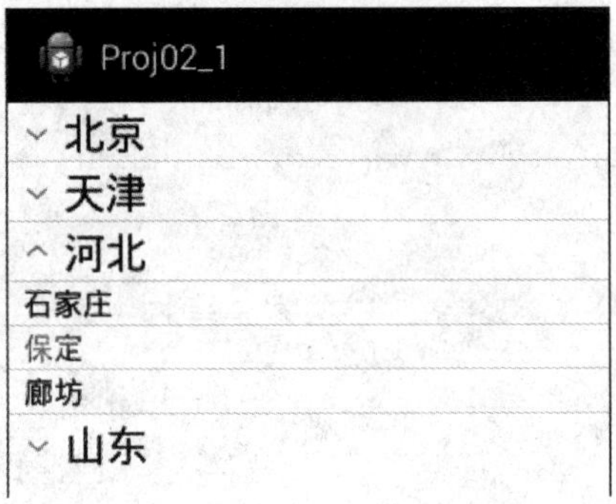

图 2.19 ExpandableListView 运行效果

2.5.6 ScrollView 与 HorizontalScrollView

垂直滚动与水平滚动控件继承 FrameLayout,具有帧布局的特性。ScrollView 只能垂直滚动,HorizontalScrollView 只能水平滚动。两个控件的共同特点是:当其内部控件在一个屏幕上显示不全的时候便会自动产生滚动效果,通过纵向或横向滚动方式来显示被遮盖的内容。

ScrollView 和 HorizontalScrollView 都只能包裹一个控件,所以一般情况下可以在 ScrollView 或 HorizontalScrollView 内部添加一个 LinearLayout 或者其他布局,然后再在这个布局内部添加各种控件,以便设计比较复杂的布局效果。

2.6 高级控件

本节介绍的控件都带有相应的视图切换、动画效果,可以用作一级或二级导航。这些控件在不同版本的 SDK 中会有变化,或随着版本的升级被其他实现类(如 Fragment)替换。它们的继承关系如图 2.20 所示。

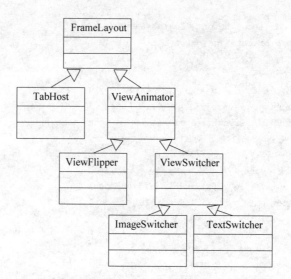

图 2.20 FrameLayout 部分子类继承关系

2.6.1 TabHost

TabHost 是 Tab(选项卡)的容器,包括两部分:TabWidget 和 FrameLayout。其中 TabWidget 就是每个 Tab 的标签区域,可以通过单击切换不同的 Tab 内容视图; FrameLayout 是 Tab 内容视图,与对应标签匹配。构建 TabHost 有两种方式,一种是在布局文件中直接使用 TabHost 控件,另一种是继承 TabAcitivty,获取 TabHost 控件。继承 TabActivity 使用 TabHost 时,TabHost 的 id 应设置为@android:id/tabhost,TabWidget 的 id 应设置为@android:id/tabs,FrameLayout 的 id 应设置为@android:id/tabcontent。直接使用 TabHost 控件,其 id 可以自定义,但标签和内容区域 id 仍需按照上面的规则命名。

下面的布局代码是在布局文件的 Graphic Layout 视图中将 TabHost 拖放到布局文件中,并对标签内容区域进行调整后的结果。TabHost 作为一种组合控件,包含两个主要的组成部分:android:id/tabs 和 android:id/tabcontent。前者对应标签区域,后者对应内容视图区域。代码中的 3 个 LinearLayout 布局位于一个 FrameLayout 布局中,意味着根据标签的切换,同一时刻只能显示一个 LinearLayout。

📖 Proj02_1 项目 activity_tabhost.xml,TabHost 控件布局

```
< LinearLayout xmlns:android = "http://schemas.android.com/apk/res/android"
    android:layout_width = "match_parent"
    android:layout_height = "match_parent"
    android:orientation = "vertical" >
    < TabHost
        android:id = "@android:id/tabhost"
        android:layout_width = "match_parent"
        android:layout_height = "match_parent" >
        < LinearLayout
            android:layout_width = "match_parent"
```

```xml
        android:layout_height = "match_parent"
        android:orientation = "vertical" >
    <!-- 标签区域 -->
    <TabWidget
        android:id = "@android:id/tabs"
        android:layout_width = "match_parent"
        android:layout_height = "wrap_content" >
    </TabWidget>
    <!-- 内容视图区域 -->
    <FrameLayout
        android:id = "@android:id/tabcontent"
        android:layout_width = "match_parent"
        android:layout_height = "match_parent" >
        <LinearLayout
            android:id = "@+id/tab1"
            android:layout_width = "match_parent"
            android:layout_height = "match_parent"
            android:background = "#CC6666"
            android:orientation = "vertical" >
            <TextView
                android:layout_width = "wrap_content"
                android:layout_height = "wrap_content"
                android:text = "标签1,内容视图"/>
        </LinearLayout>
        <LinearLayout
            android:id = "@+id/tab2"
            android:layout_width = "match_parent"
            android:layout_height = "match_parent"
            android:background = "#66CC66"
            android:orientation = "vertical" >
            <TextView
                android:layout_width = "wrap_content"
                android:layout_height = "wrap_content"
                android:text = "标签2,内容视图"/>
        </LinearLayout>
        <LinearLayout
            android:id = "@+id/tab3"
            android:layout_width = "match_parent"
            android:layout_height = "match_parent"
            android:background = "#6666CC"
            android:orientation = "vertical" >
        </LinearLayout>
    </FrameLayout>
    </LinearLayout>
</TabHost>
</LinearLayout>
```

上述布局文件直接放在 Activity 中运行没有效果。使用 TabHost 需要添加标签对象 TabHost.TabSpec,然后使用该对象的 setIndicator 方法设置标签,使用 setContent 方法设置内容视图。下面的代码演示如何创建 TabHost 的标签对象。

📖 Proj02_1 项目 MainActivity.java，TabHost 控件

```
protected void onCreate(Bundle savedInstanceState) {
    super.onCreate(savedInstanceState);
    setContentView(R.layout.activity_tabhost);
    //获取 TabHost 对象
    TabHost tabHost = (TabHost) findViewById(R.id.tabhost);
    tabHost.setup();
    //创建 Tab 标签
    TabSpec tab1 = tabHost.newTabSpec("tab1");
    tab1.setIndicator("校园新闻");              //标签标签
    tab1.setContent(R.id.tab1);                //标签内容视图
    TabSpec tab2 = tabHost.newTabSpec("tab2")
            .setIndicator("学工通信")
            .setContent(R.id.tab2);
    TabSpec tab3 = tabHost.newTabSpec("tab3")
            .setIndicator("教务通知")
            .setContent(R.id.tab3);
    //将标签添加到 TabHost
    tabHost.addTab(tab1);
    tabHost.addTab(tab2);
    tabHost.addTab(tab3);
}
```

所有的标签对象 TabSpec 必须添加到 TabHost 中，添加的顺序影响它们呈现的先后顺序。没有继承 TabActivity，直接使用 TabHost 控件，在创建标签之前，需要执行 setup 方法，否则会抛出空指针异常。项目运行效果如图 2.21 所示。

图 2.21　TabHost 运行效果

调整 activity_tabhost.xml 布局，将 TabWidget 和 Fragment 调换位置，并调整布局管理器和相应属性，可以实现将标签放置在底部的效果，感兴趣的读者可以尝试实现。

2.6.2　ViewFlipper

ViewFlipper 继承 ViewAnimator，使得其内部控件带有进入和退出动画效果。ViewFlipper 常用的属性如表 2.14 所示。

表 2.14　ViewFilpper 常用属性

属性	说明	取值
android:flipInterval	动画间隔时间	数值,单位为毫秒
android:autoStart	是否自动播放	true、false
android:inAnimation	进入动画	动画资源 R.anim.xxx
android:outAnimation	退出动画	动画资源 R.anim.xxx

下面的布局文件演示 ViewFilpper 控件属性的设置和内部元素的添加。将该布局文件设置给 Activity 后,运行项目,ViewFilpper 会自动播放动画,切换内部元素的显示,在进入和退出时带有从左到右的动画效果,该动画效果是 Android 系统动画,4.6 节会详细介绍如何自定义动画效果。

📖 Proj02_1 项目 activity_flipper.xml,ViewFlipper 控件

```xml
<LinearLayout xmlns:android = "http://schemas.android.com/apk/res/android"
    android:layout_width = "match_parent"
    android:layout_height = "match_parent"
    android:orientation = "vertical" >
    <ViewFlipper
        android:id = "@ + id/viewFlipper1"
        android:layout_width = "match_parent"
        android:layout_height = "wrap_content"
        android:autoStart = "true"
        android:flipInterval = "2000"
        android:inAnimation = "@android:anim/slide_in_left"
        android:outAnimation = "@android:anim/slide_out_right">
        <ImageView
            android:id = "@ + id/imageView1"
            android:layout_width = "wrap_content"
            android:layout_height = "wrap_content"
            android:src = "@drawable/role01" />
        <ImageView
            android:id = "@ + id/imageView2"
            android:layout_width = "wrap_content"
            android:layout_height = "wrap_content"
            android:src = "@drawable/role02" />
    </ViewFlipper>
</LinearLayout>
```

ViewFilpper 的使用并非单纯像上面代码这样,需要静态设置属性。ViewFilpper 从 ViewGroup 继承了大量方法,可以灵活地添加内部元素。ViewFilpper 还从 ViewAnimator 继承了大量方法,可以控制动画播放。本类特有的方法可以实现自动播放等功能。

2.6.3　ImageSwitcher

ImageSwitcher 是 ViewSwitcher 的子类,在切换内部元素时可以施加动画效果。ImageSwitcher 与 ImageView 的功能有很多相似之处,都可以用于显示图片,区别是前者更

容易控制图片向前、向后切换。ImageSwitcher 与 Gallery 一起使用,可以非常简单地实现幻灯片播放效果。

使用 ImageSwitcher 时需要提供 ViewSwitcher.ViewFactory 接口的实现对象,该接口只有一个方法 makeView,返回值是 View 类型。ImageSwitcher 切换图像可以通过如下 3 种方式:

setImageDrawable(Drawable drawable)

setImageResource(int resid)

setImageURI(Uri uri)

下面的布局文件包含 ImageSwitcher 和 Gallery,使用二者实现幻灯片播放。注意,Gallery 的参考位置与父容器底部对齐,并设置背景半透明,元素间距用 android:spacing 设置。

📖 Proj02_1 项目 activity_imagesw.xml,ImageSwitcher 与 Gallery 控件

```xml
<RelativeLayout xmlns:android = "http://schemas.android.com/apk/res/android"
    android:layout_width = "match_parent"
    android:layout_height = "match_parent"
    android:orientation = "vertical" >
    <ImageSwitcher
        android:id = "@ + id/imageSwitcher1"
        android:layout_width = "match_parent"
        android:layout_height = "match_parent" >
    </ImageSwitcher>
    <Gallery
        android:id = "@ + id/gallery1"
        android:layout_width = "match_parent"
        android:layout_height = "60dp"
        android:layout_alignParentBottom = "true"
        android:gravity = "center_vertical"
        android:background = "#99000000"
        android:spacing = "8dp" />
</RelativeLayout>
```

核心代码主要包括 3 个部分,onCreate 方法是对控件的初始化,准备图片资源(项目中的资源),添加监听器。内部类 MyViewFactory 是 ImageSwitcher 的工厂类,创建 ImageSwitcher 显示的视图,并设置相应参数。内部类 GalleryAdapter 是 Gallery 的适配器类,继承抽象类适配器 BaseAdapter,实现相应方法。

📖 Proj02_1 项目 MainActivity.java,核心代码

```java
@Override
protected void onCreate(Bundle savedInstanceState) {
    super.onCreate(savedInstanceState);
    setContentView(R.layout.activity_imagesw);
    //显示的图片资源
    final int pics[] = new int[]{R.drawable.sg01,R.drawable.sg02,
            R.drawable.sg03,R.drawable.sg04,R.drawable.sg05};
```

```java
        final ImageSwitcher is = (ImageSwitcher) findViewById(R.id.imageSwitcher1);
        Gallery ga   = (Gallery) findViewById(R.id.gallery1);
        //设置工厂
        is.setFactory(new MyViewFactory());
        is.setInAnimation(this,android.R.anim.slide_in_left);
        is.setOutAnimation(this, android.R.anim.slide_out_right);
        is.setImageResource(pics[0]);                //默认显示第一张
        //给 Gallery 适配数据
        ga.setAdapter(new GalleryAdapter(pics));
        //给 Gallery 添加监听器
        ga.setOnItemClickListener(new OnItemClickListener(){
            @Override
            public void onItemClick(AdapterView<?> arg0, View arg1, int arg2,long arg3) {
                is.setImageResource(pics[arg2]);       //切换图片资源
            }}
        );
    }
    //内部类,实现 SpinnerAdapter,给 Galley 提供适配器
    class GalleryAdapter extends BaseAdapter{
        int imgs[];                             //存放需要适配的数据
        public GalleryAdapter(int imgs[]){
            this.imgs = imgs;
        }
        @Override
        public int getCount() {
            return imgs.length;
        }
        @Override
        public Object getItem(int index) {
            return imgs[index];
        }
        @Override
        public long getItemId(int index) {
            return index;
        }
        @Override
        public View getView(int index, View view, ViewGroup vg) {
            //创建返回的视图,并设定参数
            ImageView iv = new ImageView(MainActivity.this);
            iv.setImageResource(imgs[index]);
            iv.setLayoutParams(new Gallery.LayoutParams(60,60));
            iv.setScaleType(ImageView.ScaleType.CENTER_CROP);
            return iv;
        }
    }
    //内部类,实现 ViewFactory 接口,创建 ImageSwitcher 的显示元素
    class MyViewFactory implements ViewFactory{
        @Override
        public View makeView() {
```

```
        ImageView iv = new ImageView(MainActivity.this);
        iv.setBackgroundColor(0xFF000000);
        iv.setScaleType(ImageView.ScaleType.FIT_CENTER);
        iv.setLayoutParams(new ImageSwitcher.LayoutParams(
                LayoutParams.MATCH_PARENT, LayoutParams.MATCH_PARENT));
        iv.setBackgroundColor(Color.WHITE);
        return iv;
    }
}
```

Gallery 监听器的作用是当用户单击某个元素时使用 ImageSwitcher.setImageResource 方法切换图片资源。项目运行效果如图 2.22 所示。读者可以尝试给 ImageSwitcher 添加触屏事件，监听左右划屏操作，这样就可以实现触屏切换图片了。

图 2.22 ImageSwitcher 和 Gallery 运行效果

2.7 日期和时间控件

本节介绍的控件位于控件面板的 Time & Date 栏，用于设置日期和时间，显示日期和时间。使用日期和时间控件，方便统一日期格式，避免用户输入无效数据。

2.7.1 DatePicker 和 TimePicker

DatePicker 和 TimePicker 都继承 FrameLayout 类，分别用于选择日期和时间。这两种控件都有对话框模式，分别是 DatePickerDialog 和 TimePickerDialog，将在 5.4.3 节介绍。DatePicker 常用属性如表 2.15 所示。日期选择控件还有 CalendarView，是 SDK 3.0 新增的控件，以日历的方式选择日期。

表 2.15 DatePicker 常用属性

属性	说明	取值
android：calendarTextColor	日历列表演示	
android：calendarViewShown	是否显示日历视图	true、false
android：datePickerMode	日历模式，低版本默认是 spinner 模式，目前默认 calendar 模式	spinner、calendar
android：endYear	允许选择的结束年份	
android：startYear	运行选择的开始年份	
android：firstDayOfWeek	设置周几作为每周的第一天	
android：maxDate	允许设置的最大日期	mm/dd/yyyy
android：minDate	允许设置的最小日期	mm/dd/yyyy

TimePicker 设置时间分 12 小时制和 24 小时制，使用 setIs24HourView 方法设置，参数是 boolean 类型，若为 false，则使用 AM/PM 区分时间。

DatePicker 使用 init(int year, int month, int day, DatePicker.OnDateChangedListener onDateChangedListener)设置监听器。TimePicker 使用 setOnTimeChangedListener(TimePicker.OnTimeChangedListener onTimeChangedListener)设置监听器。

下面的代码演示日期和时间选择控件的使用，布局文件是 activity_dt.xml，包含 DatePicker、TimePicker 和 TextView。

📖 Proj02_1 项目 MainActivity.java，核心代码

```java
public class MainActivity extends Activity {
    int year,month,day,hour,min;         //日期和时间变量
    TextView tv;
    @Override
    protected void onCreate(Bundle savedInstanceState) {
        super.onCreate(savedInstanceState);
        setContentView(R.layout.activity_dt);
        DatePicker dp = (DatePicker) findViewById(R.id.datePicker1);
        TimePicker tp = (TimePicker) findViewById(R.id.timePicker1);
        tp.setIs24HourView(true);
        tv = (TextView) findViewById(R.id.dtStr);
        //初始化日期选择器,并设置监听器
        dp.init(2017, 10, 1, new OnDateChangedListener(){
            @Override
            public void onDateChanged(DatePicker arg0, int y, int m, int d) {
                year = y;
                month = m;
                day = d;
                changeDT(year, month, day, hour, min);
            }}
        );
        //设置时间选择器的监听器
```

```
            tp.setOnTimeChangedListener(new OnTimeChangedListener(){
                @Override
                public void onTimeChanged(TimePicker arg0, int h, int m) {
                    hour = h;
                    min = m;
                    changeDT(year, month, day, hour, min);
                }}
            );
        }
        //修改日期和时间
        public void changeDT(int y,int m, int d, int h, int mm){
            tv.setText("签收日期: " + y + " - " + m + " - " + d + " " + h + ":" + mm);
        }
    }
```

代码运行效果如图 2.23 所示。当日期和时间的取值改变时,监听器方法执行,参数即为当前所设置的日期或时间。

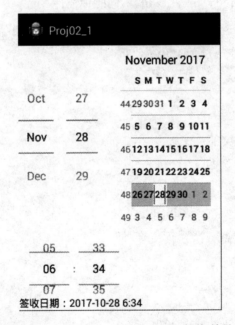

图 2.23 DatePicker 和 TimePicker 运行效果

2.7.2 Chronometer

Chronometer 是定时器类,继承 TextView,按照 MM:SS 或 H:MM:SS 格式显示时间。通过 start 和 stop 方法控制计时器的启动和停止。当时间改变时,可以使用监听器 Chronometer.OnChronometerTickListener。

2.7.3 AnalogClock 与 TextClock

AnalogClock 是模拟时钟，以指针的方式显示时间，只显示时针和分针。DigitalClock 是数字时钟，以数字形式显示时分秒。后者在 API 17 中停止使用，被 TextClock 替代。TextClock 不但可以显示时间，还可以显示日期，并可指定显示格式。AnalogClock 常用属性如表 2.16 所示。

表 2.16 AnalogClock 常用属性说明

属　　性	说　　明	取　　值
android:dial	表盘背景	drawable
android:hand_hour	时针图片	drawable
android:hand_minute	分针图片	drawable

TextClock 使用 setFormat12Hour(CharSequence)和 setFormat24Hour(CharSequence)设置显示日期和时间的格式。TextClock 在设置显示格式时会受到手机日期时间显示格式的影响，如手机的时间格式为 12 小时制，则 TextClock 设置 24 小时制将没有效果。

📖 Proj02_1 项目 activity_timer.xml，不同时钟控件

```xml
<LinearLayout xmlns:android = "http://schemas.android.com/apk/res/android"
    android:layout_width = "match_parent"
    android:layout_height = "match_parent"
    android:orientation = "vertical" >
    <AnalogClock
        android:id = "@ + id/analogClock1"
        android:layout_width = "wrap_content"
        android:layout_height = "wrap_content" />
    <DigitalClock
        android:id = "@ + id/digitalClock1"
        android:layout_width = "wrap_content"
        android:layout_height = "wrap_content"/>
    <AnalogClock
        android:id = "@ + id/analogClock2"
        android:layout_width = "wrap_content"
        android:layout_height = "wrap_content"
        android:dial = "@drawable/clock"/>
    <TextClock
        android:id = "@ + id/textClock1"
        android:layout_width = "wrap_content"
        android:layout_height = "wrap_content" />
</LinearLayout>
```

analogClock2 设置了表盘背景，textClock1 在代码部分设置了显示格式，该布局文件运行时的效果如图 2.24 所示。注意，该布局文件无法以 Graphic Layout 模式显示，会抛出空指针异常，这是由 DigitalClock 引起的。

图 2.24　时针控件运行效果

2.8　线程机制

在 Android 系统中,当程序启动时,Android 会同时启动主线程,负责处理与 UI 相关的事件,如初始化布局、响应用户的交互等,该线程又被称作 UI 线程。Android 系统默认当 UI 线程阻塞超过 20s 时将触发 ANR 异常(Application Not Responding)。在实际操作过程中,当主线程阻塞超过 5s,使用者就会表现得比较急躁,造成比较糟糕的应用体验。因此,不能在 UI 线程中执行比较耗时的操作,如更新、下载数据等。

为了避免上述现象的发生,在开发过程中应将耗时的操作放在子线程中进行。当子线程完成既定任务后,更新相应操作即可。但是,Android 系统为了避免多个线程修改 UI 可能会导致线程安全问题,规定只能由主线程(UI 线程)修改 Activity 中的组件。这就需要借助 Android 应用开发中特有的机制实现上述功能。

1. 子线程借助 Handler 修改 UI

Handler 类位于 android.os 包中,其作用是发送和处理 Message 对象、Runnable 对象。这些 Message 对象、Runnable 对象都由 MessageQueue 按照队列的特点,即先进先出的方式进行管理。MessageQueue 类被 final 修饰,没有公开的构造方法,无法直接创建,由 Looper 对象负责管理。在 UI 线程启动时,系统将会自动初始化一个 Looper 对象,因此,在 Activity 中可以直接使用 Handler 发送和处理信息。

Handler 对象运行在主线程(UI 线程)中,它与子线程通过 Message 对象来传递数据。当子线程需要更新 UI 时,使用 Handler 发送消息,并将 UI 需要显示的数据封装在 Message 对象中。Handler 常用的方法如表 2.17 所示。

表 2.17　Handler 常用方法

返回值类型	方法	说明
String	getMessageName(Message message)	获取消息的名称
void	handleMessage(Message msg)	处理消息
final boolean	hasMessages(int what)	判定消息队列中是否含有 what 指定的消息

续表

返回值类型	方法	说明
final boolean	post(Runnable r)	发送 r 到消息队列
final boolean	postDelayed(Runnable r, long delayMillis)	延迟 delayMillis 毫秒发送 r 到消息队列
final boolean	sendEmptyMessage(int what)	发送空消息,该空消息可以含有 what 属性值
final boolean	sendEmptyMessageDelayed(int what, long delayMillis)	延迟 delayMillis 毫秒发送空消息
final boolean	sendMessage(Message msg)	发送消息
final boolean	sendMessageDelayed(Message msg, long delayMillis)	延迟 delayMillis 毫秒发送消息

在 Android 应用中创建线程的语法与 J2SE 一致,可以选择继承 Thread 类或实现 Runnable 接口。

下面的代码演示子线程使用 Handler 更新 UI,修改进度条的取值。布局文件 activity_progressbar 包含一个水平进度条,id 是 progressBar4。

📖 Proj02_1 项目 MainActivity.java,核心代码

```java
public class MainActivity extends Activity {
    //声明 Handler 引用,并创建对象
    private Handler handler = new Handler(){
        public void handleMessage(android.os.Message msg) {
            switch(msg.what){
                case 0x100:
                    pb.setProgress(pb.getProgress() + 2);
                    break;
                case 0x200:
                    Toast.makeText(MainActivity.this, "数据更新完成",
                            Toast.LENGTH_SHORT).show();
            }
        }
    };
    ProgressBar pb;              //进度条控件
    @Override
    protected void onCreate(Bundle savedInstanceState) {
        super.onCreate(savedInstanceState);
        setContentView(R.layout.activity_progressbar);
        //获取水平进度条
        pb = (ProgressBar) findViewById(R.id.progressBar4);
        //创建并启动线程
        new Thread(new Runnable(){
            @Override
            public void run() {
                //当进度条取值未达最大时
                while(pb.getProgress()< pb.getMax()){
                    //使用 Handler 发送空消息,数值 0x100 表示更新进度值
                    handler.sendEmptyMessage(0x100);
                    try {
```

```
                    Thread.sleep(300);
                } catch (InterruptedException e) {
                    e.printStackTrace();
                }
            }
            //发送空消息,数值 0x200 是自定义的,表示进度条更新完成
            handler.sendEmptyMessage(0x200);
        }
    }).start();
}
```

子线程使用 Handler 发送两种类型的空消息,分别表示修改进度值和进度条完成,因此对空消息 what 取值进行了标识。0x100 和 0x200 是自定义取值,标识该空消息的含义,以便 Handler 对象区分不同消息所对应的功能。

由于 handler 对象位于 Activity 中,属于主线程(UI 线程)中的对象,因此可以直接获得 Looper 对象中的 MessageQueue,可以直接发送和接收消息。根据空消息的标识(即 Message 类中 what 属性的取值)修改 UI 中控件的取值。

上述代码的功能也可以使用 Message 类的属性实现。Message 类的属性 arg1、arg2 和 obj 都用 public 修饰,可以直接赋值使用,前两个属性是 int 类型,obj 是 Object 类型,除此之外 setData(Bundle data)方法可以向 Message 对象中设置复杂数据。

2. 开启异步任务 AsyncTask 修改 UI

AsyncTask 是抽象类,位于 android.os 包中,它在不需要借助线程和 Handler 机制的前提下完成轻量级应用,修改 UI 显示。该类定义了 3 个泛型:AsyncTask < Params, Progress, Result >。Params 是任务启动时传入的参数,Progress 是后台任务完成的进度值类型,Result 是执行完任务的返回值类型。AsyncTask 常用的方法如表 2.18 所示。

表 2.18 AsyncTask 常用方法说明

返回值类型	方法	说明
abstract Result	doInBackground(Params ... params)	必须重写,需要完成的业务逻辑
void	onPreExecute()	任务开始前执行
void	onPostExecute(Result result)	任务完成后执行
void	onProgressUpdate(Progress ... values)	更新 UI
final void	publishProgress(Progress ... values)	任务执行过程中调用该方法,会触发更新 UI 的方法 onProgressUpdate
static void	execute(Runnable runnable)	执行一个 Runnable 对象
final AsyncTask< Params, Progress, Result >	execute(Params ... params)	根据传入的参数执行任务

下面的代码使用 AsyncTask 实现进度条更新,布局文件 activity_progressbar.xml 中含有进度条控件,id 是 progressBar4。创建和启动任务应在主线程(UI 线程)中进行,调用 execute 方法后任务即开始执行。

📖 Proj02_1 项目 MainActivity.java,核心代码

```java
public class MainActivity extends Activity {
    ProgressBar pb;
    @Override
    protected void onCreate(Bundle savedInstanceState) {
        super.onCreate(savedInstanceState);
        setContentView(R.layout.activity_progressbar);
        pb = (ProgressBar) findViewById(R.id.progressBar4);
        new PbTask().execute(pb);                    //创建并执行任务
    }
    //实现任务类,重写任务方法
    class PbTask extends AsyncTask< Object, Integer, Integer >{
        @Override
        protected Integer doInBackground(Object... arg0) {
            while(pb.getProgress()< pb.getMax()){
                pb.setProgress(pb.getProgress() + 2);
                publishProgress(pb.getProgress());   //更新 UI
                try {
                    Thread.sleep(300);
                } catch (InterruptedException e) {
                    e.printStackTrace();
                }
            }
            return null;
        }
        @Override
        protected void onPostExecute(Integer result) {
            super.onPostExecute(result);
            Toast.makeText(MainActivity.this, "数据更新完成!", Toast.LENGTH_SHORT).show();
        }
    }
}
```

在上面有关进度条演示的两段代码中,为了控制进度条的现实,在更新进度条时执行了线程休眠操作,但在实际项目中请勿执行线程休眠。

2.9 习　　题

1. 选择题

(1) android：layout_width 属性可以被赋予值,不包括(　　)。

　　A. fill_parent　　　　　　　　　　B. wrap_content

　　C. match_parent　　　　　　　　　D. @drawable 引用

(2) EditText 控件属性 android：inputType 设置为密码框的取值是(　　)。

　　A. textPassword　　　　　　　　　B. Phone

　　C. textCapSentences　　　　　　　D. number

(3) RelativeLayout 布局管理中元素所特有的属性是(　　)。
　　A. android：orientation　　　　　　B. android：grivity
　　C. layout_below　　　　　　　　　D. android：weight
(4) 适配器 ArrayAdapter(Context context，int resource，int textViewResourceId，T [] objects)中参数 resource 的作用是(　　)。
　　A. 布局资源　　　B. 数据资源　　　C. 颜色资源　　　D. 控件主键
(5) 下列关于 Android 应用开发中线程机制的说法无误的是(　　)。
　　A. Android 应用中不支持使用多线程机制，只能使用 UI 线程
　　B. Android 应用中 UI 线程是主线程，可以修改 UI
　　C. Android 应用中子线程可以直接修改 UI，保证了线程安全
　　D. Android 应用中开启线程必须声明权限

2. 简答题

(1) Android 应用开发中常用的布局管理器有哪些？各有何特点？
(2) Android 应用开发中数据适配器的作用是什么？常用数据适配器有哪些？
(3) 列举常用适配器的构造方法，并说明参数含义。
(4) 在 Android 应用开发中，如何实现在子线程中修改 UI？
(5) 请设计如下布局：

第 3 章　Activity 与 Intent

本章学习目标
- 理解 Activity 的生命周期。
- 掌握 Activity 的创建与不同启动模式。
- 掌握 Intent 对象的使用。
- 掌握 Fragment 的创建与使用。
- 了解使用 Intent 启动手机组件的方法。

Activity 与 Intent 涉及 Activity 的创建和管理、Fragment 的创建以及在 Activity 中的使用。Activity 是 Android 的重要组件之一，它提供了一个用户界面，而 Fragment 可以作为 Activity 中的一个部分，可以将 Fragment 组合起来构成一个 Activity。Intent 是一个消息对象，可以使用 Intent 启动应用程序的另一个组件并传递消息。

3.1　Activity 的创建与管理

Activity 是一个应用程序组件，它负责 UI 的显示以及处理各种输入事件。Activity 提供了一个界面供用户进行交互操作，如拨打电话、拍照、查看地图等。一个应用程序通常由多个存在关联的 Activity 组成。通常情况下，在应用程序中存在一个主 Activity，该 Activity 在第一次启动该应用程序时呈现给用户。当一个新的 Activity 开始时，它的生命周期回调方法将根据状态变化而被调用。

3.1.1　创建 Activity 与配置信息

创建一个 Activity 时，必须定义一个类继承 Activity（或者 Activity 的派生类），并且在自定义的类中重写一些所需要的生命周期回调方法。这些生命周期方法在 Activity 的状态切换时将会被自动调用，例如 Activity 创建时会回调生命周期方法 onCreate。

　📖 Proj3_1 项目 SecondActivity.java 文件，创建一个 Activity

```
public class SecondActivity extends Activity {
    @Override
    protected void onCreate(Bundle savedInstanceState) {
        super.onCreate(savedInstanceState);
        setContentView(R.layout.activity_second);
```

```
    }
    ...
}
```

　　一个 Activity 的用户界面部分是由 View 类的视图对象提供的。每个 View 控件在界面中占一个矩形区域,并且可以在这个矩形区域响应用户的交互。比如一个 View 控件——按钮,当用户在这个按钮的矩形区域单击时,可以启动一个动作执行相应的逻辑。Android 提供了一个 XML 文件(定义在 res/layout/目录下)用来设计和排布 Activity 中需要展示的 View 控件,这样可以把界面设计从源代码中分离出来,单独进行用户界面的设计。当设置完界面后,可以在 Activity 的 onCreate 方法中通过 setContentView 方法引入该 XML 文件。

　　📖 Proj3_1 项目 activity_second.xml 文件,描述 SecondActivity 的用户界面部分

```xml
<LinearLayout xmlns:android = "http://schemas.android.com/apk/res/android"
    xmlns:tools = "http://schemas.android.com/tools"
    android:layout_width = "match_parent"
    android:layout_height = "match_parent"
    android:paddingBottom = "@dimen/activity_vertical_margin"
    android:paddingLeft = "@dimen/activity_horizontal_margin"
    android:paddingRight = "@dimen/activity_horizontal_margin"
    android:paddingTop = "@dimen/activity_vertical_margin"
    tools:context = ".SecondActivity" >
    <TextView
        android:id = "@ + id/second_textview_show"
        android:layout_width = "wrap_content"
        android:layout_height = "wrap_content"
        android:text = "@string/show" />
    <Button
        android:id = "@ + id/second_button_sure"
        android:layout_width = "wrap_content"
        android:layout_height = "wrap_content"
        android:text = "@string/back" />
</LinearLayout>
```

　　当创建完 Activity 之后,还必须在 AndroidManifest.xml 文件中声明该 Activity。打开 AndroidManifest.xml 文件,找到 application 元素,在该元素内声明 Activity。

　　📖 Proj3_1 项目 AndroidManifest.xml 文件中 SecondActivity 的声明

```xml
<?xml version = "1.0" encoding = "utf - 8"?>
<manifest ... >
    ...
    <application ... >
        ...
        <activity android:name = "com.example.proj_3_1.SecondActivity">
```

```
        </activity>
        ...
    </application>
    ...
</manifest>
```

在AndroidManifest.xml文件中声明Activity时,必须添加android：name属性,表明该Activity类所在的位置。Activity配置的常用属性如表3.1所示。

表3.1 Activity的常用配置属性

名 称	常用属性值	说 明
android：alwaysRetainTaskState	"true"|"false"	设置Activity所属的任务是否始终保持原来的状态。如果设置为true,则由系统来维护并保持状态,设置为false,那么在某些情况下系统会允许重设任务的初始状态。默认值是false。这个属性只对任务根节点的Activity有意义,其他所有的Activity都会被忽略
android：clearTaskOnLaunch	"true"|"false"	是否从任务清除除根Activity之外的所有Activity,true表示清除,false表示不清除,默认值是false。同样,这个属性也只对根Activity起作用,其他的Activity都会被忽略
android：enabled	"true"|"false"	设置activity是否能被系统初始化,true表示可以,false表示不可以,默认值是false
android：excludeFromRecents	"true"|"false"	设置这个任务是否会列在最近使用的应用列表中。当一个Activity是一个任务的根Activity时,这个属性将决定这个任务是否会显示在最近使用的应用列表中。如果设置为true,这个任务将不会显示在最近使用的应用列表中,如果设置为false,则会显示。这个属性的默认值是false
android：exported	"true"|"false"	设置Activity是否能被其他应用程序中的组件启动。设置为true,表示可以;如果设置为false则表示不可以,此时的Actvity只能被一个应用或者相同user ID的组件启动。这个属性的默认值依赖于Activity是否设置了intent filters
android：finishOnTaskLaunch	"true"|"false"	用于设置当用户重新启动一个任务(重新单击桌面的应用图标),Activity是否会被结束。设置为true会被结束,false则不会。如果这个属性和allowTaskReparenting属性同时被设置为true,这个属性的级别要高于后者

续表

名　称	常用属性值	说　明
android:hardwareAccelerated	"true" | "false"	用于设置 Activity 能否被硬件加速渲染，true 表示能，false 表示不能，默认值是 false
android:icon	drawable resource	设置 Activity 的图标
android:label	string resource	设置用户可读的标签
android:launchMode	"standard" | "singleTop" | "singleTask" | "singleInstance"	设置 Activity 的启动模式
android:multiprocess	"true" | "false"	一个 Activity 实例是否允许运行于多个应用的进程中，true 表示可以，false 表示不可以，默认值是 false
android:name	string	描述 Activity 的名称。Activity 的名称描述有两种，一种是全路径的，例如：com.example.project.ExtracurricularActivity；另一种是相对路径，即在当前包的路径后面加上类的名称，例如".ExtracurricularActivity"
android:noHistory	"true" | "false"	当用户离开这个 Activity，即此 Activity 在屏幕上不可见时，当前 Activity 是否会从 Activity 栈中移除。true 表示会被移除并结束，false 则不会，默认值是 false
android:parentActivityName	string	逻辑父 Activity 的类名。这个名字必须与<activity>元素中的 android:name 属性中的值相匹配。如果设置了这个属性，当用户在 actionbar 中单击了 Up button 时，系统会读取这个逻辑父 Activity 并启动它
android:permission	string	一个允许客户端必须启动该 Activity 或以其他方式来响应一个 Intent 的权限
android:process	string	Activity 运行的进程的名称
android:screenOrientation	"landscape" | "portrait"	Activity 在屏幕中显示的方向
android:stateNotNeeded	"true" | "false"	用于设置 Activity 在被杀死和成功地再次启动之前是否保存它的状态。如果设置为 true，它在重新启动后不会关联到死之前的状态；如果是 false 则会关联。默认值是 false
android:taskAffinity	string	与 Activity 有亲和力的任务。有相同亲和力的 Activity 属于同一个任务。根 Activity 的亲和力取决于所在的这个任务的亲和力。默认情况下，一个应用中的所有 Activity 都有相同的亲和力。可以通过设置这个属性对 Activity 分组，甚至可以在不同的应用中将它们的 Activity 定义到同一个任务中。也可以设置一个空的字符串来指定 Activity 对任何任务都没有亲和力

续表

名　称	常用属性值	说　明
android:theme	resource 或 theme	设定 Activity 的主题。如果这个属性没有设置，Activity 会默认继承应用的主题。如果应用的主题也没有设置，Activity 会默认使用系统的主题
android:uiOptions	" none " | " splitActionBar-WhenNarrow "	Activity UI 的扩展选项。 • "none"：没有多余的 UI 选项，这是默认选项 • "splitActionBarWhenNarrow"：当手机处于横屏状态下时，在屏幕底部添加 bar 来显示 ActionBar 的操作项目

除了在 activity 元素中为它设置属性以外，还可以在 activity 元素内部嵌套 intent-filter 元素，以便 Intent 启动该 Activity 时进行筛选并且知道如何启动它。比如，在创建一个 Android 项目时，开发工具会默认创建一个具有优先运行权的主 Activity，该 Activity 在 AndroidManifest.xml 文件中的声明如下：

```
<activity
    android:name="com.example.proj_3_1.MainActivity"
    android:label="@string/app_name" >
    <intent-filter>
        <action android:name="android.intent.action.MAIN" />
        <category android:name="android.intent.category.LAUNCHER" />
    </intent-filter>
</activity>
```

3.1.2　Activity 的生命周期

管理 Activity 的生命周期可以通过重写它的生命周期回调方法实现。Activity 的生命周期和其他 Activity、应用程序、后台任务有着密切的联系。

在 Activity 的生命周期图中，可以将 Activity 表现为 3 种状态：

- 激活态。当 Activity 位于屏幕最前端，并可以获取用户焦点，接收用户输入时，这种状态称为激活态，也可以称为运行态。
- 暂停态。当 Activity 在运行时被另一个 Activity 所遮挡并获取焦点，此时 Activity 仍然可见，也就是说另一个 Activity 是部分遮挡的，或者另一个 Activity 是透明或者半透明的，此时的 Activity 处于暂停态。
- 停止态。当 Activity 被另一个 Activity 完全遮挡不可见时处于停止态。这个 Activity 仍然是存在的，它保留在内存中并保持所有状态和成员信息。但是当设备内存不足时该 Activity 可能会被系统杀掉以释放其占用的内存空间。再次打开该 Activity 时，它会被重新创建。

当 Activity 在以上 3 种状态间切换时，会回调相应的生命周期方法。因此可以在 Activity 中重写相应的生命周期方法来实现某种状态下要执行的逻辑。这些生命周期方法如下：

- onCreate()，当 Activity 被创建时调用。
- onStart()，当 Activity 被创建后即将可见时调用。
- onResume()，当 Activity 位于设备最前端，对用户可见时调用。
- onPause()，当另外一个 Activity 遮挡当前 Activity，当前 Activity 被切换到后台时调用。
- onRestart()，当 Activity 执行完 onStop 方法，又被用户打开时调用。
- onStop()，如果另一个 Activity 完全遮挡了当前 Activity 时，该方法被调用。
- onDestroy()，当 Activity 被销毁时调用。

以上生命周期方法并不需要描述全部，只需要根据具体需求去重写相应的方法，状态转换关系如图 3.1 所示。

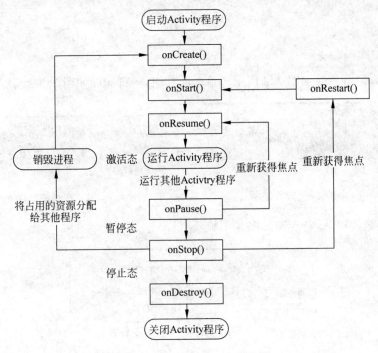

图 3.1　Activity 的生命周期图

3.1.3　Activity 启动模式

Activity 的启动模式有 4 种，分别是 standard、singleTop、singleTask 和 singleInstance，在配置文件中使用 android：launchMode 属性配置。下面分别介绍这种启动模式。

（1）standard：Activity 的默认启动模式。该模式每次启动都会在任务中新建一个 Activity 实例。

📖 Proj3_2 项目 MainActivity.java 文件代码

```
public class MainActivity extends Activity implements OnClickListener{
    private TextView tv;
    @Override
```

```java
protected void onCreate(Bundle savedInstanceState) {
    super.onCreate(savedInstanceState);
    setContentView(R.layout.activity_main);
    tv = (TextView) findViewById(R.id.main_tv);
    tv.setText(this.toString());
}
@Override
public void onClick(View v) {
    switch(v.getId()) {
    case R.id.main_btn:
        Intent intent = new Intent(MainActivity.this, MainActivity.class);
        startActivity(intent);
        break;
    }
}
```

启动项目,运行界面如图 3.2 所示。单击此按钮,运行界面如图 3.3 所示。

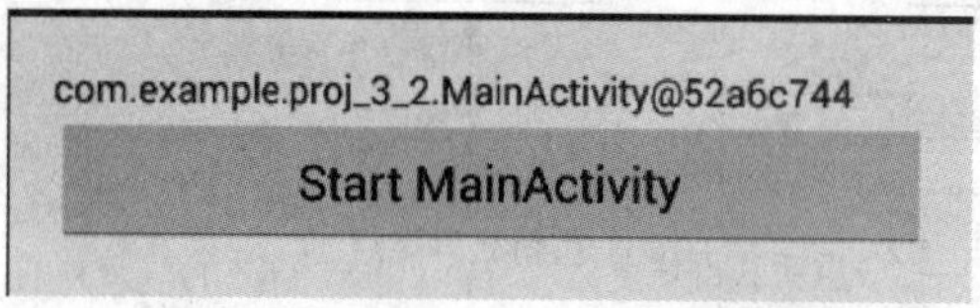

图 3.2　Activity 首次启动界面

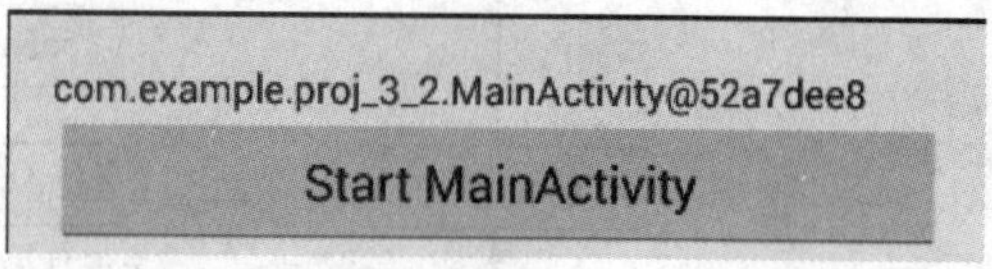

图 3.3　第二次创建 Activity 界面

再次单击此按钮,运行界面如图 3.4 所示。

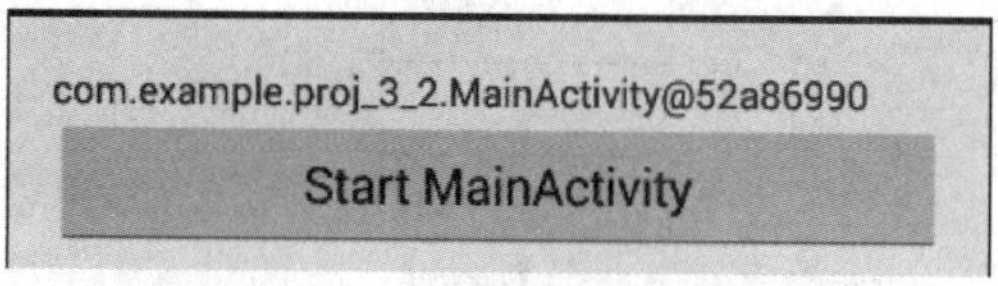

图 3.4　第三次创建 Activity 界面

可以发现,运行界面中显示的 MainActivity 的序列号是不一样的,每次单击该按钮都产生一个新的 Activity 实例,而且需要两次按下手机上的返回键才能回到第一个界面。

(2) singleTop:如果该 Activity 实例位于栈顶则不重新创建。

将上方 MainActivity 在 AndroidManifest.xml 文件中的 launchMode 设置为 singleTop,然后运行应用程序。启动项目,运行界面如图 3.5 所示。单击此按钮,运行界面如图 3.6 所示。

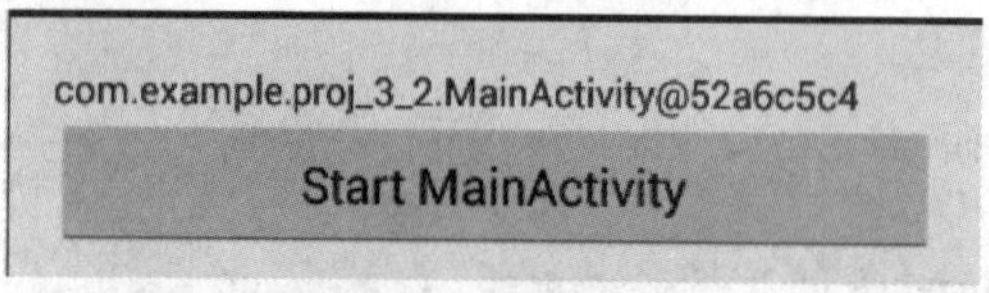

图 3.5　SingleTop 模式首次启动

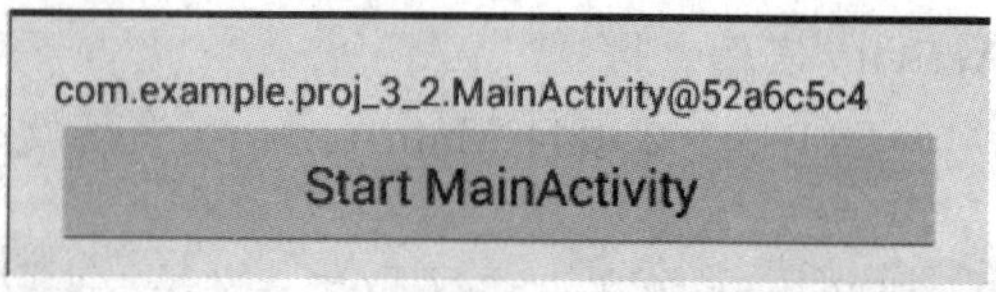

图 3.6　SingleTop 模式第二次启动

再次单击此按钮,运行界面如图 3.7 所示。

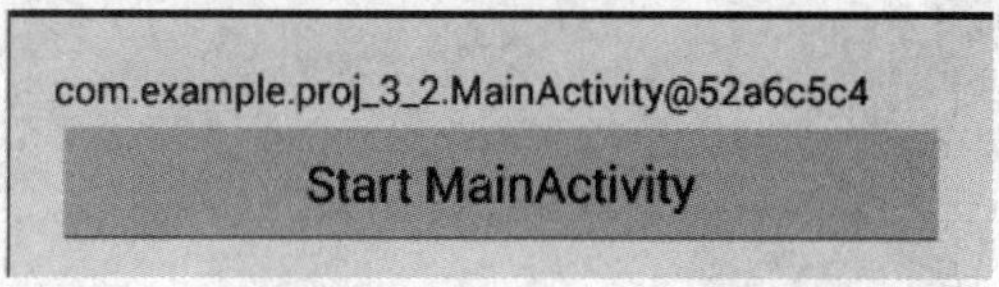

图 3.7 SingleTop 第三次启动

此次运行显示的 MainActivity 的序列号是一样的,也就是每次单击按钮都使用的是已有的 MainActivity 实例,按下返回键时程序退出了,因为运行程序时 MainActivity 实例在任务顶部,当需要 MainActivity 实例时,并不产生新的实例,而是调用任务顶部的实例,任务中只有一个 MainActivity 实例。

如果启动 MainActivity 时任务中存在 MainActivity 实例,但不位于栈顶,会怎样呢?修改项目 Proj_3_2MainActivity 代码。

📖 Proj3_2 项目 MainActivity.java 文件代码

```java
public class MainActivity extends Activity implements OnClickListener{
    private TextView tv;
    @Override
    protected void onCreate(Bundle savedInstanceState) {
        super.onCreate(savedInstanceState);
        setContentView(R.layout.activity_main);
        tv = (TextView) findViewById(R.id.main_tv) ;
        tv.setText(this.toString()) ;
    }
    @Override
    public void onClick(View v) {
        switch(v.getId()) {
            case R.id.main_btn:
                Intent intent = new Intent(MainActivity.this , SecondActivity.class) ;
                startActivity(intent) ;
                break;
        }
    }
}
```

📖 Proj3_2 项目 SecondActivity.java 文件代码

```java
public class SecondActivity extends Activity implements OnClickListener{
    private TextView tv;
    @Override
    protected void onCreate(Bundle savedInstanceState) {
        super.onCreate(savedInstanceState);
        setContentView(R.layout.activity_second);
        tv = (TextView) findViewById(R.id.second_tv) ;
        tv.setText(this.toString()) ;
    }
```

```
@Override
public void onClick(View v) {
    switch(v.getId()) {
        case R.id.second_btn:
            Intent intent = new Intent(SecondActivity.this, MainActivity.class);
            startActivity(intent);
            break;
    }
}
```

启动项目,运行界面如图 3.8 所示。单击按钮,启动 SecondActivity,运行界面如图 3.9 所示。

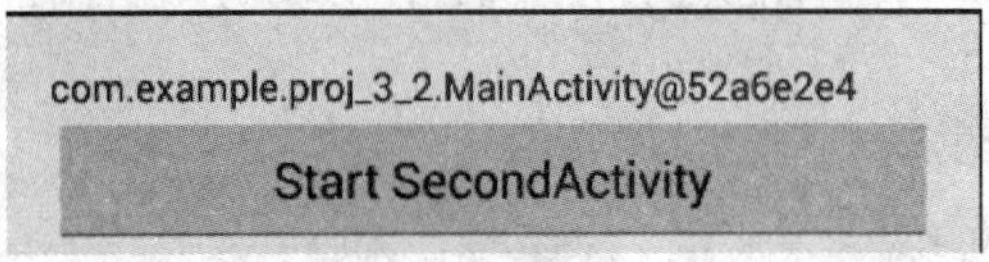

图 3.8 SingleTop 首次启动 MainActivity 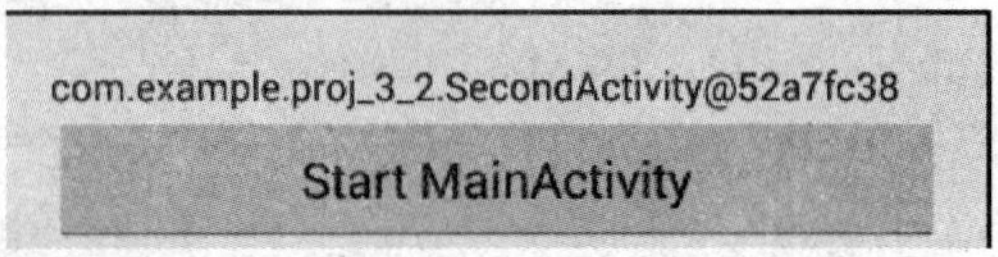 　　　图 3.9 启动 SecondActivity

单击 SecondActivity 中的按钮,启动 MainActivity,运行界面如图 3.10 所示。

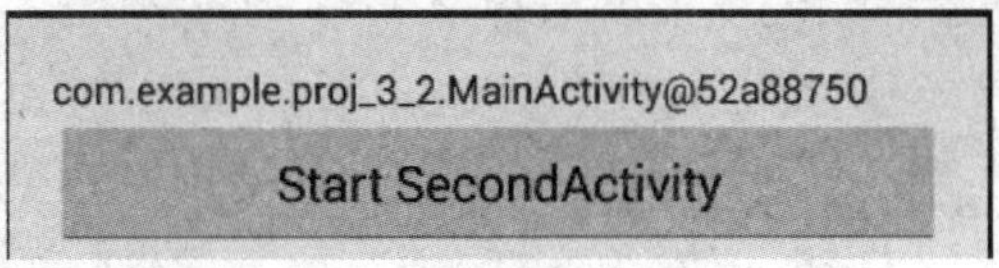

图 3.10 SingleTop 第二次启动 MainActivity

从运行结果可以看出,最开始启动的 MainActivity 的实例序列号和从 SecondActivity 启动的 MainActivity 的实例序列号是不一样的,因为当从 SecondActivity 界面中启动 MainActivity 时,已有的 MainActivity 实例并不位于任务顶部,所以产生了一个新的运行实例。

从前后对比可以看出,singleTop 模式,如果 Activity 实例位于栈顶则不产生新的实例,如果不位于栈顶则会重新产生一个新的运行实例。

(3) singleTask:如果栈中有该 Activity 实例,则直接启动,中间的 Activity 实例将被关闭。关闭顺序与启动顺序相同。

将上方 MainaActivity 在 AndroidManifest.xml 文件中的 launchMode 设置为 singleTask,然后运行应用程序。

启动项目,运行界面如图 3.11 所示,单击按钮,启动 SecondActivity,运行界面如图 3.12 所示。

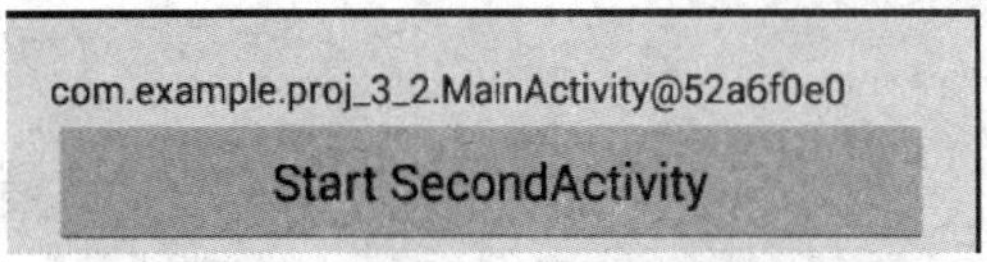

图 3.11 SingleTask 首次启动 MainActivity 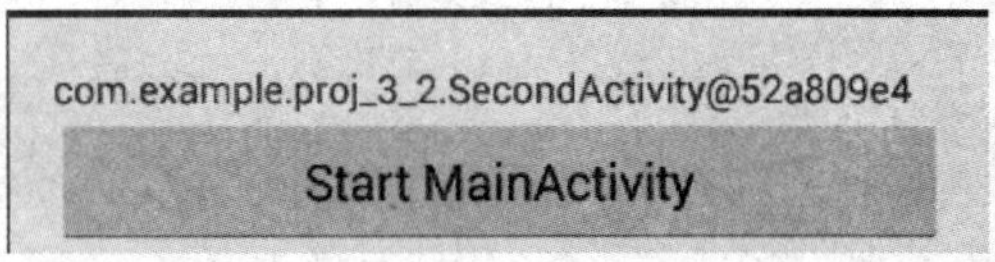 　　　图 3.12 SingleTask 启动 SecondActivity

单击 SecondActivity 中的按钮，运行界面如图 3.13 所示。单击 MainActivity 中的按钮，再次启动 SecondActivity，运行界面如图 3.14 所示。

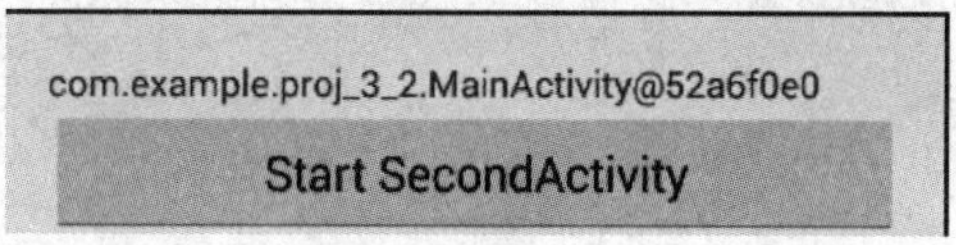

图 3.13　SingleTask 再次启动 MainActivity 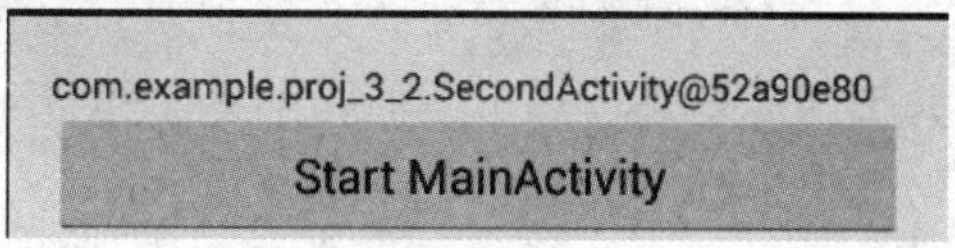　　　图 3.14　SingleTask 再次启动 SecondActivity

从运行结果可以看出，两次启动 MainActivity 的实例序列号是一样的，而两次启动 SecondActivity 的实例序列号是不一样的。最后按下两次返回键即可退出应用程序，因为中间启动的 SecondActivity 实例已经被销毁了。

（4）singleInstance：启用一个新的任务，将该 Activity 实例放置在这个任务中，并且该任务中不保存其他的 Activity 实例。

将上方 MainaActivity 在 AndroidManifest.xml 文件中的 launchMode 设置为 standard，将 SecondActivity 的 lanchMode 设置为 singleInstance。并且在两个 Activity 各自的 TextView 中显示 taskId。

启动项目，运行界面如图 3.15 所示。单击 MainActivity 中的按钮，启动 SecondActivity，运行界面如图 3.16 所示。

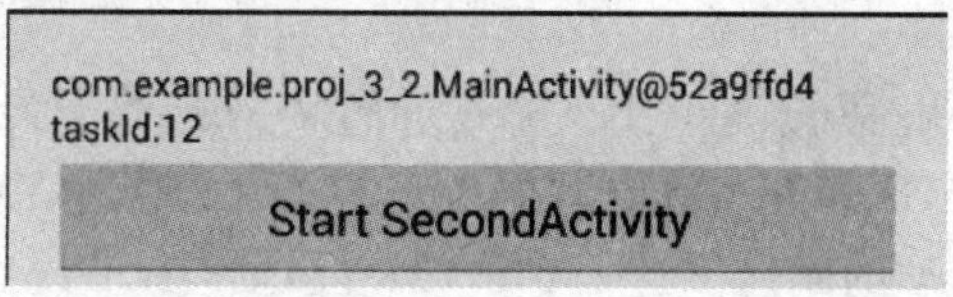

图 3.15　首次启动 MainActivity 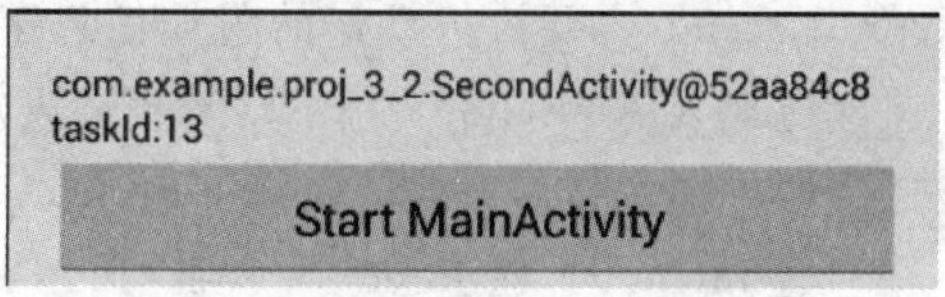　　　图 3.16　首次启动 SecondActivity

单击 SecondActivity 中的按钮，重新启动 MainActivity，运行界面如图 3.17 所示。
单击 MainActivity 中的按钮，再次启动 SecondActivity，运行界面如图 3.18 所示。

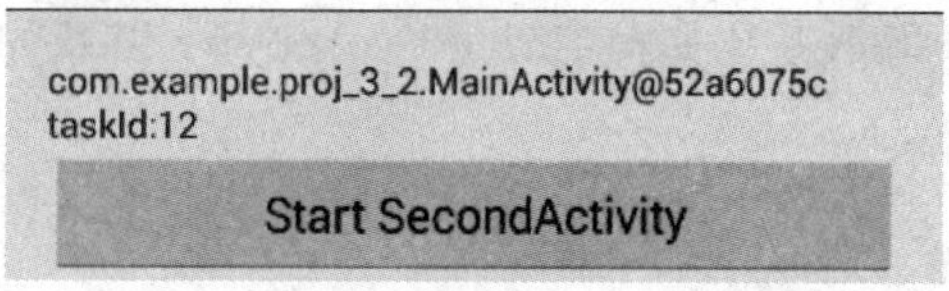

图 3.17　再次启动 MainActivity 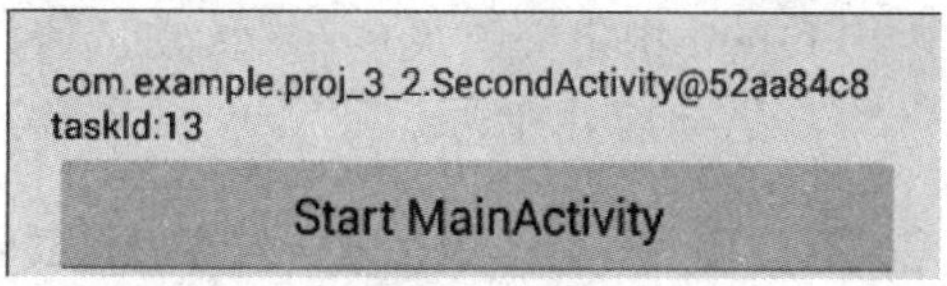　　　图 3.18　再次启动 SecondActivity

从运行结果可以看出，两次启动 MainActivity 产生了两个运行实例，都是位于同一个任务中，而前后两次启动 SecondActivity 都是同一个运行实例，而且单独开启了一个任务来存放该实例；而且按下返回键三次便可退出程序，因为有两个 MainActivity 和一个 SecondActivity 运行实例。

3.2　Intent 对象

Intent 是一个消息对象，并且可以通过它来启动应用程序的其他组件。通常通过 Intent 在组件之间传递消息有 3 种情况：启动 Activity，启动 Service，发送 Broadcast。

3.2.1 创建 Intent 对象

一个 Intent 对象携带了很多信息，Android 系统通过这些信息会决定启动哪个组件，并且会把相关数据传递给要启动的那个组件。

一个 Intent 中的主要信息如下：

- ComponentName，要启动的组件的名称。
  ComponentName 类中包含两个 String 成员：mPackage 与 mClass，分别用于指定报名和类名。Intent 一旦使用该属性，则不再通过其他属性查找要启动的组件。
- Action，指定要执行的操作的字符串。
  在 Broadcast Intent 中，这一属性非常重要，它是正在发生的行为。Intent 类中定义了很多的 action 常量，如 ACTION_VIEW、ACTION_SEND，这些特定的 action 字符串包含了一些系统中具体的功能模块意图。也可以在应用程序中为自己的组件定义特定的 action 字符串，这样可以通过设定该属性来启动这一组件。
- Data，Uri 类型的对象。
  该 Uri 对象用于引用数据的作用和/或数据的 MIME 类型。例如，如果 action 是 Intent.ACTION_CALL，则 Data 中应包含要拨打电话的 Uri 信息，如 Uri.parse("tel:电话号码")。
- Category，要执行动作的附加信息字符串。
  比如，CATEGORY_BROWSER 表示要启动的 Activity 将通过网络浏览器来显示要传递的链接数据；CATEGORY_HOME 则表示放回到 Home 界面。当程序中创建 Intent 时，该 Intent 默认启动的 Category 的属性值为 CATEGORY_DEFAULT。
- Extra，执行所需操作附加信息的键-值对数据。
  可以通过 putExtras 方法添加数据，该方法需要两个参数，分别是键和值。它可以放置多组键-值对，这些键-值对会被放置到一个 Bundle 对象中在组件间传递数据。
- Flags，在 Intent 中定义的标志值，它是一个 int 类型的值。
  Intent 中定义了很多 Flags 常量值，它将指示系统如何启动一个 Activity 以及如何在启动后处理它。

3.2.2 使用 Intent 启动 Activity

启动 Activity 可以分别使用 startActivity(Intent intent)方法和 startActivityForResult (Intent intent, int requestCode)方法。前者直接启动另一个 Activity，不等待返回；后者启动另一个 Activity，并等待回传结果，如果要启动的 Activity 存在，requestCode 的值将会传递到 onActivityResult 方法中。

📖 Proj3_3 项目 MainActivity.java 文件代码

```
public class MainActivity extends Activity implements OnClickListener{
    private TextView tv;
    @Override
    protected void onCreate(Bundle savedInstanceState) {
        super.onCreate(savedInstanceState);
```

```java
            setContentView(R.layout.activity_main);
            tv = (TextView) findViewById(R.id.main_tv);
        }
        @Override
        public void onClick(View v) {
            switch (v.getId()) {
            case R.id.main_btn1:
                Intent intent1 = new Intent(MainActivity.this, SecondActivity.class);
                intent1.putExtra("from", "main");
                intent1.putExtra("result", false);
                startActivity(intent1);
                break;
            case R.id.main_btn2:
                Intent intent2 = new Intent(MainActivity.this, SecondActivity.class);
                Bundle b = new Bundle();
                b.putString("from", "main");
                b.putBoolean("result", true);
                intent2.putExtras(b);
                startActivityForResult(intent2, 1);
                break;
            case R.id.main_btn3:
                Intent intent3 = new Intent(MainActivity.this, ThirdActivity.class);
                Bundle bundle = new Bundle();
                bundle.putString("from", "main");
                bundle.putLong("time", System.currentTimeMillis());
                intent3.putExtras(bundle);
                startActivityForResult(intent3, 2);
                break;
            default:
                break;
            }
        }
        @Override
        protected void onActivityResult(int requestCode, int resultCode, Intent data) {
            if(resultCode == RESULT_OK) {
                switch (requestCode) {
                case 1:
                        tv.setText(" SecondActivity result: " + data.getStringExtra("resultData"));
                    break;
                case 2:
                    tv.setText("ThirdActivity result: " + data.getStringExtra("result"));
                    break;
                }
            }
        }
    }
```

📖 Proj3_3 项目 SecondActivity.java 文件代码

```java
public class SecondActivity extends Activity {
    private TextView tv ;
    private Button btn ;
    private boolean needResult ;
    @Override
    protected void onCreate(Bundle savedInstanceState) {
        super.onCreate(savedInstanceState);
        setContentView(R.layout.activity_second);
        tv = (TextView) findViewById(R.id.second_tv) ;
        btn = (Button) findViewById(R.id.second_btn_back) ;
        // 获取 Intent 传递的数据
        Intent intent = getIntent() ;
        tv.append("\nfrom:" + intent.getStringExtra("from")) ;
        tv.append("\nresult:" + intent.getBooleanExtra("result", false)) ;
        needResult = intent.getBooleanExtra("result", false);
        btn.setOnClickListener(new OnClickListener() {
            @Override
            public void onClick(View v) {
                if(needResult) {
                    Intent dataIntent = new Intent() ;
                    dataIntent.putExtra("resultData", "second result data.") ;
                    setResult(RESULT_OK, dataIntent) ;
                }
                finish();
            }
        }) ;
    }
}
```

📖 Proj3_3 项目 ThirdActivity.java 文件代码

```java
public class ThirdActivity extends Activity {
    private TextView tv ;
    private Button btn ;
    @Override
    protected void onCreate(Bundle savedInstanceState) {
        super.onCreate(savedInstanceState);
        setContentView(R.layout.activity_third);
        tv = (TextView) findViewById(R.id.third_tv) ;
        btn = (Button) findViewById(R.id.third_btn_back) ;
        // 获取 Intent 传递的数据
        Intent intent = getIntent() ;
        tv.append("\nfrom:" + intent.getStringExtra("from")) ;
        tv.append("\ntime:" + intent.getLongExtra("time", 0)) ;
        btn.setOnClickListener(new OnClickListener() {
```

```
            @Override
            public void onClick(View v) {
                Intent dataIntent = new Intent();
                dataIntent.putExtra("result", "ThirdActivity 返回了数据。");
                setResult(RESULT_OK, dataIntent);
                finish();
            }
        });
    }
}
```

运行应用程序,初始 MainActivity 界面如图 3.19 所示。单击"启动 Activity"按钮,启动 SecondActivity,界面如图 3.20 所示。

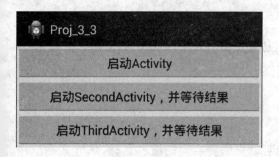

图 3.19　MainActivity 运行界面

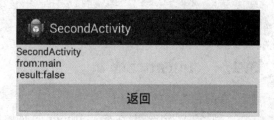

图 3.20　SecondActivity 界面

在 SecondActivity 界面单击"返回"按钮,回到 MainActivity,界面并没有任何变化。单击第二个按钮"启动 SecondActivity,并等待结果",启动 SecondActivity,界面如图 3.21 所示。

SecondActivity 接收到的 result 参数的值是不同的,单击 SecondActivity 中的"返回"按钮,MainActivity 将接收到 SecondActivity 返回的数据,此时运行界面如图 3.22 所示。

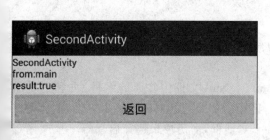

图 3.21　返回模式下启动 SecondActivity

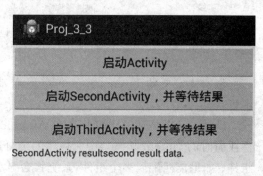

图 3.22　返回 MainActivity 界面

如果启动 ThirdActivity 后单击 ThirdActivity 中的"返回"按钮,则 MainActivity 的运行界面如图 3.23 所示。

从上面的案例可以看出,如果在一个 Activity 中要分别启动不同的 Activity 并等待回传结果数据,这就需要靠 startActivityForResult 的参数 requestCode 进行区分了。

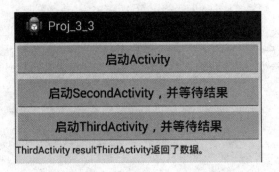

图 3.23 启动 ThirdActivity 并返回的效果

3.2.3 使用 Intent 传递数据

使用 Intent 在启动组件的时候传递数据，如上例所示，既可以直接使用 Intent 类的 putExtra 方法，也可以通过 Bundle 来传递。但是无论哪个方法，传递的数据都是以键-值的形式表达的。

3.2.4 Intent 过滤器

有两种类型的 Intent：

（1）显式 Intent。在构造 Intent 的时候就明确指明要启动的组件名称，该方式一般用于应用程序内部启动组件。例如：

```
Intent intent = new Intent(MainActivity.this , SecondActivity.class);
startActivity(intent);
```

（2）隐式 Intent。构造 Intent 时没有明确指出要启动的组件，而是定义一个要执行的操作。这样系统会通过查找组件注册时的 Intent Filter 来筛选适合的组件。例如：

```
Intent intent = new Intent() ;
intent.setAction("com.example.proj33.action.second");
startActivity(intent);
```

像这样定义的 Intent，并不知道要启动的组件具体是哪个，这样就需要通过 Intent 解析。

Intent 解析是通过比较所有在 AndroidManifest.xml 文件中注册的组件，去和这些组件注册时的 <intent-filter> 中所描述的部分进行匹配。如果只能匹配到一个组件，则启动该组件；如果能匹配到多个组件，系统将会显示一个对话框，由用户来选择启动哪一个应用程序的组件。

Intent 在解析时是通过 Intent 的 action、type、category 来进行判断的。

如上面所定义的 Intent，将会匹配到如下设置的 Activity：

```
< activity
        android:name = "com.example.proj_3_3.SecondActivity"
```

```
            android:label = "@string/title_activity_second" >
        < intent - filter >
            < action android:name = "com.example.proj33.action.second"/>
            < category android:name = "android.intent.category.DEFAULT"/>
        </intent - filter >
    </activity>
```

3.2.5 使用 Intent 启动手机组件

Intent 中定了很多 action 常量,用来描述所启动组件的标识字符串。Android 系统中常用的一些 action 字符串,如拨打电话 Activity、发送短信 Activity 等,在 Intent 中均已定义,具体如下。

- 呼叫指定的电话号码:

```
Intent intent = new Intent();
intent.setAction(Intent.ACTION_CALL);
intent.setData(Uri.parse("tel:电话号码"));
startActivity(intent);
```

- 调用拨打电话界面:

```
Intent intent = new Intent();
intent.setAction(Intent.ACTION_DIAL);
intent.setData(Uri.parse("tel:电话号码"));
startActivity(intent);
```

- 调用发送短消息界面:

```
Intent intent = new Intent();
intent.setAction(Intent.ACTION_SENDTO);
intent.setData(Uri.parse("smsto:接收者号码"));
startActivity(intent);
```

- 打开浏览器浏览网页:

```
Intent intent = new Intent();
intent.setAction(Intent.ACTION_VIEW);
intent.setData(Uri.parse("url"));
startActivity();
```

- 返回系统 HOME 桌面:

```
Intent intent = new Intent();
intent.setAction(Intent.ACTION_MAIN);
intent.addCategory(Intent.CATEGORY_HOME);
intent1.setFlags(Intent.FLAG_ACTIVITY_NEW_TASK);
startActivity(intent);
```

- 安装 apk：

```
Intent intent = new Intent();
intent.setAction("android.intent.action.VIEW");
intent.setDataAndType(apkUri, "application/vnd.android.package-archive");
intent.setFlags(Intent.FLAG_ACTIVITY_NEW_TASK);
startActivity(intent);
```

- 卸载 apk：

```
Intent intent = new Intent();
intent.setAction(Intent.ACTION_DELETE);
intent.setData(Uri.fromParts("package", "要卸载 apk 的 packageName", null));
startActivity(intent);
```

以上只是列举了一些常用的启动系统的 Activity。更加详细的 Intent 的 Action 设置请参考官方 API。

3.3 Activity 与 Fragment

Android 3.0 引入了 Fragment 技术，Fragment 被译为"碎片、片段"。使用 Fragment 目的是为了解决不同屏幕分辨率动态和灵活的 UI 设计。Fragment 可以将 Activity 分割成多个可重用的组件，每个部分都有它自己的生命周期和 UI。在 Activity 中可以通过 FragmentManager 来添加、移除和管理所加入的 Fragment。

当 Activity 中需要加入 Fragment 时，可以通过两种方式，一是在 Activity 的 Layout 文件中声明 Fragment，二是通过代码将 Fragment 添加到一个已存在的 ViewGroup。

3.3.1 Fragment 生命周期

每个 Fragment 有自己布局、生命周期，交互事件处理。但是由于 Fragment 是嵌入到 Activity 的，所以 Fragment 的生命周期又和 Activity 的生命周期有密切的关联。如果 Activity 是暂停状态，其中所有的 Fragment 都是暂停状态；如果 Activity 是 stopped 状态，这个 Activity 中所有的 Fragment 都不能被启动；如果 Activity 被销毁，那么其中的所有 Fragment 都会被销毁。但是，当 Activity 在活动状态时，可以独立控制 Fragment 的状态，比如添加或者移除 Fragment。

其中 Fragment 的生命周期方法如下：

- onAttach，当 Fragment 和 Activity 产生关联时被调用。
- onCreate，当 Fragment 被创建时调用。
- onCreateView，创建并返回与 Fragment 相关的 View 界面。
- onActivityCreated，通知 Fragment，它所关联的 Activity 已经完成了 onCreate 的调用。
- onStart，让 Fragment 准备可以被用户所见，该方法和 Activity 的 onStart 相关联。
- onResume，Fragment 可见，可以和用户交互，该方法和 Activity 的 onResume 相关

联。当一个 Fragment 不被使用时，会调用下面一系列的生命周期方法：
- onPause,：当用户离开 Fragment 时调用该方法，此操作是由于它所在的 Activity 被遮挡或者是在 Activity 中的一个 Fragment 操作所引起的。
- onStop，对用户而言，Fragment 不再可见时调用该方法，此操作是由于它所在的 Activity 不再可见或者是在 Activity 中的一个 Fragment 操作所引起的。
- onDestroyView，Fragment 清理和它的 View 相关的资源。
- onDestroy，最终清理 Fragment 的状态。
- onDetach，Fragment 与 Activity 不再产生关联。

Fragment 的生命周期如图 3.24 所示。

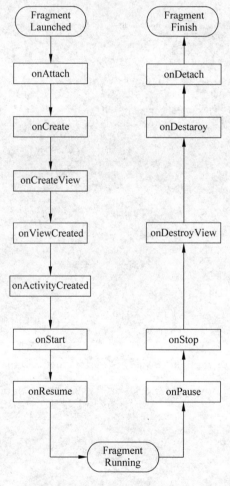

图 3.24 Fragment 的生命周期

📖 Proj3_4 项目 DemoFragment.java 文件代码

```java
public class DemoFragment extends Fragment {
    static final String TAG = "DemoFragment";
    @Override
    public void onCreate(Bundle savedInstanceState) {
```

```java
        super.onCreate(savedInstanceState);
        Log.v(TAG, "onCreate") ;
    }
    @Override
    public View onCreateView(LayoutInflater inflater, ViewGroup container,
            Bundle savedInstanceState) {
        Log.v(TAG, "onCreateView");
        View view = inflater.inflate(R.layout.fragment_demo, null) ;
        return view ;
    }
    @Override
    public void onResume() {
        Log.v(TAG, "onResume") ;
        super.onResume();
    }
    @Override
    public void onPause() {
        Log.v(TAG, "onPause");
        super.onPause();
    }
    @Override
    public void onStop() {
        Log.v(TAG, "onStop") ;
        super.onStop();
    }
    @Override
    public void onDestroy() {
        Log.v(TAG, "onDestroy");
        super.onDestroy();
    }
}
```

 📖 Proj3_4 项目 DemoFragment 的布局文件 fragment_demo.xml 文件代码

```xml
<?xml version = "1.0" encoding = "utf-8"?>
<LinearLayout xmlns:android = "http://schemas.android.com/apk/res/android"
    android:layout_width = "match_parent"
    android:layout_height = "match_parent"
    android:orientation = "vertical"
    android:background = "#ccc">
    <TextView
        android:layout_width = "match_parent"
        android:layout_height = "wrap_content"
        android:textSize = "18sp"
        android:text = "我来自一个 Fragment"/>
</LinearLayout>
```

📖 Proj3_4 项目 DemoFragment2.java 文件代码

```java
public class DemoFragment2 extends Fragment {
    static final String TAG = "DemoFragment2";
    @Override
    public void onCreate(Bundle savedInstanceState) {
        super.onCreate(savedInstanceState);
        Log.v(TAG, "onCreate") ;
    }
    @Override
    public View onCreateView(LayoutInflater inflater, ViewGroup container,
            Bundle savedInstanceState) {
        Log.v(TAG, "onCreateView");
        View view = inflater.inflate(R.layout.fragment_demo, null) ;
        return view ;
    }
    @Override
    public void onResume() {
        Log.v(TAG, "onResume") ;
        super.onResume();
    }
}
```

📖 Proj3_4 项目 MainActivity.java 文件代码

```java
public class MainActivity extends FragmentActivity implements OnClickListener {
    static final String TAG = "MainActivity";
    private DemoFragment fragmentDemo;
    @Override
    protected void onCreate(Bundle savedInstanceState) {
        super.onCreate(savedInstanceState);
        Log.v(TAG, "onCreate");
        setContentView(R.layout.activity_main);
    }
    @Override
    protected void onResume() {
        super.onResume();
        Log.v(TAG, "onResume");
    }
    @Override
    protected void onPause() {
        super.onPause();
        Log.v(TAG, "onPause");
    }
    @Override
    protected void onStop() {
        super.onStop();
        Log.v(TAG, "onStop");
    }
```

```java
    @Override
    protected void onDestroy() {
        super.onDestroy();
        Log.v(TAG, "onDestroy");
    }
    @Override
    public void onClick(View v) {
        switch (v.getId()) {
        case R.id.main_btn_addfragment:
            if (null == fragmentDemo) {
                fragmentDemo = new DemoFragment();
                // 添加 Fragment
                getSupportFragmentManager().beginTransaction()
                        .add(R.id.main_frame_container, fragmentDemo).commit();
            }
            break;
        case R.id.main_btn_startactivity:
            startActivity(new Intent(this, SecondActivity.class));
            break;
        }
    }
}
```

 Proj3_4 项目 MainActivity 的布局文件 activity_main.xml 文件代码

```xml
<LinearLayout xmlns:android = "http://schemas.android.com/apk/res/android"
    xmlns:tools = "http://schemas.android.com/tools"
    android:layout_width = "match_parent"
    android:layout_height = "match_parent"
    android:orientation = "vertical" >
    <fragment
        android:id = "@ + id/fragment1"
        android:name = "com.example.proj_3_4.fragment.DemoFragment2"
        android:layout_width = "match_parent"
        android:layout_height = "150dp"/>
    <Button
        android:id = "@ + id/main_btn_addfragment"
        android:layout_width = "match_parent"
        android:layout_height = "wrap_content"
        android:onClick = "onClick"
        android:text = "添加 Fragment" />
    <Button
        android:id = "@ + id/main_btn_startactivity"
        android:layout_width = "match_parent"
        android:layout_height = "wrap_content"
        android:onClick = "onClick"
        android:text = "切换 Activity" />
    <FrameLayout
        android:id = "@ + id/main_frame_container"
```

```
        android:layout_width = "match_parent"
        android:layout_height = "0dp"
        android:layout_weight = "1">
</FrameLayout>
</LinearLayout>
```

运行应用程序,界面如图 3.25 所示。

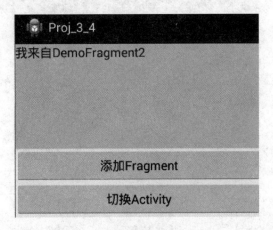

图 3.25　Fragment 运行效果

LogCat 打印日志如图 3.26 所示。

图 3.26　Log 日志输出信息

在 MainActivity 的布局文件中加载了 DemoFragment2,所以当 Activity 加载时,DemoFragment2 也被加载了,当 MainActivity 可见时调用了 onResume 方法,随之 DemoFragment2 的 onResume 方法也被调用了。

单击 MainActivity 中的"添加 Fragment"按钮,在 Java 代码中完成 Fragment 的添加。运行界面如图 3.27 所示。

LogCat 新增加的打印日志如图 3.28 所示。

单击 MainActivity 中的"切换 Activity"按钮,此时 MainActivity 将会被 SecondActivity 所遮挡,运行界面如图 3.29 所示。

由于 MainActivity 被遮挡,不能再和用户产生交互,也不再可见,所以 MainActivity 中的两个 Fragment 也是如此,因此 LogCat 打印内容如图 3.30 所示。

当单击 SecondActivity 的 back 按钮,MainActivity 又重新可见,两个 Fragment 也重新可见,LogCat 打印内容如图 3.31 所示。

最后按下返回键,退出应用程序,MainActivity 将被销毁,则两个 Fragment 也将被销毁,LogCat 打印内容如图 3.32 所示。

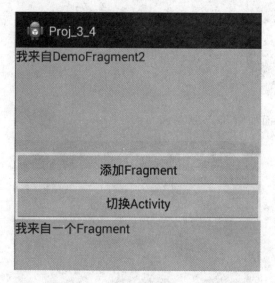

图 3.27 动态增加 Fragment 效果

com.example.proj_3_4	DemoFragment	onCreate
com.example.proj_3_4	DemoFragment	onCreateView
com.example.proj_3_4	DemoFragment	onResume

图 3.28 Log 日志输出信息

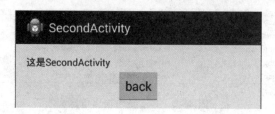

图 3.29 SecondActivity 运行效果

com.example.proj_3_4	DemoFragment2	onStop
com.example.proj_3_4	DemoFragment	onStop
com.example.proj_3_4	MainActivity	onStop

图 3.30 Fragment 停止日志

com.example.proj_3_4	MainActivity	onResume
com.example.proj_3_4	DemoFragment2	onResume
com.example.proj_3_4	DemoFragment	onResume

图 3.31 Fragment 恢复活动日志

由示例可以看出，Fragment 有自己的生命周期，但又和它所在的 Activity 是密切相关的。

com.example.proj_3_4	DemoFragment2	onPause
com.example.proj_3_4	DemoFragment	onPause
com.example.proj_3_4	MainActivity	onPause
com.example.proj_3_4	DemoFragment2	onStop
com.example.proj_3_4	DemoFragment	onStop
com.example.proj_3_4	MainActivity	onStop
com.example.proj_3_4	DemoFragment2	onDestroy
com.example.proj_3_4	DemoFragment	onDestroy
com.example.proj_3_4	MainActivity	onDestroy

图 3.32 Fragment 销毁日志

3.3.2 Fragment 传递数据

当 Activity 加载 Fragment 的时候，有时需要往内部传递参数。官方推荐 Fragment.setArguments(Bundle bundle)这种方式来传递参数，而不推荐通过构造方法直接传递参数。因为当 Activity 重新创建时，会重新构建它所管理的 Fragment，原先的 Fragment 的字段值将会全部丢失，但是通过 Fragment.setArguments(Bundle bundle)方法设置的 bundle 会保留下来。所以尽量使用 Fragment.setArguments(Bundle bundle)方式来传递参数。

📖 Proj3_5 项目 NewsFragment.java 文件代码

```java
public class NewsFragment extends Fragment {
    private String title;
    private ListView listview;
    private String[] data;
    // 从 Fragment 的 Bundle 中取数据
    public void getTitle() {
        // 获取 Bundle 对象
        Bundle b = getArguments();
        if (null != b) {
            // 从 Bundle 对象中取出存入的数据,并赋值给成员变量 title
            title = b.getString("title");
        }
    }
    @Override
    public View onCreateView(LayoutInflater inflater, ViewGroup container,
            Bundle savedInstanceState) {
        View view = inflater.inflate(R.layout.fragment_hot, null);
        initView(view);
        return view;
    }
    private void initView(View view) {
        getTitle();
        listview = (ListView) view.findViewById(R.id.hot_listview);
        // 根据 title 变量的值不同,初始化不同的 ListView 的数据源
        if ("头条".equals(title)) {
            data = new String[] { "头条新闻 1", "头条新闻 2", "头条新闻 3",
                    "头条新闻 4", "头条新闻 5" };
        } else if ("娱乐".equals("title")) {
```

```java
            data = new String[] { "娱乐新闻1", "娱乐新闻2", "娱乐新闻3",
                    "娱乐新闻4", "娱乐新闻5" };
        } else {
            data = new String[] {};
        }
        ArrayAdapter<String> adapter = new ArrayAdapter<String>(getActivity(),
                android.R.layout.simple_list_item_1, android.R.id.text1, data);
        listview.setAdapter(adapter);
    }
}
```

📖 Proj3_5 项目 MainActivity 的布局文件 activity_main.xml 代码

```xml
<LinearLayout xmlns:android="http://schemas.android.com/apk/res/android"
    xmlns:tools="http://schemas.android.com/tools"
    android:layout_width="match_parent"
    android:layout_height="match_parent"
    android:orientation="vertical"
    android:paddingBottom="@dimen/activity_vertical_margin"
    android:paddingLeft="@dimen/activity_horizontal_margin"
    android:paddingRight="@dimen/activity_horizontal_margin"
    android:paddingTop="@dimen/activity_vertical_margin" >
    <LinearLayout
        android:layout_width="match_parent"
        android:layout_height="wrap_content"
        android:orientation="horizontal" >
        <Button
            android:id="@+id/main_btn_topline"
            android:layout_width="0dp"
            android:layout_height="wrap_content"
            android:layout_weight="1"
            android:text="头条"
            android:onClick="onClick"/>
        <Button
            android:id="@+id/main_btn_entertainment"
            android:layout_width="0dp"
            android:layout_height="wrap_content"
            android:layout_weight="1"
            android:layout_marginLeft="8dp"
            android:text="娱乐"
            android:onClick="onClick"/>
    </LinearLayout>
    <FrameLayout
        android:id="@+id/main_container"
        android:layout_width="match_parent"
        android:layout_height="match_parent">
    </FrameLayout>
</LinearLayout>
```

📖 **Proj3_5 项目 MainActivity.java 文件部分代码**

```java
public class MainActivity extends FragmentActivity implements OnClickListener {
    private Fragment tFragment, eFragment;
    private Button btnTopline, btnEntertainment;
    @Override
    protected void onCreate(Bundle savedInstanceState) {
        super.onCreate(savedInstanceState);
        setContentView(R.layout.activity_main);
        initView();
    }
    private void initView() {
        btnTopline = (Button) findViewById(R.id.main_btn_topline);
        btnEntertainment = (Button) findViewById(R.id.main_btn_entertainment);
        btnTopline.setEnabled(false);
        // 初始化 HotFragment 对象 tFragment
        tFragment = new NewsFragment();
        // 为 tFragment 传递数据
        Bundle b = new Bundle();
        b.putString("title", "头条");
        tFragment.setArguments(b);
        // 初始化 HotFragment 对象 eFragment 并为它传递数据
        eFragment = new NewsFragment();
        Bundle b2 = new Bundle();
        b2.putString("title", "娱乐");
        eFragment.setArguments(b2);
        // 将 Fragment 加载到 MainActivity 所设置的容器
        getSupportFragmentManager().beginTransaction()
            .add(R.id.main_container, tFragment)
            .add(R.id.main_container, eFragment)
            .hide(eFragment).commit();
    }
    @Override
    public void onClick(View v) {
        switch (v.getId()) {
        case R.id.main_btn_topline:
            getSupportFragmentManager().beginTransaction()
            .show(tFragment)
            .hide(eFragment).commit();
            btnTopline.setEnabled(false);
            btnEntertainment.setEnabled(true);
            break;
        case R.id.main_btn_entertainment:
            getSupportFragmentManager().beginTransaction()
            .show(eFragment)
            .hide(tFragment).commit();
            btnTopline.setEnabled(true);
            btnEntertainment.setEnabled(false);
            break;
        }
    }
}
```

从以上代码可以看出，当初始化自定义 Fragment 对象后，可将所有要传递的参数放置到 Bundle 中，然后调用 Fragment 的 setArguments 来保存参数。在自定义的 Fragment 类中，可以调用 getArgments 方法获取 Bundle 对象，从而取出传递到 Fragment 中的数据。运行界面如图 3.33 所示。

当单击"娱乐"按钮时，界面下方的 ListView 将切换为娱乐相关数据，运行界面如图 3.34 所示。

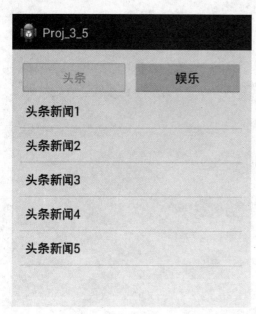
图 3.33 "头条"列表运行效果

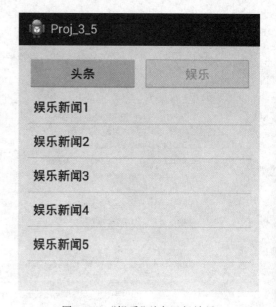
图 3.34 "娱乐"列表运行效果

3.3.3 管理 Fragment

要管理 Fragment，需要使用 FragmentManager，要获取它，需要在 Activity 中调用 getFragmentManager 方法（导入 android.app.FragmentManager 包时调用该方法）或者 getSupportFragmentManager 方法（导入 android.support.v4.app.FragmentActivity 包时调用该方法）。

1. FragmentManager

使用 FragmentManager 对象，可以调用如下主要方法：
- findFragmentById 或 findFragmentByTag，获取 Activity 中已存在的 Fragment。
- getFragments，获取所有加入 Activity 中的 Fragment。
- beginTransaction，获取一个 FragmentTransaction 对象，用来执行 Fragment 的事务。
- popBackStack，从 Activity 的后退栈中弹出 Fragment。
- addOnBackStackChangedListerner，注册一个侦听器以监视后退栈的变化。

2. FragmentTransaction

在 Activity 中对 Fragment 进行添加、删除、替换以及执行其他的动作将引起 Fragment

的变化，叫做一个事务。事务通过 FragmentTransaction 来执行，可以用 add、remove、replace、show、hide()等方法构成事务，最后使用 commit 方法提交事务。

如 3.3.2 节 MainActivity.java 代码中：

```
getSupportFragmentManager().beginTransaction()
            .add(R.id.main_container, tFragment)
            .add(R.id.main_container, eFragment)
            .hide(eFragment).commit();
```

当向事务添加了多个动作，比如多次调用了 add、hide 等方法，那么所有的在 commit 之前调用的方法都被作为一个事务。

事务中动作的执行顺序可随意，但要注意以下两点：
- 必须最后调用 commit。
- 如果添加了多个 Fragment，那么它们的显示顺序与添加顺序一致（后显示的覆盖前面的）。
- 如果一个 Fragment 已经添加到 Activity 指定的容器中，则不能重复添加。
- 调用 replace 方法时，会先把容器中的所有 Fragment 清空，然后再添加 Fragment。

📖 Proj3_6 项目 MainActivity.java 文件代码

```java
public class MainActivity extends FragmentActivity implements OnClickListener {
    final String FRAGMENT1_TAG = "fragment1";
    final String FRAGMENT2_TAG = "fragment2";
    private Fragment fragment1, fragment2;
    @Override
    protected void onCreate(Bundle savedInstanceState) {
        super.onCreate(savedInstanceState);
        setContentView(R.layout.activity_main);
        fragment1 = new Fragment1();
        fragment2 = new Fragment2();
    }
    @Override
    public void onClick(View v) {
        switch (v.getId()) {
            case R.id.main_btn_add_frag1:
                if (null == getSupportFragmentManager().findFragmentByTag(
                        FRAGMENT1_TAG)) {
                    getSupportFragmentManager()
                            .beginTransaction()
                            .add(R.id.main_fragment_container, fragment1,
                                    FRAGMENT1_TAG).commit();
                }
                break;
            case R.id.main_btn_add_frag2:
                if (null == getSupportFragmentManager().findFragmentByTag(
                        FRAGMENT2_TAG)) {
                    getSupportFragmentManager()
                            .beginTransaction()
                            .add(R.id.main_fragment_container, fragment2,
```

```java
                                    FRAGMENT2_TAG).commit();
            }
            break;
        case R.id.main_btn_remove_frag2:
            FragmentManager fMan = getSupportFragmentManager();
            Fragment f;
            f = fMan.findFragmentByTag(FRAGMENT1_TAG) == null;
            if (null != f) {
                fMan.beginTransaction().remove(f).commit();
            }
            break;
        case R.id.main_btn_repalce_frag1:
            getSupportFragmentManager()
                    .beginTransaction()
                    .replace(R.id.main_fragment_container, fragment2,
                            FRAGMENT2_TAG).commit();
            break;
    }
  }
}
```

运行界面如图 3.35 所示。

依次单击 4 个按钮,运行界面如图 3.36 至图 3.39 所示。

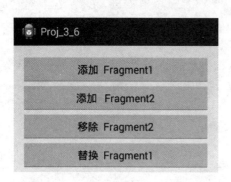

图 3.35　MainActivity 初始界面

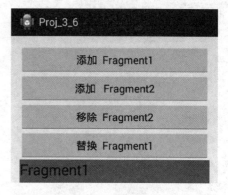

图 3.36　添加 Fragment1

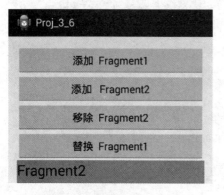

图 3.37　添加 Fragment2

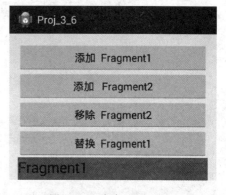

图 3.38　移除 Fragment2

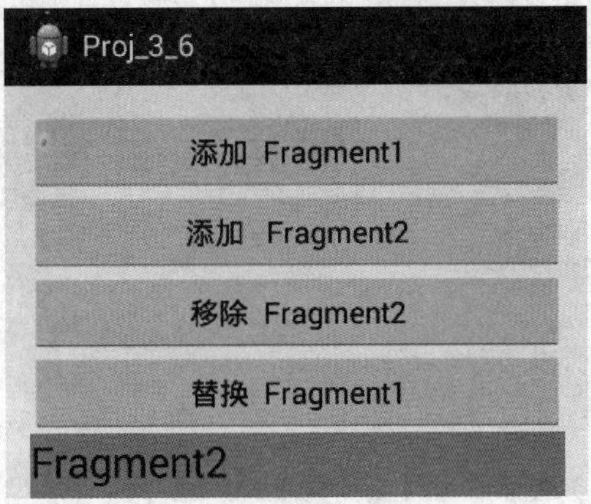

图 3.39 替换 Fragment1

3.4 习　　题

1. 选择题

(1) 结束一个 Activity 的方法是(　　)。
　　A. finish()　　　　B. close()　　　　C. destroy()　　　　D. shutdown()
(2) 对 Activity 布局文件的绑定应该在生命周期的(　　)函数中进行。
　　A. onPause()　　　B. onStart()　　　C. onCreate()　　　D. onResume()
(3) 在 Android 中,属于 Intent 的作用的是(　　)。
　　A. 实现应用程序间的数据共享
　　B. 没有用户界面的程序可以保持应用在后台运行
　　C. 可以实现界面间的切换,可以包含动作和动作数据
　　D. 处理一个应用程序整体性的工作
(4) Fragment 是(　　)。
　　A. 用户界面
　　B. 后台执行任务
　　C. 独立的一个部分,有自己的生命周期,但是必须加入 Activity 才能表现出来
　　D. 用于实现组件跳转
(5) Intent 包括以下(　　)部分。(多选)
　　A. action　　　　B. data　　　　C. extra　　　　D. flag

2. 简答题

(1) Activity 有哪几个基本状态?
(2) 请描述 Activity 生命周期函数以及调用的先后顺序。
(3) 请简述什么是 Intent 及其在 Android 中的作用。
(4) 使用 Fragment 实现不同频道新闻消息列表的显示。

第 4 章 使用项目资源

本章学习目标
- 掌握 Android 中各类资源的创建与引用。
- 掌握 NinePatch(9.png)图片的制作。
- 掌握动画资源的设置与使用。
- 了解主题资源的设置。

Android 应用程序也会使用各种资源,例如图片、字符串等,会把它们放入源码的相应文件夹下面,如/res/drawable、/res/xml、/res/values/、/res/raw、/res/layout 和/assets。Android 也支持并鼓励开发者把 UI 相关的布局和元素用资源来实现。

4.1 Android 资源类型

Android 中的资源可以分为两大类:可直接访问的资源以及无法直接访问的原生资源。
- 直接访问资源。这些资源可以使用 R 类进行访问,都保存在 res 目录下,在编译时会自动生成 R.java 资源索引文件。本章重点介绍该类资源。
- 原生资源。这些资源存放在 assets 下,不能使用 R 类进行访问,只能通过 AssetManager 以二进制流形式读取。

4.1.1 资源的创建与引用

1. assets 目录

assets 目录下保存的文件不能通过存取资源的方式在代码中访问。访问 assets 目录下文件的方式类似于打开文件,例如:

```
AssetsManager assetsMan = getAssets();
InputStream in = assetsMan.open("filepath");
```

在 assets 目录下还可以再创建目录,没有限制。但需要注意打开该资源时需要指出路径。在 assets 目录下可以放置任意类型的文件。这些文件会原封不动地被打包进 apk 中。

2. res 目录

res 目录下有固定的子目录,不同的子目录存放不同类别的资源文件。应用程序编译后会自动生产一个 R.java 文件,该文件中包含了 res 下所有定义的资源的 ID。

在 R 类中有很多的内部子类,每个子类对应一种类型的资源。这些资源不仅可以在 Java 代码中引用,也可以在资源文件 xml 中进行引用。

在 Java 代码中使用资源,可以通过 R 类来进行调用,如 R.drawable.photo 引用图片资源,R.dimen.lineheight 引用尺寸资源。

具体引用方式为

```
[<package_name>.]R.<resource_type>.<resource_name>
```

其中 package_name 是被引用的资源所在的包名。如果引用应用程序中的自定义资源,并且该 Java 类和 R 类不在同一个包,则需要导入 R 类所在包。例如:

```
ImageView imageView = (ImageView) findViewById(R.id.myimageview);
imageView.setImageResource(R.drawable.myimage);
```

如果引用 Android 系统定义的资源,则一般不导入包,而是在 R 类名前添加包名,包名为 android。例如:

```
setListAdapter(new ArrayAdapter<String>(this, android.R.layout.simple_list_item_1, myarray));
```

如果需要直接获取资源,可以通过 Resources 类来实现。Resources 类的实例可以通过 Context 类的 getResources 方法获得。例如:

```
Drawable drawPhoto = getResources().getDrawable(R.drawable.photo);
Color colorBg = getResources().getColor(R.color.bg_gray);
```

在 XML 文件中使用资源不需要借助 R 类,直接以引用的方式调用即可。调用的一般形式为

```
@[<package_name>:]<resource_type>/<resource_name>
```

例如:

```
<Button
    android:layout_width="fill_parent"
    android:layout_height="wrap_content"
    android:text="@string/submit" />
```

如果引用 Android 系统定义的资源,则需要在@后面使用包名。例如:

```
<EditText
    android:layout_width="match_parent"
    android:layout_height="wrap_content"
    android:textColor="@android:color/secondary_text_dark"
    android:text="@string/hello" />
```

除了引用某种资源外,在 xml 文件中还可以引用 style 资源的属性定义。引用方式为

?[<package_name>:][<resource_type>/]<resource_name>

其中 resource_type 是资源的类型,这里使用的是属性,一般都是 attr 这一项,所以一般省略。例如:

```
<EditText id = "text"
    android:layout_width = "fill_parent"
    android:layout_height = "wrap_content"
    android:textColor = "?android:textColorSecondary"
    android:text = "@string/hello_world" />
```

4.1.2 资源的分类

Android 中定义了如下的资源类型:

(1) 布局资源。定义应用程序中 UI 布局的 xml 文件,保存在 res/layout 目录下,通过 R.layout 类访问。

(2) 菜单资源。定义应用程序菜单的内容,保存在 res/menu 目录下,通过 R.menu 类访问。

(3) "值"资源。应用程序中所需要的字符串、尺寸、颜色等均可在 res/values 目录下定义相应的 xml 文件,通过 R.<resource_type>类来访问。

(4) 图片资源。定义各种图片或者 xml 的图片资源,保存在 res/drawable 目录下,通过 R.drawable 类访问。

(5) 动画资源。定义预设动画,保存在 res/anim 目录下,通过 R.anim 类访问。

(6) 样式资源。定义用户界面元素的外观和格式,保存在 res /values/styles.xml 文件中,通过 R.style 类访问。

4.2 布局资源

布局资源定义了一个 Activity 的 UI 结构或者一个组件的 UI。该资源以 xml 文件的形式定义在 res/layout 目录下,该文件名称将作为资源的标识 ID。

布局资源的访问方式有以下两种:

- Java 代码: R.layout.filename。
- Xml 代码: @[packagename]layout/filename。

xml 布局文件允许嵌套使用。例如,在 activity_main 中嵌套布局文件 menu.xml,那么在 activity_main.xml 中要使用<include layout="@layout/menu" />来引入资源。

布局资源可以被实例化为 View 类。布局文件的一般形式如下:

```
<?xml version = "1.0" encoding = "utf-8"?>
<ViewGroup xmlns:android = "http://schemas.android.com/apk/res/android"
    android:id = "@[ + ][package:]id/resource_name"
```

```
        android:layout_height = ["dimension" | "match_parent" | "wrap_content"]
        android:layout_width = ["dimension" | "match_parent" | "wrap_content"]
        [ViewGroup - specific attributes] >
    < View
            android:id = "@[ + ][package:]id/resource_name"
            android:layout_height = ["dimension" | "match_parent" | "wrap_content"]
            android:layout_width = ["dimension" | "match_parent" | "wrap_content"]
            [View - specific attributes] >
        < requestFocus/>
    </View >
    < ViewGroup >
        < View />
    </ViewGroup >
    < include layout = "@layout/layout_resource"/>
</ViewGroup >
```

例如，res/layout/activity_main.xml 代码如下：

```xml
<?xml version = "1.0" encoding = "utf - 8"?>
< LinearLayout xmlns:android = "http://schemas.android.com/apk/res/android"
              android:layout_width = "match_parent"
              android:layout_height = "match_parent"
              android:orientation = "vertical" >
    < TextView android:id = "@ + id/text"
              android:layout_width = "wrap_content"
              android:layout_height = "wrap_content"
              android:text = "Hello, I am a TextView" />
    < Button android:id = "@ + id/button"
            android:layout_width = "wrap_content"
            android:layout_height = "wrap_content"
            android:text = "Hello, I am a Button" />
</LinearLayout >
```

在应用程序的 MainActivity.java 的 onCreate 方法中可以引用该布局资源，代码如下：

```java
public void onCreate(Bundle savedInstanceState) {
    super.onCreate(savedInstanceState);
    setContentView(R.layout.main_activity);
}
```

4.3 菜单资源

Android 系统中有 options menu 与 context menu。不论任何类型的菜单资源都可以被定义在 res/menu 目录下，该文件名称将作为资源的标识 ID。

菜单资源只能通过 Java 代码访问，访问方式为 R.menu.filename。

菜单资源可以被实例化为 Menu 类。菜单文件的一般形式如下：

```xml
<?xml version = "1.0" encoding = "utf-8"?>
<menu xmlns:android = "http://schemas.android.com/apk/res/android">
    <item android:id = "@[+][package:]id/resource_name"
        android:title = "string"
        android:titleCondensed = "string"
        android:icon = "@[package:]drawable/drawable_resource_name"
        android:onClick = "method name"
        android:showAsAction = ["ifRoom" | "never" | "withText" | "always" |
                                "collapseActionView"]
        android:actionLayout = "@[package:]layout/layout_resource_name"
        android:actionViewClass = "class name"
        android:actionProviderClass = "class name"
        android:alphabeticShortcut = "string"
        android:numericShortcut = "string"
        android:checkable = ["true" | "false"]
        android:visible = ["true" | "false"]
        android:enabled = ["true" | "false"]
        android:menuCategory = ["container" | "system" | "secondary" | "alternative"]
        android:orderInCategory = "integer" />
    <group android:id = "@[+][package:]id/resource name"
        android:checkableBehavior = ["none" | "all" | "single"]
        android:visible = ["true" | "false"]
        android:enabled = ["true" | "false"]
        android:menuCategory = ["container" | "system" | "secondary" | "alternative"]
        android:orderInCategory = "integer" >
        <item />
    </group>
    <item >
        <menu>
            <item />
        </menu>
    </item>
</menu>
```

4.3.1 普通菜单

普通菜单有 3 种类型：options menu、context menu 和 sub menu。

1. OptionsMenu

OptionsMenu 用于在 Activity 中单击 Menu 显示出来的菜单选项。OptionMenu 默认最多显示 6 个选项，当多于 6 个时就会自动显示为"更多/More"，单击时可展开其他菜单选项。

options menu 创建步骤如下：

(1) 创建菜单资源 xml 文件。

(2) 重写 Activity 的 onCreateOptionsMenu 方法，实现创建菜单功能。

(3) 当菜单项(MenuItem)被选中时，重写 Activity 的 onOptionsMenuSelected 方法响应事件。

📖 Proj4_1 项目中创建 MainActivity 的菜单资源 menu.xml 文件

```xml
<menu xmlns:android = "http://schemas.android.com/apk/res/android" >
    <item
        android:id = "@ + id/item1"
        android:icon = "@drawable/group_item1_icon"
        android:title = "@string/item1"/>
    <group android:id = "@ + id/group" >
        <item
            android:id = "@ + id/group_item1"
            android:icon = "@drawable/group_item1_icon"
            android:checkable = "true"
            android:title = "@string/group_item1"/>
        <item
            android:id = "@ + id/group_item2"
            android:icon = "@drawable/group_item2_icon"
            android:checkable = "true"
            android:title = "@string/group_item2"/>
    </group>
    <item
        android:id = "@ + id/submenu"
        android:title = "@string/submenu_title">
        <menu>
            <item
                android:id = "@ + id/submenu_item1"
                android:title = "@string/submenu_item1"/>
        </menu>
    </item>
</menu>
```

📖 Proj4_1 项目中 MainActivity.java 中创建菜单和处理菜单选项逻辑部分代码

```java
/**
 * 创建 Activity 的菜单
 */
@Override
public boolean onCreateOptionsMenu(Menu menu) {
    getMenuInflater().inflate(R.menu.main, menu);
    return true;
}
/**
 * 处理 Activity 的菜单选项被选中时的逻辑
 */
@Override
public boolean onOptionsItemSelected(MenuItem item) {
    switch(item.getItemId()) {
        case R.id.item1 :
            Toast.makeText(this, "item1", Toast.LENGTH_LONG).show() ;
            break;
        case R.id.group_item1 :
            Toast.makeText(this, "group_item1", Toast.LENGTH_LONG).show() ;
            break;
```

```
            case R.id.group_item2 :
                Toast.makeText(this, "group_item2", Toast.LENGTH_LONG).show() ;
                break;
            case R.id.submenu_item1 :
                Toast.makeText(this, "submenu_item1", Toast.LENGTH_LONG).show() ;
                break;
        }
        return false;
    }
```

启动项目,运行界面如图 4.1 和图 4.2 所示。

图 4.1　单击菜单弹出菜单选项

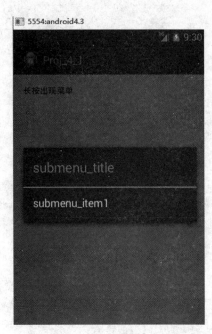

图 4.2　子菜单效果

2. SubMenu

Menu 的菜单项可以是 MenuItem,也可以是 SubMenu。MenuItem 直接显示在 Menu 上;SubMenu 可以包含二级菜单,可将功能类似的一组菜单项组合在一起形成多级显示。在上面的 menu.xml 代码中,最下方 item 内部嵌套的 Menu 项就是 SubMenu。

3. ContextMenu

ContextMenu 是一个浮动菜单,当用户在 Activity 的某一个控件元素上长按时出现该菜单。ContextMenu 的创建步骤如下:

(1) 创建菜单资源 xml 文件。

(2) 重写 Activity 的 onCreateOptionsMenu 方法,实现创建菜单功能。

(3) 当菜单项(MenuItem)被选中时,重写 Activity 的 onOptionsMenuSelected 方法响应事件。

(4) 在 Activity 中调用 registerForContextMenu 为控件注册 ContextMenu。

📖 Proj4_1 项目中 MainActivity.java 主要代码

```java
public class MainActivity extends Activity {
    private TextView tv;
    @Override
    protected void onCreate(Bundle savedInstanceState) {
        super.onCreate(savedInstanceState);
        setContentView(R.layout.activity_main);
        tv = (TextView) findViewById(R.id.main_tv);
        registerForContextMenu(tv);
    }
    @Override
    public void onCreateContextMenu(ContextMenu menu, View v,
            ContextMenuInfo menuInfo) {
        if(v == tv) {
            getMenuInflater().inflate(R.menu.main_context, menu);
        }
    }
    @Override
    public boolean onContextItemSelected(MenuItem item) {
        switch(item.getItemId()) {
        case R.id.context_item1 :
            Toast.makeText(this, "submenu_item1", Toast.LENGTH_LONG).show();
            break;
        case R.id.context_item2 :
            Toast.makeText(this, "submenu_item1", Toast.LENGTH_LONG).show();
            break;
        }
        return false;
    }
}
```

在项目 Proj_4_1 项目中,长按 TextView 控件出现 ContextMenu,运行界面如图 4.3 所示。

4.3.2 ActionBar 中的菜单

ActionBar 是 Android 3.0 新增的内容,取代了原有的标题栏组件,为用户提供菜单和导航。当 ActionBar 不能显示全部菜单项时,可以将其隐藏起来,变为二级菜单。用户可以单击 ActionBar 右侧的一级菜单(或设备的菜单按钮,如果可用的话),打开菜单列表,选择二级菜单。

当在 xml 资源文件中添加 ActionItem 时,只需要为原有的 item 设置 android:showAsAction 属性,该属性有 5 个不同的取值,如果需要取两个值,值中间使用"|"分隔。该属性的 5 个值含义如下:

- never,该菜单项不显示在 ActionBar 上。
- ifRoom,当 ActionBar 上有足够的空间时,显

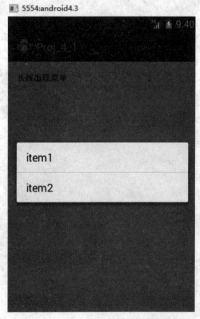

图 4.3 弹出式菜单

示该菜单项。
- always,一直显示该菜单项。
- withText,菜单项的图标和文本信息都显示在 ActionBar 上。
- collapseActionView,将 ActionView 折叠为普通的菜单项。

📖 Proj_4_2 项目中菜单资源 main.xml 代码

```
< menu xmlns:android = "http://schemas.android.com/apk/res/android" >
    < item
        android:id = "@ + id/search"
        android:icon = "@android:drawable/ic_menu_search"
        android:showAsAction = "always|withText"
        android:title = "搜索"/>
    < item
        android:id = "@ + id/edit"
        android:icon = "@android:drawable/ic_menu_edit"
        android:showAsAction = "always|withText"
        android:title = "编辑"/>
    < item
        android:id = "@ + id/search"
        android:icon = "@android:drawable/ic_menu_delete"
        android:showAsAction = "always|withText"
        android:title = "删除"/>
    < item
        android:id = "@ + id/exit"
        android:id = "@ + id/exit"
        android:icon = "@android:drawable/ic_menu_close_clear_cancel"
        android:showAsAction = "always|withText"
        android:title = "退出"/>
</menu >
```

运行时如果是竖屏显示,界面如图 4.4 所示;如果是横屏显示,界面如图 4.5 所示。

图 4.4　竖屏菜单效果

图 4.5　横屏菜单效果

4.4 "值"资源

在 Android 应用程序中,字符串、颜色、尺寸、样式等均可在 xml 文件中定义,这些文件被设置在 res/values 目录下,按照类别分别被保存为独立的 xml 文件,并且这些 xml 文件的根元素均是 resource。

4.4.1 字符串

字符串资源存放在/res/values/strings.xml 中。在 xml 中访问时,格式为@string/*；在 Java 代码中获取 id 的格式为 R.string.<string_name>,获取字符串的格式为 getResource().getString()。例如：

```
<string name = "info">从字符串资源引用</string>
<string name = "show">来自 values/strings.xml 文件</string>
<string name = "show2">字符串资源</string>
```

在 xml 文件中引用字符串资源：

```
<TextView
    android:layout_width = "wrap_content"
    android:layout_height = "wrap_content"
    android:text = "@string/info" />
```

在 Java 代码中引用字符串资源：

```
TextView tv = (TextView) findViewById(R.id.main_tv);
tv.setText(R.string.show);
String str = getResources().getString(R.string.show2);
TextView tv2 = (TextView) findViewById(R.id.main_tv);
tv2.setText(str);
```

4.4.2 颜色资源

颜色资源放置在/res/values/colors.xml,xml 中的引用方式为@color/*；在 Java 代码中获取 id 的格式为 R.color.<color_name>,获取颜色值的格式为 getResource().getColor()。例如：

```
<color name = "red">#f00</color>
<color name = "green">#0f0</color>
<color name = "blue">#00f</color>
```

在 xml 文件中引用颜色资源：

```
<TextView
    android:layout_width = "wrap_content"
```

```
    android:layout_height = "wrap_content"
    android:textColor = "@color/blue"
    android:text = "@string/info" />
```

在 Java 代码中引用字符串资源:

```
TextView tv = (TextView) findViewById(R.id.main_tv);
tv.setTextColor(getResources().getColor(R.color.red));
```

4.4.3 尺寸资源

尺寸资源放置在 /res/values/dimens.xml 中。在 xml 中的引用方式是 @dimen/*;在 Java 代码中获取 id 的格式为 R.dimen.<dimen_name>,获取尺寸值的格式为 getResource().getDimen()。例如:

```
<dimen name = "line_height">40dp</dimen>
<dimen name = "textsize">16sp</dimen>
```

在 xml 文件中引用尺寸资源:

```
<TextView
    android:id = "@+id/main_tv2"
    android:layout_width = "wrap_content"
    android:layout_height = "@dimen/line_height"
    android:background = "@color/red" />
```

在 Java 代码中引用尺寸资源:

```
TextView tv = (TextView) findViewById(R.id.main_tv);
tv.setTextSize(getResources().getDimension(R.dimen.textsize));
```

4.5 可绘制资源

一个 drawable 资源可以来自一个图片文件资源,也可以来自 xml 文件中的定义。可以使用 android:drawable 或者 android:icon 属性把它们应用到 XML 配置文件中。

4.5.1 Android 中的图片类型

在 Android 应用程序中,目前支持的图片类型有如下 3 种:
- JPG。是图片基本格式,也是照片的标准格式,图片颜色丰富,但是不支持透明。大小比较适中,只有几十千字节。
- GIF。被限制在 256 色,因此对于大面积纯色和简单图像的显示效果好。GIF 支持透明,但是会产生二维图形中锯齿边缘的效果。Android 目前不支持图形的动态效果。

- PNG。是 Android 推荐使用的图片格式。具有 JPG 格式的图片质量和 GIF 格式的透明度,且无锯齿缺陷。一般 PNG 格式的大小是几百千字节,比 JPG 格式的要大。

4.5.2 NinePatch 图片格式

NinePatch 是一种 PNG 图片资源,这种格式的图片资源可以定义拉伸区域。如果将 View 对象的一个尺寸设置为 wrap_content,当 View 对象根据容纳的内容增长时,NinePatch 图片也会根据 View 对象的大小被缩放。例如,使用 NinePatch 图片资源作为按钮的背景,背景图片可以随着按钮的大小而伸缩。

NinePatch 资源是一个标准的 PNG 图片外加 1 像素的边框,必须保存为 9.PNG 格式,而且该资源放置在 drawable 目录下。9.PNG 的 1 像素的边框用来定义图形的拉伸和静态区域,如图 4.6 所示,left(左)和 top(上)边的交叉部分是可拉伸部分,未选中部分是静态区域部分。如图 4.7 所示,right(右)和 bottom(下)边框的交叉部分则是内容部分。

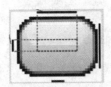
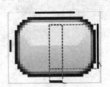

图 4.6　拉伸区域　　　　　图 4.7　内容区域

无论是 left 和 top 还是 right 和 bottom,都是把图片分成 9 块(边角处的 4 块是不能缩放的,其他 5 块则是允许缩放的),所以叫做 9.PNG。

9.PNG 图片的编辑可以使用 SDK 自带的工具 draw9patch(位于 SDK 的 tools 目录下)。双击运行该批处理文件,等待一会儿将会弹出一个编辑窗体,如图 4.8 所示。

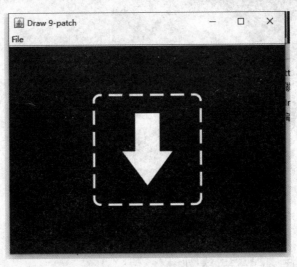

图 4.8　9.png 编辑器

单击 File 按钮,打开一张普通的 PNG 图片,如图 4.9 所示。

可以看到 PNG 图片边缘有一圈透明像素,这是用来标记拉伸区域在横向和纵向上的对应区域位置的。默认的拉伸是整体拉伸,在实际应用中,一般圆角部分并不拉伸,所以要

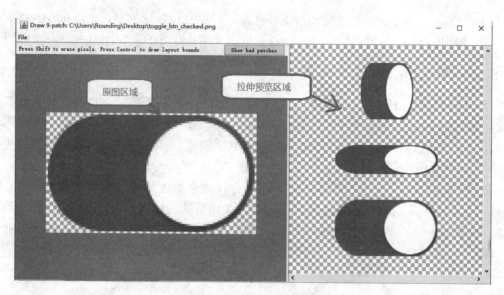

图 4.9 未设置拉伸区域的图片

编辑拉伸区域。将鼠标放在白色边框，按住左键不放拖动，会出现黑色线条，如图 4.10 所示。

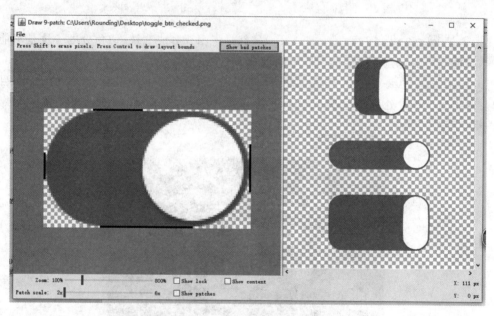

图 4.10 设置拉伸区域

单击图片窗口右上角 Show bad patches 按钮，将会出现两个矩形，并有一部分交叉区域，交叉区域即为拉伸时的拉伸区域，如图 4.11 所示。

编辑区域拉伸设置作用如下：
- 左侧黑色条位置向右覆盖的区域表示图片纵向拉伸时只拉伸该区域。
- 上方黑色条位置向下覆盖的区域表示图片横向拉伸时只拉伸该区域。
- 右侧黑色条位置向左覆盖的区域表示图片纵向显示内容的区域。

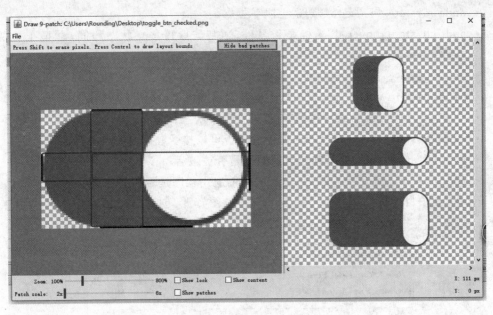

图 4.11　显示拉伸区域

- 下方黑色条位置向上覆盖的区域表示图片横向显示内容的区域。
- 没有黑色条的位置覆盖的区域是图片拉伸时保持不变。

如果操作失误多选了拉伸部分,可按住 Shift 键,单击黑色条将其去掉。图 4.11 右侧的 3 个图片表示拉伸后的预览效果。

选择好区域后,单击左上角 File,然后单击下拉菜单中的 Save 9-patch 保存图片。对比为编辑的 PNG 图片,9.PNG 格式的图略有不同,如图 4.12 所示。

图 4.12　PNG 与 9.PNG 对比

使用编辑好的 9.PNG 图片做 View 背景时,当 View 大小和图片大小不一致时,图片会按照上方预览效果进行拉伸,而不是最初严重变形的拉伸效果。

最后要说明的是,9.PNG 图片和其他图片一样放置在 res/drawable 目录下,使用@drawable/ * 来引用。

4.5.3　selector 资源

在开发过程中往往需要针对一个 View 在不同的状态使用不同的背景图片或者颜色,例如一个按钮希望按下状态使用 A 背景图,其他状态使用 B 背景图,这时可以使用 StateListDrawable 来实现效果。

selector 资源可以分别针对不同状态设置不同的颜色或者不同的图片,保存位置分别为 res/color/ * .xml 与 res/drawable/ * .xml。

selector 资源定义的 xml 文件,根元素选择 selector,在该 xml 文件中每种状态可以使用<item>元素来描述。

Item 中描述状态的常用属性如下:

- android:state_pressed,判断一个按钮被触摸或被点击状态。

- android:state_focused,判断是否取得焦点。
- android:state_hovered,光标是否悬停,通常与取得焦点状态相同,它是4.0的新特性。
- android:state_selected,判断是否被选中。
- android:state_checkable,判断当前组件是否被选中,多用于复选框中。
- android:state_checked,判断当前控件被选中状态。
- android:state_enabled,判断当前控件是否可以接受触摸或者点击事件。
- android:state_activated,判断当前控件是否被激活。
- android:state_window_focused,判断当前应用程序是否在前台显示。

Proj_4_3项目为按钮创建一个selector背景图资源,创建时资源类型选择如图4.13所示。

图4.13 创建Selector资源

📖 Proj_4_3项目中文件/res/drawable/btn_red_selector.xml代码

```
<?xml version = "1.0" encoding = "utf - 8"?>
<selector xmlns:android = "http://schemas.android.com/apk/res/android">
    <!-- 设置按钮被按下时的背景图 -->
    <item android:drawable = "@drawable/common_btn_red_pressed"
          android:state_pressed = "true"/>
    <!-- 设置按钮不可用时的背景图 -->
    <item android:drawable = "@drawable/common_btn_unabled"
```

```
            android:state_enabled = "false" />
    <!-- 设置按钮默认状态的背景图 -->
    <item android:drawable = "@drawable/common_btn_red"></item>
</selector>
```

📖 Proj_4_3 项目中文件 activity_main 将以上资源用于按钮的相关代码

```
<Button
    android:id = "@+id/btn1"
    android:layout_width = "wrap_content"
    android:layout_height = "wrap_content"
    android:background = "@drawable/btn_red_selector"
    android:text = "selector 资源"
    android:layout_marginTop = "16dp" />
<Button
    android:id = "@+id/btn2"
    android:layout_width = "wrap_content"
    android:layout_height = "wrap_content"
    android:background = "@drawable/btn_red_selector"
    android:enabled = "false"
    android:text = "selector 资源"
    android:textColor = "#000"
    android:layout_marginTop = "16dp" />
```

运行时的界面效果如图 4.14 所示,当按钮被按下时效果如图 4.15 所示。

图 4.14　默认状态效果　　　　　　图 4.15　按钮被按下时的效果

如果希望按钮在不同状态下文字颜色也有所不同,则在/res/color/下定义 xml 文件来进行设置。

📖 Proj_4_3 项目中文件/res/color/btn_redbg_textcolor_selector.xml 代码

```
<?xml version = "1.0" encoding = "utf-8"?>
<selector xmlns:android = "http://schemas.android.com/apk/res/android">
    <!-- 设置被按下时颜色为红色 -->
    <item android:state_pressed = "true" android:color = "#fff"/>
    <!-- 设置被按下时颜色为灰色 -->
    <item android:state_enabled = "false" android:color = "#ccc"/>
    <!-- 设置默认状态颜色为黑色 -->
    <item android:color = "#000"/>
</selector>
```

📖 Proj_4_3 项目中文件 activity_main 将以上资源用于按钮的相关代码

```
<Button
    android:id = "@ + id/btn3"
    android:layout_width = "wrap_content"
    android:layout_height = "wrap_content"
    android:background = "@drawable/btn_red_selector"
    android:text = "selector 资源"
    android:textColor = "@color/btn_redbg_textcolor_selector"
    android:layout_marginTop = "16dp"/>
<Button
    android:id = "@ + id/btn4"
    android:layout_width = "wrap_content"
    android:layout_height = "wrap_content"
    android:background = "@drawable/btn_red_selector"
    android:enabled = "false"
    android:text = "selector 资源"
    android:textColor = "@color/btn_redbg_textcolor_selector"
    android:layout_marginTop = "16dp"/>
```

运行时的默认界面效果如图 4.16 所示,单击按钮时的效果如图 4.17 所示。

图 4.16　默认状态时的文字颜色效果　　　　图 4.17　按钮被按下时的文字颜色效果

4.5.4　shape 资源

shape 资源用于设定形状,可以在 selector、layout 等里面使用,有 6 个子标签,各属性如下:

```
<?xml version = "1.0" encoding = "utf - 8"?>
<shape xmlns:android = http://schemas.android.com/apk/res/android
    android:shape = ["rectangle" | "oval" | "line" | "ring"]>
    <!-- 圆角 -->
    <corners
        android:radius = "integer"
        android:topLeftRadius = "integer"
        android:topRightRadius = "integer"
        android:bottomLeftRadius = "integer"
        android:bottomRightRadius = "integer"/>
    <!-- 渐变 -->
    <gradient
        android:angle = "integer"
        android:centerX = "integer "
        android:centerY = "integer "
```

```
        android:gradientRadius = "90"
        android:startColor = "color"
        android:centerColor = "color"
        android:endColor = "color"
        android:type = ["linear" | "radial" | "sweep"]
        android:useLevel = ["true" | "false" />
<!-- 间隔 -->
<padding
        android:left = "integer"
        android:top = "integer"
        android:right = "integer"
        android:bottom = "integer"/>
<!-- 大小 -->
<size
        android:width = "integer"
        android:height = "integer"/>
<!-- 填充 -->
<solid
        android:color = "color"/>
<!-- 描边 -->
<stroke
        android:width = "integer"
        android:color = "color"
        android:dashWidth = "integer"
        android:dashGap = "integer"/>
</shape>
```

在 shape 资源的 xml 文件中必须使用< shape >作为根元素。android：shape = ["rectangle" | "oval" | "line" | "ring"]用来设置形状,其中 rectagle 为矩形,oval 为椭圆,line 为水平直线,ring 为环形。

< corner >设置圆角。android：radius 设置圆角的半径,值越大角越圆;android：topRightRadius 设置右上圆角半径;android：bottomLeftRadius 设置左下圆角角半径;android：topLeftRadius 设置左上圆角半径;android：bottomRightRadius 设置右下圆角半径。如果同时设置五个属性,则 radius 属性无效。

< gradient >设置渐变。android：startColor 设置起始颜色;android：endColor 设置结束颜色;android：angle 设置渐变角度,0 表示从上到下,90 表示从左到右,数值必须为 45 的整数倍,默认为 0,该设置仅在 type＝"linear"时有效;android：type 设置渐变的样式,linear 为线性渐变,radial 为环形渐变,sweep 为扫描线渐变。

< padding >设置 4 个方向上的填充间隔。

< size >设置大小。

< solid >设置填充的颜色。

< stroke >设置描边。dashWidth 和 dashGap 属性只要其中一个设置为 0dp,则边框为实线边框;android：width 设置边框的宽度;android：color 设置边框的颜色;android：dashWidth 设置虚线的宽度;android：dashGap 设置虚线的间隔宽度。

📖 Proj_4_3 项目中文件/res/drawable/textview_bg.xml 代码

```xml
<?xml version = "1.0" encoding = "utf-8"?>
<shape xmlns:android = "http://schemas.android.com/apk/res/android"
    android:shape = "oval">
    <!-- 从左下角到右上角绘制渐变色  -->
    <gradient android:startColor = "#FFFF0000" android:endColor = "#80FF00FF"
        android:angle = "45" />
    <!-- 定义控件内容到边界的距离,到4条边界的距离都是8像素  -->
    <padding android:left = "8dp" android:top = "8dp" android:right = "8dp"
        android:bottom = "8dp" />
    <!-- 边框线宽度是2dp,颜色为白色的边框线  -->
    <stroke android:width = "2dp" android:color = "#fff" />
    <!-- 圆角半径是8dp  -->
    <corners android:radius = "8dp" />
</shape>
```

📖 Proj_4_3 项目中将以上资源用于 TextView 的 activity_main.xml 相关代码

```xml
<TextView
    android:layout_width = "wrap_content"
    android:layout_height = "wrap_content"
    android:text = "Shape Resource"
    android:layout_margin = "8dp"
    android:background = "@drawable/textview_bg" />
```

使用了上述 shape 后,TextView 控件界面运行效果如图 4.18 所示。

图 4.18 渐变背景

4.6 动画资源

Android 中对动画的支持非常强大,可以使用资源文件定义动画,也可以在 Java 代码中创建动画,(在此重点介绍使用资源文件定义动画)。Android 中的动画主要分为两大类: Tween Animation 与 Frame Animation。

- Tween Animation(补间动画)通过对显示对象的内容执行连续的简单变化来创建一个动画,它提供了旋转、移动、伸展和淡入淡出效果。
- Frame Animation(帧动画)按顺序播放事先规划好的一系列图像来产生动画的效果,如同卡通电影的带子。

4.6.1 Tween Animation

Tween Animation 也称为属性动画,通过控制透明度、位置、角度和放缩属性实现动画效果。资源定义位置为/res/anim/ * .xml,访问方式为 R.anim. * 。Tween Animation 共

提供了 4 种动画效果:

(1) AlphaAnimation,透明度动画效果,使用< alpha >作为 xml 中的元素。
(2) RotateAnimation,旋转动画效果,使用< rotate >作为 xml 中的元素。
(3) ScaleAnimation,缩放动画效果,使用< scale >作为 xml 中的元素。
(4) TranslateAnimation,位移动画效果,使用< translate >作为 xml 中的元素。

以上 4 种动画在设置时有一些共同的属性,如下:

- duration,long 类型,设置动画显示的时间,以毫秒为单位。
- fillAfter,设置为 true 时,动画最终停留在结束后。
- fillBefore,设置为 true 时,动画最终停留在开始时。
- repeatCount,int 类型,动画重复次数。
- repeatMode,设置动画重复模式,restart 为重新播放,reverse 为反向播放。
- interpolator,设置动画的内插件。
- zAdjustment,设置动画的 Z 轴模式,0 为保持不变,1 为保持在最上层,−1 为保持在最下层。
- startOffSet,int 类型,设置动画播放的延迟时间。

Tween Animation 的设置一般为如下形式:

```
<?xml version = "1.0" encoding = "utf-8"?>
< set xmlns:android = "http://schemas.android.com/apk/res/android"
    android:interpolator = "@[package:]anim/interpolator_resource"
    android:shareInterpolator = ["true" | "false"] >
    < alpha
        android:fromAlpha = "float"
        android:toAlpha = "float" />
    < scale
        android:fromXScale = "float"
        android:toXScale = "float"
        android:fromYScale = "float"
        android:toYScale = "float"
        android:pivotX = "float"
        android:pivotY = "float" />
    < translate
        android:fromXDelta = "float"
        android:toXDelta = "float"
        android:fromYDelta = "float"
        android:toYDelta = "float" />
    < rotate
        android:fromDegrees = "float"
        android:toDegrees = "float"
        android:pivotX = "float"
        android:pivotY = "float" />
    < set >
        …
    </ set >
</ set >
```

1. alpha 动画

属性 fromAlpha 与 toAlpha 分别表示动画开始时和结束时的透明度,取值为 0.0~1.0。

📖 Proj_4_4 项目中文件/res/anim/alpha.xml 代码

```xml
<?xml version = "1.0" encoding = "utf-8"?>
<set xmlns:android = "http://schemas.android.com/apk/res/android">
    <alpha
        android:fromAlpha = "1.0"
        android:toAlpha = "0.1"
        android:fillAfter = "true"
        android:duration = "2000" />
</set>
```

2. translate 动画

属性 fromXDelta、toXDelta 表示动画开始和结束时的 X 坐标位置,属性 fromYDelta、toYDelta 表示动画开始和结束时的 Y 坐标位置。

注意,采用坐标值表示时有 3 种设置方法:int 表示以自身为起点的绝对坐标,％表示相对自己大小的百分比计算值,％p 表示父控件的百分比计算值。

📖 Proj_4_4 项目中文件/res/anim/translate.xml 代码

```xml
<?xml version = "1.0" encoding = "utf-8"?>
<set xmlns:android = "http://schemas.android.com/apk/res/android" >
    <translate
        android:fromXDelta = "0%"
        android:fromYDelta = "0%"
        android:toXDelta = "50%p"
        android:toYDelta = "50%p"
        android:duration = "2000" />
</set>
```

3. scale 动画

属性 fromXScale,toXScale 表示动画开始和结束时的 X 坐标上的尺寸,值从 0.0 到 1.0 表示缩小,值大于 1 表示放大,值为负数会产生水平镜像;属性 fromYScale,toYScale 表示动画开始和结束时的 Y 坐标上的尺寸,值从 0.0 到 1.0 表示缩小,值大于 1 表示放大,值为负数会产生竖直镜像;属性 pivotX、pivotY 表示动画执行时 X、Y 坐标的开始位置。

📖 Proj_4_4 项目中文件/res/anim/scale.xml 代码

```xml
<?xml version = "1.0" encoding = "utf-8"?>
<set xmlns:android = "http://schemas.android.com/apk/res/android" >
    <scale
        android:fromXScale = "1"
        android:fromYScale = "1"
        android:toXScale = "0.1"
        android:toYScale = "0.1"
        android:pivotX = "50%"
        android:pivotY = "50%"
        android:duration = "2000"/>
</set>
```

4. rotate 动画

属性 fromDegrees、toDegrees 表示动画开始和结束时的对象角度，属性 pivotX、pivotY 表示动画执行时 X、Y 坐标的开始位置。

📖 Proj_4_4 项目中文件/res/anim/rotate.xml 代码

```
<?xml version = "1.0" encoding = "utf-8"?>
<set xmlns:android = "http://schemas.android.com/apk/res/android">
    <rotate
        android:fromDegrees = "0"
        android:toDegrees = "280"
        android:pivotX = "50%"
        android:pivotY = "50%"
        android:duration = "2000"
        android:repeatCount = "1"
        android:repeatMode = "reverse" />
</set>
```

以上 4 种动画可以单独设置，也可以将多个单独动画叠加使用，也就是在< set >内部设置多种动画，让这些动画一起演示。

📖 Proj_4_4 项目中文件/res/anim/set.xml 代码

```
<?xml version = "1.0" encoding = "utf-8"?>
<set xmlns:android = "http://schemas.android.com/apk/res/android">
    <alpha
        android:fromAlpha = "1.0"
        android:toAlpha = "0.1"
        android:duration = "2000" />
    <rotate
        android:fromDegrees = "0"
        android:toDegrees = "280"
        android:pivotX = "50%"
        android:pivotY = "50%"
        android:duration = "2000"
        android:repeatCount = "1"
        android:repeatMode = "reverse"/>
    <translate
        android:fromXDelta = "0%"
        android:fromYDelta = "0%"
        android:toXDelta = "90%p"
        android:toYDelta = "90%p"
        android:duration = "2000" />
    <scale
        android:fromXScale = "1"
        android:fromYScale = "1"
        android:toXScale = "0.1"
        android:toYScale = "0.1"
        android:pivotX = "50%"
        android:pivotY = "50%"
        android:duration = "2000" />
</set>
```

当设置好动画资源后，可以使用 AnimationUtils 的静态方法 loadAnimation 载入动画。控件可以通过调用 startAnimation 方法开启动画。

📖 Proj_4_4 项目中文件 MainActivity.java 代码

```java
public class MainActivity extends Activity implements OnClickListener {
    private ImageView iv;
    // 动画对象
    private Animation anim;
    @Override
    protected void onCreate(Bundle savedInstanceState) {
        super.onCreate(savedInstanceState);
        setContentView(R.layout.activity_main);
        iv = (ImageView) findViewById(R.id.iv);
    }
    @Override
    public void onClick(View v) {
        switch (v.getId()) {
            case R.id.btn_alpha:
                // 使用 AnimationUtils 的静态方法 loadAnimation 载入动画
                anim = AnimationUtils.loadAnimation(this, R.anim.alpha);
                break;
            case R.id.btn_rotate:
                anim = AnimationUtils.loadAnimation(this, R.anim.rotate);
                break;
            case R.id.btn_scale:
                anim = AnimationUtils.loadAnimation(this, R.anim.scale);
                break;
            case R.id.btn_translate:
                anim = AnimationUtils.loadAnimation(this, R.anim.translate);
                break;
            case R.id.btn_set:
                anim = AnimationUtils.loadAnimation(this, R.anim.set);
                break;
            default:
                iv = null;
                break;
        }
        // 控件加载动画
        iv.startAnimation(anim);
    }
}
```

项目运行时，界面的初始效果如图 4.19 所示。

单击 alpha 按钮，图形会改变透明度，如图 4.20 所示。

单击 translate 按钮，图形会发生位移，如图 4.21 所示。

单击 rotate 按钮,图形旋转,如图 4.22 所示。

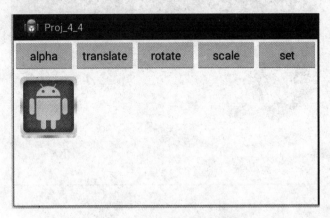

图 4.19　动画效果初始界面

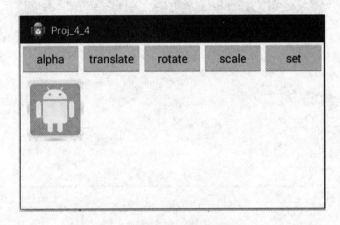

图 4.20　alpha 动画效果

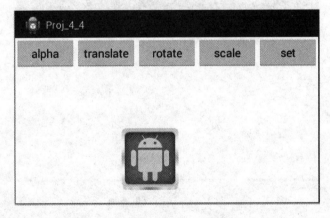

图 4.21　translate 动画效果

单击 scale 按钮,图形会缩放,如图 4.23 所示。
单击 set 按钮,图形会叠加演示几种动画,如图 4.24 所示。

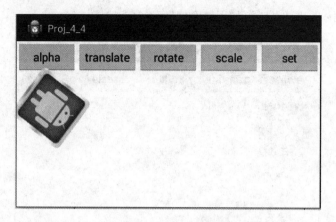

图 4.22 rotate 动画效果

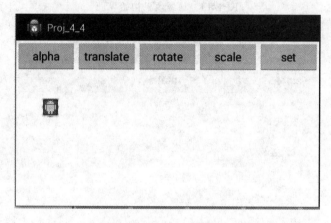

图 4.23 rotate 动画效果

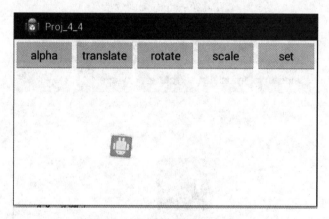

图 4.24 动画效果叠加

4.6.2 Frame Animation

Frame Animation 也称为关键帧动画,在很短的时间内连续呈现多张内容相关的图片,利用视觉暂留效应产生动画效果。资源定义位置为/res/drawable/ * .xml,访问方式为 R.

drawable.*。

设置代码如下：

```xml
<?xml version = "1.0" encoding = "utf-8"?>
<animation-list xmlns:android = "http://schemas.android.com/apk/res/android"
    android:oneshot = ["true" | "false"] >
    <item
        android:drawable = "@[package:]drawable/drawable_resource_name"
        android:duration = "integer" />
</animation-list>
```

其中根元素必须使用<animation-list>，内部可以根据图片的数量放置若干<item>元素，android：oneshot 属性设置为 true 表示动画仅播放一次，为 false 表示循环播放。

使用动画时，可以将 FrameAnimation 资源设置为控件背景，但是动画并不播放，FrameAnimation 动画由 AnimationDrawable 管理，通过调用该对象的 start 方法开始播放，调用 stop 方法停止播放。

📖 Proj_4_4 项目中文件/res/drawable/frame.xml 代码

```xml
<?xml version = "1.0" encoding = "utf-8"?>
<animation-list xmlns:android = "http://schemas.android.com/apk/res/android"
    android:oneshot = "true" >
    <item
        android:drawable = "@drawable/progress_1"
        android:duration = "200"/>
    <item
        android:drawable = "@drawable/progress_2"
        android:duration = "200"/>
    <item
        android:drawable = "@drawable/progress_3"
        android:duration = "200"/>
    <item
        android:drawable = "@drawable/progress_4"
        android:duration = "200"/>
    <item
        android:drawable = "@drawable/progress_5"
        android:duration = "200"/>
    <item
        android:drawable = "@drawable/progress_6"
        android:duration = "200"/>
    <item
        android:drawable = "@drawable/progress_7"
        android:duration = "200"/>
    <item
        android:drawable = "@drawable/progress_8"
        android:duration = "200"/>
</animation-list>
```

其中图片 progress_1 到 progress_8 如图 4.25 所示。

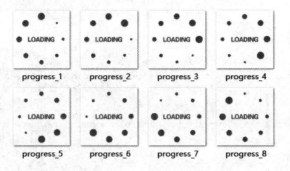

图 4.25 帧动画原图

📖 Proj_4_4 项目中文件 activity_main.xml 中相关代码

```
< ImageView
    android:id = "@ + id/iv2"
    android:layout_width = "wrap_content"
    android:layout_height = "wrap_content"/>
< LinearLayout
    android:layout_width = "match_parent"
    android:layout_height = "wrap_content"
    android:orientation = "horizontal" >
    < Button
        android:id = "@ + id/btn_start"
        android:layout_width = "0dp"
        android:layout_height = "wrap_content"
        android:layout_weight = "1"
        android:text = "start"
        android:onClick = "onClick2" />
    < Button
        android:id = "@ + id/btn_stop"
        android:layout_width = "0dp"
        android:layout_height = "wrap_content"
        android:layout_weight = "1"
        android:text = "stop"
        android:onClick = "onClick2" />
</LinearLayout >
```

📖 Proj_4_4 项目中文件 MainActivity.java 中为 ImageView 设置背景和启动、停止动画的相关代码

```
private AnimationDrawable animDraw ;
    @Override
    protected void onCreate(Bundle savedInstanceState) {
        super.onCreate(savedInstanceState);
        setContentView(R.layout.activity_main);
        ...
        iv2 = (ImageView) findViewById(R.id.iv2) ;
```

```
            iv2.setBackgroundResource(R.drawable.frame);
            animDraw = (AnimationDrawable) iv2.getBackground();
    }
    public void onClick2(View v) {
        if(v.getId() == R.id.btn_start) {
            animDraw.start();
        }
        else {
            animDraw.stop();
        }
    }
    …
}
```

4.7 样式与主题资源

4.7.1 样式资源

Style(样式)资源是一个 View 或者 Window 的若干外观、格式的属性的集合。一个 Style 资源中可以定义若干属性,如高度、填充、字体颜色、字体大小、背景颜色等,然后 View 或者 Window 就可以引用该资源,能够实现外观的统一和代码的重用。

资源定义位置为/res/values/styles.xml,访问方式为@style/＊。

设置代码如下:

```
<resources>
    …
    <style name = "…" parent = "…">
        <item name = "…">…</item>
        …
    </style>
</resources>
```

styles.xml 文件的根元素必须是 resources。当创建样式资源时,在其内部添加 style 元素,每一个 style 元素将被转换为一个 style 资源。其中,name 属性的值就是该 style 资源的标识；style 的 parent 属性是可选的,它指定了另一个资源标识,如果设置该属性将继承这一资源。

📖 Proj_4_5 项目中文件/res/values/styles.xml 中定义两个 style 资源的相关代码

```
<!-- 定义文本默认样式 -->
<style name = "textDefault">
    <item name = "android:textSize">16sp</item>
    <item name = "android:textColor">#000</item>
    <item name = "android:textStyle">bold</item>
    <item name = "android:layout_height">50dp</item>
```

```xml
        <item name = "android:gravity">center_vertical</item>
        <item name = "android:layout_marginLeft">16dp</item>
    </style>
    <!-- 定义间隔线样式 -->
    <style name = "lineGray">
        <item name = "android:layout_width">match_parent</item>
        <item name = "android:layout_height">0.5dp</item>
        <item name = "android:background">#ccc</item>
    </style>
```

📖 Proj_4_5 项目中 activity_main.xml 引用该资源的相关代码

```xml
<TextView
    style = "@style/textDefault"
    android:layout_width = "wrap_content"
    android:text = "姓名" />
<View style = "@style/lineGray" />
<TextView
    style = "@style/textDefault"
    android:layout_width = "wrap_content"
    android:text = "身高" />
<View style = "@style/lineGray" />
<TextView
    style = "@style/textDefault"
    android:layout_width = "wrap_content"
    android:text = "体重" />
<View style = "@style/lineGray" />
```

运行时,界面效果如图 4.26 所示。

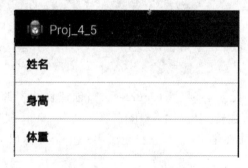

图 4.26 使用样式的界面效果

4.7.2 主题资源

Theme(主题)资源是将 style 样式应用到 Application 或者 Activity。主题资源依然在 style 中设置,也依然使用同样的方法引用。

Android 系统的 themes.xml 和 style.xml 中包含了很多系统定义好的 style,如果想自定义主题，可以继承其中一个，然后再重新定义其中的某个属性。

当设置主题时，可以在 AndroidManifest.xml 中编辑 application 标签，让其包含 android：theme 属性，值是一个主题的名字。如果只是想让某个 Activity 使用一个主题，则编辑 activity 标签，设置 android：theme 属性。

📖 Proj_4_5 项目中 AndroidManifest.xml 为 application、SecondActivity、ThirdActivity 分别设置 theme 属性的相关代码

```xml
<application
    android:allowBackup = "true"
    android:hasCode = "true"
    android:icon = "@drawable/ic_launcher"
    android:label = "@string/app_name"
    <!-- 使用系统无标题栏主题 -->
    android:theme = "@android:style/Theme.Light.NoTitleBar" >
    <activity
        android:name = "com.example.proj_4_5.MainActivity"
        android:label = "@string/app_name" >
        <intent-filter>
            <action android:name = "android.intent.action.MAIN" />
            <category android:name = "android.intent.category.LAUNCHER" />
        </intent-filter>
    </activity>
    <activity
        android:name = "com.example.proj_4_5.SecondActivity"
        android:label = "@string/title_activity_second"
        <!-- 使用系统对话框风格主题 -->
        android:theme = "@android:style/Theme.Dialog" >
    </activity>
    <activity
        android:name = "com.example.proj_4_5.ThirdActivity"
        <!-- 使用自定义主题 -->
        android:theme = "@style/third_window"
        android:label = "@string/title_activity_third" >
    </activity>
</application>
```

📖 Proj_4_5 项目中文件/res/values/styles.xml 自定义主题的相关代码

```xml
<style name = "third_window" parent = "android:Theme.Light">
    <item name = "android:windowTitleSize">50dp</item>
</style>
```

运行项目，该应用程序初始化界面如图 4.27 所示，没有标题栏。打开 SecondActivity，界面如图 4.28 所示，以对话框模式显示。打开 ThirdActivity，显示效果如图 4.29 所示，以自定义标题栏高度显示。

图 4.27 无标题栏主题　　　　图 4.28 对话框主题

图 4.29 自定义标题栏主题

4.8 习　　题

1. 选择题

(1) res 目录下的文件名可以使用（　　）字符。

　　A. 小写字母　　　　B. 大写字母　　　　C. 数字　　　　D. 下画线

(2) （　　）将 a.jpg 与 a.png 两张图片同时放入 drawable-hdpi。

　　A. 能　　　　　　　B. 不能

(3) res/values 目录下允许使用（　　）文件。

　　A. strings.xml　　B. colors.xml　　C. string.xml　　D. color.xml

(4) 在一个 layout 文件中想要包含 menu.xml 这个 layout 文件，可以使用（　　）方式实现。

　　A. < includes layout="@layout/menu.xml" />

　　B. < include layout="@layout/menu.xml" />

　　C. < include layout="@layout/menu" />

　　D. < include layout="menu" />

(5) 在 Android 中使用 GIF 图片资源时，（　　）显示出动画效果。

　　A. 能　　　　　　　B. 不能

2. 简答题

(1) 完成一个注册界面，其中界面控件外观样式设置为样式资源并调用。

(2) 将以上界面中所有使用到的字符串定义在 strings.xml 中并调用。

(3) 将以上界面中所有控件的尺寸定义在 dimens.xml 中并调用。

(4) 在以上注册界面中添加一个默认头像 ImageView 控件，单击该控件时加载一个自定义动画集。

第 5 章 使用系统组件

本章学习目标
- 掌握 Android 中菜单的创建与使用。
- 掌握 ActionBar 的使用。
- 掌握对话框的创建和使用。
- 掌握通知的创建与发送。

Android 应用程序除了前面介绍的系统控件外，还有菜单组件、ActionBar 组件、简单消息提示、对话框、通知消息等非常重要的系统组件。

5.1 菜单的使用

当用户单击设备上的 Menu 按钮时，应用程序界面底部将弹出一个菜单。使用资源 xml 文件创建菜单，在前面已经详细介绍过，本章将重点介绍如何使用 Java 代码来创建菜单。

5.1.1 创建菜单

创建菜单，只需要重写 Activity 的 onCreateOptionsMenu 方法，便可实现创建菜单功能。其中相关类为 Menu，Menu 类提供 add 方法来添加菜单项。

📖 Proj_5_1 项目中文件 MainActivity.java 创建菜单相关方法的代码

```java
public boolean onCreateOptionsMenu (Menu menu) {
    menu.add(Menu.NONE, Menu.FIRST, 1, "新建")
        .setIcon(android.R.drawable.ic_menu_add);
    menu.add(Menu.NONE, Menu.FIRST + 1, 2, "编辑")
        .setIcon(android.R.drawable.ic_menu_edit);
    menu.add(Menu.NONE, Menu.FIRST + 2, 3, "删除")
        .setIcon(android.R.drawable.ic_menu_delete);
    menu.add(Menu.NONE, Menu.FIRST + 3, 4, "帮助")
        .setIcon(android.R.drawable.ic_menu_help);
    return true;
}
```

Menu 类的 add 方法共有 4 个重载，最常用的如下：
public MenuItem add(int groupId, int itemId, int order, CharSequence title);
其中，groupId 为菜单组的组号，为一组的菜单项可以用来批处理状态变化，如果不需要设置组可以使用 Menu.NONE；itemId 是该菜单项的唯一标识；order 用于设置顺序；title 用于设置菜单项文本。add 方法的返回值是 MenuItem 类型，调用 setIcon 设置菜单项的图标。

5.1.2 监听菜单选中

监听菜单选中需要重写 Activity 的 onOptionsItemSelected 方法。

📖 Proj_5_1 项目中文件 MainActivity.java 处理菜单项选中相关方法的代码

```java
public boolean onOptionsItemSelected(MenuItem item) {
    switch(item.getItemId()) {
        case Menu.FIRST :
            tv.append("MenuItem-新建被选中\n") ;
            break;
        case Menu.FIRST + 1 :
            tv.append("MenuItem-编辑被选中\n") ;
            break;
        case Menu.FIRST + 2 :
            tv.append("MenuItem-删除被选中\n") ;
            break;
        case Menu.FIRST + 3 :
            tv.append("MenuItem-帮助被选中\n") ;
            break;
    }
    return false ;
}
```

其中，参数 MenuItem 为选中的菜单项，在方法体中通过判断选中 MenuItem 的 id 值来判断用户选中的是哪一个菜单项，其中 MenuItem 的 id 值在 onCreateOptionsMenu 方法体中添加菜单项时设置。

项目运行时的效果如图 5.1 所示，菜单项在 Android 2.X 和 Android 4.X 下外观不一样，这里的演示效果为 Android 4.4 下的效果。选中菜单项时，执行 onOptionsItemSelected 方法，显示效果如图 5.2 所示。

图 5.1　菜单运行界面

图 5.2　选中菜单项

5.1.3 子菜单与弹出菜单

在为 Menu 添加菜单项时,可以将若干个功能相近的菜单项放置在一起,设置为一个子菜单中的若干项。也就是说可以为 Menu 添加子菜单,然后再为该子菜单添加若干菜单项。

Menu 添加子菜单使用 addSubMenu 方法,该方法将返回一个 SubMenu 对象。SubMenu 继承自 Menu,但是 SubMenu 中添加的菜单项不支持图标设置。

修改项目 Proj_5_1 中 MainActivity.java 的 onOptionsItemSelected 方法,在其中添加代码如下:

```
SubMenu subMore = menu.addSubMenu(Menu.NONE, Menu.FIRST + 4, 5, "更多");
subMore.setHeaderTitle("更多");
subMore.setHeaderIcon(android.R.drawable.ic_menu_more);
subMore.add(Menu.NONE, Menu.FIRST + 5, 6, "版本检查");
subMore.add(Menu.NONE, Menu.FIRST + 6, 7, "关于我们");
```

运行效果图如图 5.3 所示,最后一个菜单项"更多"包含子菜单项。点击该菜单项,将弹出子菜单,如图 5.4 所示。

图 5.3 带有子菜单效果   图 5.4 子菜单显示效果

子菜单仍然是 Menu 的一部分,其显示效果类似于列表。弹出菜单与菜单不同,是和控件绑定在一起的,可以通过在某个 View 控件上长按而触发从而显示该菜单。一个 Activity 中只有一个菜单,而弹出菜单可以有多个,不同的 View 使用不同的弹出菜单。

弹出菜单相关类为 ContextMenu,ContextMenu 也继承自 Menu,但是 ContextMenu 不支持图标和快捷键设置。

创建弹出菜单需要在 Activity 中重写 onCreateContextMenu 方法来创建菜单,重写

onContextItemSelected 方法来响应菜单项的选中事件,需要调用 registerForContextMenu 方法为 View 控件注册菜单。在项目 Proj_5_1 中分别为两个 TextView 控件添加弹出菜单。

📖 Proj_5_1 项目中文件 MainActivity.java 中相关方法的代码

```java
protected void onCreate(Bundle savedInstanceState) {
    super.onCreate(savedInstanceState);
    setContentView(R.layout.activity_main);
    tv = (TextView) findViewById(R.id.tv);
    tvShare = (TextView) findViewById(R.id.tv_share);
    tvMore = (TextView) findViewById(R.id.tv_more);
    // 为控件注册 ContextMenu
    registerForContextMenu(tvShare);
    registerForContextMenu(tvMore);
}
@Override
public void onCreateContextMenu(ContextMenu menu, View v,ContextMenuInfo menuInfo) {
    menu.setHeaderIcon(android.R.drawable.ic_menu_more);
    menu.setHeaderTitle("请选择:");
    if(v == tvShare) {           // 为控件 TextView"分享"设置弹出菜单子项
        menu.add(Menu.NONE, Menu.FIRST + 7, 8, "短信分享");
        menu.add(Menu.NONE, Menu.FIRST + 8, 9, "QQ分享");
    }
    else if (v == tvMore) {      // 为控件 TextView"更多操作"设置弹出菜单子项
        menu.add(Menu.NONE, Menu.FIRST + 9, 10, "删除");
        menu.add(Menu.NONE, Menu.FIRST + 10, 11, "更新");
    }
}
// ContextMenu 中响应菜单项选中的逻辑处理
@Override
public boolean onContextItemSelected(MenuItem item) {
    switch(item.getItemId()) {
        case 8:  tv.append("ContextMenu-短信分享被选中\n");break;
        case 9:  tv.append("ContextMenu-QQ分享被选中\n");break;
        case 10: tv.append("ContextMenu-删除被选中\n");break;
        case 11: tv.append("ContextMenu-更新被选中\n");break;
    }
    return false;
}
```

项目运行效果如图 5.5 所示。长按"分享"控件,会弹出菜单,效果如图 5.6 所示。弹出菜单以对话框的形式显示在 Activity 之上,如果不需要选中菜单项,点击返回键或菜单之外的区域,即可关闭弹出式菜单。

长按"更多操作",弹出第二个弹出菜单,效果如图 5.7 所示。单击"更新"菜单项时,显示效果如图 5.8 所示。

图 5.5 初始界面图

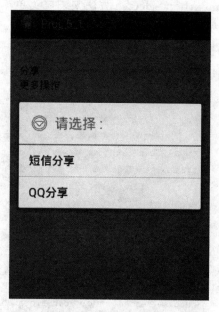

图 5.6 "分享"弹出菜单

图 5.7 "更多操作"弹出菜单

图 5.8 选中"更新"菜单效果

5.2 ActionBar 的使用

前面已经介绍了使用 ActionBar 完成菜单功能的方法,但是 ActionBar 的功能不仅限于此,ActionBar 还可以提供 ActionBar 图标导航和提供多种方式的导航功能以便于在 Fragment 间进行切换。

5.2.1 导航菜单

ActionBar 的导航图标默认为未开启,可以通过设置显示该图标,并提供导航功能。例如在 A 界面执行某个操作跳转到 B 界面,B 界面可以开启导航图标从而实现到 A 界面的跳转。要显示导航按钮,需要进行如下配置:

```
ActionBar actionBar = getActionBar();
actionBar.setDisplayHomeAsUpEnabled(true);
```

如果考虑版本向下兼容到 Android 2.X,则可以使用 support-v7 包中的 getSupportActionBar 方法来获取 ActionBar 对象。

当设置 setDisplayHomeAsUpEnabled(true) 后,可以发现当前 Activity 界面标题栏中最左侧会出现一个返回箭头图标,如图 5.9 所示。

开启该图标后,还需要为导航添加事件。导航应用的 id 值为 android.R.id.home,这是 Android 系统规定的 id 值。要监听该菜单项选中,可以在 onOptionsItemSelected 方法中实现相应逻辑,代码如下:

图 5.9 启用返回箭头图标

```
public boolean onOptionsItemSelected(MenuItem item) {
    switch(item.getItemId()) {
    case android.R.id.home :
        finish();
        break;
    }
    return true;
}
```

以上为最简单的导航功能处理,和返回按键的功能一样。但是 ActionBar 的导航功能并不是这样简单。例如分别可以从 MainActivity 和 ThirdActivity 跳转到 SecondActivity,在 SecondActivity 按下返回键则可能返回 MainActivity,也可能返回 ThirdActivity,完全取决于是从哪个 Activity 跳转到 SecondActivity 的。如果希望从 SecondActivity 返回时总是返回到 MainActivity,就需要使用 ActionBar 的导航图标实现上述要求。首先仍然需要调用 setDisplayHomeAsUpEnabled(true) 开启导航,另外需要在 AndroidManifest.xml 文件中对 Activity 进行相应的设置。

📖 项目 Proj_5_2 中 AndroidManifest.xml 设置 SecondActivity 的相关代码

```
<activity android:name = "com.example.proj_5_2.SecondActivity" >
    <meta-data
        android:name = "android.support.PARENT_ACTIVITY"
        android:value = "com.example.proj_5_2.MainActivity" />
</activity>
```

在 onOptionsItemSelected 方法中进行如下逻辑描述:

```java
public boolean onOptionsItemSelected(MenuItem item) {
    switch(item.getItemId()) {
    case android.R.id.home :
        Intent upIntent = NavUtils.getParentActivityIntent(this);
        if (NavUtils.shouldUpRecreateTask(this, upIntent)) {
            TaskStackBuilder.create(this)
                    .addNextIntentWithParentStack(upIntent)
                    .startActivities();
        } else {
            upIntent.addFlags(Intent.FLAG_ACTIVITY_CLEAR_TOP);
            NavUtils.navigateUpTo(this, upIntent);
        }
        break;
    }
    return true;
}
```

5.2.2 导航模式

ActionBar 除以上导航菜单功能外,还可以实现导航标签(tab)功能。首先设置 ActionBar 导航模式为 Tab,然后为 ActionBar 添加导航标签。不过需要注意的是,在为 ActionBar 添加导航标签时需要设置 TabListener。

  项目 Proj_5_3 中 MainActivity.java 代码

```java
public class MainActivity extends Activity {
    private ActionBar actionBar;
    private TabListener tabListener;
    @Override
    protected void onCreate(Bundle savedInstanceState) {
        super.onCreate(savedInstanceState);
        setContentView(R.layout.activity_main);
        actionBar = getActionBar();
        actionBar.setNavigationMode(ActionBar.NAVIGATION_MODE_TABS);
        tabListener = new MyTabListener();
        actionBar.addTab(actionBar.newTab().setText("头条").setTabListener(tabListener)) ;
        actionBar.addTab(actionBar.newTab().setText("娱乐").setTabListener(tabListener)) ;
        actionBar.addTab(actionBar.newTab().setText("体育").setTabListener(tabListener)) ;
    }
    class MyTabListener implements TabListener {
        @Override
        public void onTabSelected(Tab tab, FragmentTransaction ft) { }
        @Override
        public void onTabUnselected(Tab tab, FragmentTransaction ft) { }
        @Override
        public void onTabReselected(Tab tab, FragmentTransaction ft) { }
    }
}
```

项目运行,界面如图 5.10 所示,3 个导航标签可以切换。当手机由竖屏切换为横屏时,导航标签的效果会发生变化,感兴趣的读者可以尝试一下。

图 5.10 导航标签效果

5.2.3 Actionbar 与 Fragment

在以上示例中实现了 ActionBar 的导航标签功能,但是仅出现了导航标签,没有内容区域切换与之配合。如果希望单击标签部分实现下方内容的切换,可以结合 Fragment 来实现。

首先设置标签内容 Fragment,然后在 TabListener 的 onTabSelected 方法中实现当标签页选中时添加 Fragment,并在 onTabUnSelected 方法中实现当标签页未选中时分离 Fragment。

📖 项目 Proj_5_3 中 NewsFragment.java 代码

```java
public class NewsFragment extends Fragment {
    private String type;
    public void getType() {
        Bundle bundle = getArguments();
        if(null != bundle){
            type = bundle.getString("type");
        }
    }
    @Override
    public View onCreateView(LayoutInflater inflater, ViewGroup container,
            Bundle savedInstanceState) {
        TextView tv = new TextView(getActivity());
        getType();
        tv.setText(type == null ? "Fragment" : type);
        tv.setBackgroundColor(Color.WHITE);
        tv.setTextSize(40);
        return tv;
    }
}
```

📖 修改项目 Proj_5_3 中 MainActivity.java 代码

```java
public class MainActivity extends Activity {
    private ActionBar actionBar;
    private TabListener tabListener;
```

```java
    @Override
    protected void onCreate(Bundle savedInstanceState) {
        super.onCreate(savedInstanceState);
        setContentView(R.layout.activity_main);
        actionBar = getActionBar();
        actionBar.setNavigationMode(ActionBar.NAVIGATION_MODE_TABS);
        tabListener = new MyTabListener();
        actionBar.addTab(actionBar.newTab().setText("头条").setTabListener(tabListener));
        actionBar.addTab(actionBar.newTab().setText("娱乐").setTabListener(tabListener));
        actionBar.addTab(actionBar.newTab().setText("体育").setTabListener(tabListener));
    }
    class MyTabListener implements TabListener {
        Fragment hotF , entertainmentF , sportsF;
        @Override
        public void onTabSelected(Tab tab, FragmentTransaction ft) {
            Bundle bundle = new Bundle();
            switch(tab.getPosition()) {
            case 0 :
                if(null == hotF) {
                    hotF = new NewsFragment();
                    bundle.putString("type", "头条");
                    hotF.setArguments(bundle);
                    ft.add(android.R.id.content, hotF);
                }
                else {
                    ft.attach(hotF);
                }
                break;
            case 1 :
                if(null == entertainmentF) {
                    entertainmentF = new NewsFragment();
                    bundle.putString("type", "娱乐");
                    entertainmentF.setArguments(bundle);
                    ft.add(android.R.id.content, entertainmentF);
                }
                else {
                    ft.attach(entertainmentF);
                }
                break;
            case 2:
                if(null == sportsF) {
                    sportsF = new NewsFragment();
                    bundle.putString("type", "体育");
                    sportsF.setArguments(bundle);
                    ft.add(android.R.id.content, sportsF);
                }
                else {
                    ft.attach(sportsF);
                }
                break;
```

```java
            }
        }
        @Override
        public void onTabUnselected(Tab tab, FragmentTransaction ft) {
            switch(tab.getPosition()) {
            case 0:
                if(null != entertainmentF) {
                    ft.detach(entertainmentF);
                }
                if(null != sportsF) {
                    ft.detach(sportsF);
                }
                break;
            case 1:
                if(null != hotF) {
                    ft.detach(hotF);
                }
                if(null != sportsF) {
                    ft.detach(sportsF);
                }
                break;
            case 2:
                if(null != entertainmentF) {
                    ft.detach(entertainmentF);
                }
                if(null != hotF) {
                    ft.detach(hotF);
                }
                break;
            }
        }
        @Override
        public void onTabReselected(Tab tab, FragmentTransaction ft) { }
    }
}
```

运行项目,默认界面如图 5.11 所示,切换导航标签,下方的内容区域会自动切换,效果如图 5.12 所示。

图 5.11 默认运行界面

图 5.12 "娱乐"导航标签

5.3 Toast 与 Notification

在应用程序运行过程中,如果需要给出用户一些提醒或通知,但又不希望影响用户当前的操作,可以使用 Android 中提供的 Toast 和 Notification。

5.3.1 创建并显示 Toast

Toast 主要用于在屏幕下方位置弹出一条提示消息,Toast 显示的消息在屏幕显示短暂时间后会自动消失,不会中断用户操作。

常用的简单 Toast 可以使用 Toast 的静态方法 makeText 来创建,并调用 show 方法显示。makeText 方法原型如下:

public static Toast makeText(Context context, CharSequence text, int duration)

其中 duration 表示 Toast 显示时长,Toast 提供了常量 LENGTH_SHORT 与 LENGTH_LONG,也可以自定义以毫秒为单位的时长。Toast 默认显示效果如图 5.13 所示。

图 5.13 Toast 显示效果

5.3.2 自定义 Toast

上面介绍的 Toast 是最简单也最常用的方式,如果想添加一些个性化设置,可以自定义 Toast 来完成。Toast 也可以自定义 layout 布局文件,然后调用 Toast 的 setView 方法来加载该布局对应的 View,从而完成个性化设置。

📖 项目 Proj_5_4 中 toast.xml 代码

```xml
<?xml version = "1.0" encoding = "utf-8"?>
<LinearLayout xmlns:android = "http://schemas.android.com/apk/res/android"
    android:layout_width = "match_parent"
    android:layout_height = "match_parent"
    android:orientation = "vertical"
    android:background = "#000">
    <ImageView
        android:layout_width = "wrap_content"
        android:layout_height = "wrap_content"
        android:src = "@drawable/ic_launcher"/>
    <TextView
        android:layout_width = "match_parent"
        android:layout_height = "wrap_content"
        android:gravity = "center"
        android:padding = "4dp"
        android:text = "My Toast"
        android:textColor = "#fff"/>
</LinearLayout>
```

📖 项目 Proj_5_4 中 MainActivity.java 创建 Toast 的代码

```
Toast toast = new Toast(this);
toast.setView(LayoutInflater.from(this).inflate(R.layout.toast, null));
toast.setGravity(Gravity.CENTER, 0, 0);
toast.setDuration(Toast.LENGTH_LONG);
toast.show();
```

自定义 Toast 组件运行效果如图 5.14 所示。

图 5.14 自定义 Toast 效果

Toast 的主要方法如下：
- makeText，最常用的静态方法，是直接创建 Toast 对象的方法。
- setDuration，设置停留的时间。
- setGravity，设置屏幕上的位置。
- setText，设置显示的文本信息。
- setView，设置所显示的视图。
- show，显示。

5.3.3 创建并发出通知

以上介绍的 Toast 显示消息会有一定时长的显示，不管用户是否看到该消息，显示过后 Toast 都会消失，而如果希望发送一些全局的消息，并且可以通过声音等方式提醒用户查看，则可以使用 Notification（通知）来完成。

通知一般是通过 NotificationManager 服务来发送一个 Notification 对象完成。NotificationManager 是一个重要的系统级服务，该对象位于应用程序的框架层中，应用程序可以通过它向系统发送全局的通知。NotificationManager 类是一个通知管理类，该类对象是由系统维护的服务。可以通过调用 getSystemService(String) 方法获取 NotificationManager

对象，getSystemService(String)方法可以通过 Android 系统级服务方法返回相应的对象。因为需要返回 NotificationManager，所以直接传递 Context.NOTIFICATION_SERVICE 即可。

Notification 对象用于承载通知的内容。一般在实际使用过程中不会直接构建 Notification 对象，而是使用它的一个内部类 NotificationCompat.Builder 来实例化一个对象（Android 3.0 版本之后使用 Notification.Builder），并设置通知的属性，最后通过 NotificationCompat.Builder.build 方法得到一个 Notification 对象。当获得这个对象之后，可以使用 NotificationManager.notify 方法发送通知。

📖 项目 Proj_5_5 中 MainActivity.java 创建通知部分的代码

```java
private void createNotify() {
    //封装一个 Intent,用于设置单击通知后所跳转到的 Activity
    Intent resultIntent = new Intent(this, WeatherActivity.class);
    resultIntent.putExtra("weather", "天气晴朗,微风1-2级,适当户外散步!");
    PendingIntent resultPendingIntent = PendingIntent.getActivity(
        MainActivity.this, 0, resultIntent, PendingIntent.FLAG_UPDATE_CURRENT);
    // 设置通知详情
    Notification notify = new NotificationCompat.Builder(this)
        .setSmallIcon(R.drawable.sunny)                    // 设置小图标
        .setContentTitle("天气")                            // 设置通知标题
        .setContentText("天气晴朗,微风1-2级")                // 设置通知内容
        .setTicker("天气信息")                              // 设置消息到达时的提示文字
        .setLargeIcon(BitmapFactory.decodeResource(getResources(), R.drawable.weather))
                                                           // 设置大图标
        .setWhen(System.currentTimeMillis())               // 设置通知的时间
        .setAutoCancel(true)                               // 设置通知查看后消失
        .setContentIntent(resultPendingIntent)             //查看通知时所启动的组件
        .build();
    // 发送通知
    NotificationManager nMan =
        (NotificationManager) getSystemService(Context.NOTIFICATION_SERVICE);
    nMan.notify(0x100, notify);
}
```

运行项目，通知到达手机之后的效果如图 5.15 所示，该通知会直接显示在手机的标题栏中。当打开手机通知栏后，会显示通知的详细信息，如图 5.16 所示。单击该通知时，可以跳转到预设的目标组件。

图 5.15 通知到达时的效果

图 5.16 显示通知的详细信息

在上例中，使用了 NotificationCompat.Builder 的一系列 set 方法组来创建了通知的相关设置内容，主要方法如下：

- setSmallIcon,设置小图标。
- setContentTitle,设置通知标题。
- setContentText,设置通知内容。
- setTicker,设置消息到达时的提示文字。
- setLargeIcon,设置大图标。
- setWhen,设置通知的时间,一般为系统时间。
- setAutoCancel,设置通知查看后消失。
- setContentIntent,设置查看通知详情时所启动的组件。
- build,创建 Notification。

5.4 对话框的使用

在 Android UI 界面开发中,对话框是非常重要的一个组成部分。和 Toast 相比,对话框可以阻塞 UI 主线程的执行,可以获取用户焦点,并且对话框中不仅显示信息,而且可以提供选择供用户进行下一步操作。

5.4.1 普通对话框的创建

AlertDialog 是最常用的普通对话框,要创建它一般使用它的内部类 AlertDialog.Builder 来完成。Builder 对象的常用方法如下:

- setTitle,设置对话框标题。
- setIcon,设置对话框图标。
- setMessage,设置对话框信息。
- setView,设置对话框自定义一个 View 视图对象。
- setItems,设置对话框要显示的一个集合 items 信息。
- setMultiChoiceItems,设置对话框显示列表前对应的复选框。
- setSingleChoiceItems,设置对话框显示列表为单选按钮模式。
- setNeutralButton,设置对话框中间按钮。
- setPositiveButton,添加对话框"确定"按钮。
- setNegativeButton,添加对话框"取消"按钮。
- create,创建对话框。
- show,显示对话框。

对话框组件中可以添加 PositiveButton、NegativeButton 和 NeutralButton 3 个按钮控件,也可以不添加,或有选择地添加。添加按钮时使用 setXXXButton 方法,其中第一个参数是按钮的文本,第二个参数是按钮的监听器。一般按钮的监听器和对话框中按钮的监听器都是 OnClickListener,但两个监听器不同。一般按钮的监听器是 android.view.View.OnClickListener,对话框中按钮的监听器是 android.content.DialogInterface.OnClickListener。Android 系统中有很多同名类,在编写代码时可以借助"包名.类名"的方法确认需要使用的类。

📖 项目 Proj_5_5 中 MainActivity.java 创建普通对话框部分的代码

```java
private void commonDialog() {
    new AlertDialog.Builder(this)
        .setIcon(android.R.drawable.ic_menu_close_clear_cancel)
        .setTitle("退出")
        .setMessage("确认退出?")
        .setNegativeButton("取消", null)
        .setPositiveButton("退出", new DialogInterface.OnClickListener() {
            public void onClick(DialogInterface dialog, int which) {
                finish();
            }
        })
        .create()
        .show();
}
```

运行项目,打开对话框的效果如图 5.17 所示。

图 5.17 一般对话框

5.4.2 选择对话框

可以在对话框中设置一系列的文本内容,提供一个列表供用户进行选择,带有列表选择的对话框称为选择对话框。在上例中使用 setItems 方法来替换 setMessage 方法,便可创建一个选择对话框。

📖 项目 Proj_5_5 中 MainActivity.java 创建选择对话框部分的代码

```java
private void choiceDialog() {
    final String[] lan = new String[]{"Java", "C", "C#", "C++"};
    new AlertDialog.Builder(this)
        .setIcon(R.drawable.star)
        .setTitle("选择你所掌握的语言")
        .setItems(lan, new DialogInterface.OnClickListener() {
            public void onClick(DialogInterface dialog, int which) {
                Toast.makeText(MainActivity.this, "your choose:" + lan[which],
                        Toast.LENGTH_SHORT).show();
            }
        })
```

```
        .create()
        .show();
}
```

运行项目,效果如图 5.18 所示。当用户选中某一个列表项后,会出现 Toast 提示,同时对话框会自动关闭。

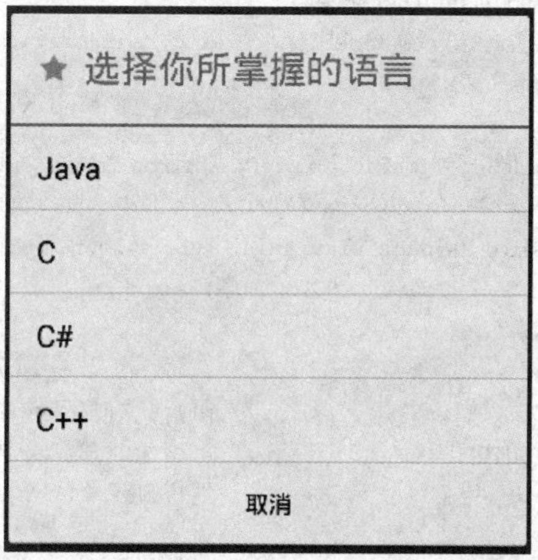

图 5.18 列表对话框

5.4.3 日期与时间对话框

日期与时间对话框 DatePickerDialog 与 TimePickerDialog 都继承自 AlertDialog,但是各自内置了日期和时间控件。可以利用带参数构造函数来创建日期对话框 DatePickerDialog。

📖 项目 Proj_5_5 中 MainActivity.java 创建日期对话框部分的代码

```
private void dateDialog() {
    Calendar now = Calendar.getInstance();
    DatePickerDialog dialog = new DatePickerDialog(this, new OnDateSetListener() {
        @Override
        public void onDateSet(DatePicker view, int year, int monthOfYear,int dayOfMonth) {
            String info = String.format("%4d-%02d-%02d", year ,
                                        (1+monthOfYear) , dayOfMonth) ;
            Toast.makeText(MainActivity.this, "the selected date is : " + info,
                            Toast.LENGTH_LONG).show();
        }
    },
    now.get(Calendar.YEAR),              //初始化时显示年份
    now.get(Calendar.MONTH),             //初始化时显示月份
    now.get(Calendar.DAY_OF_MONTH)       //初始化时显示日期
    );
    dialog.show();
}
```

运行项目，效果如图 5.19 所示。选择日期，单击"完成"按钮，OnDateSetListener 监听器中的 onDateSet 方法开始执行，通过参数就以读取用户设置的日期。注意，在 Java 程序中月份对应的数值是从 0 开始的。

  📖 项目 Proj_5_5 中 MainActivity.java 中创建时间对话框部分代码如下：

```java
private void timeDialog() {
    Calendar now = Calendar.getInstance();
    TimePickerDialog dialog = new TimePickerDialog(MainActivity.this,
     new OnTimeSetListener() {
        @Override
        public void onTimeSet(TimePicker view, int hourOfDay, int minute) {
            String info = String.format("%02d:%02d", hourOfDay, minute);
            Toast.makeText(MainActivity.this, "the selected time is : " + info,
                    Toast.LENGTH_LONG).show();
        }
    },
    now.get(Calendar.HOUR_OF_DAY),      // 初始化时显示小时
    now.get(Calendar.MINUTE),            // 初始化时显示分钟
    true                                 // 是否显示 24 小时制
    );
    dialog.show();
}
```

运行效果如图 5.20 所示。选择时间，单击"完成"按钮，监听器 OnTimeSetListener 中的 onTimeSet 方法开始执行，通过参数即可读取用户设置的时间。

图 5.19　日期对话框 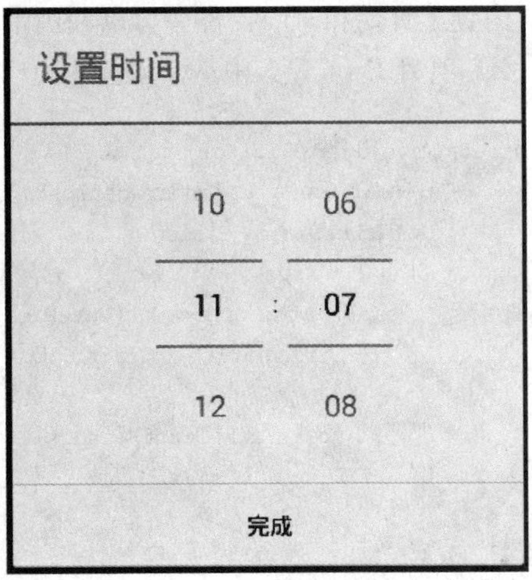　　　　　　　　图 5.20　时间对话框

5.4.4　进度条对话框

  进度条对话框 ProgressDialog 继承自 AlertDialog，其中内置了控进度条件 Progress 可以直接通过构造方法创建进度条对话框。

📖 项目 Proj_5_5 中 MainActivity.java 创建环形进度条对话框部分的代码

```
private void progressDialog1() {
    ProgressDialog dialog = new ProgressDialog(MainActivity.this);
    dialog.setIcon(android.R.drawable.ic_menu_upload);
    dialog.setTitle("上传");
    dialog.setMessage("文件上传中,请稍候……");
    dialog.setMax(100);
    dialog.show();
}
```

运行效果如图 5.21 所示。进度条对话框不会自动消失,如果相应的业务逻辑代码完成,调用 dismiss 方法直接结束对话框即可。

ProgressDialog 中的进度条默认显示为图 5.21 所示的环形。如果想显示为水平进度条,可以调用 setProgressStyle(ProgressDialog.STYLE_HORIZONTAL)来完成。如果需要显示进度,则要开启一个子线程,利用 Handle 接口修改主界面的 UI。

📖 项目 Proj_5_5 中 MainActivity.java 创建水平进度条对话框部分的代码

```
private void progressDialog2() {
    ProgressDialog dialog = new ProgressDialog(MainActivity.this);
    dialog.setIcon(android.R.drawable.ic_menu_upload);
    dialog.setTitle("上传");
    dialog.setMessage("文件上传中,请稍候……");
    dialog.setProgressStyle(ProgressDialog.STYLE_HORIZONTAL);
    dialog.setMax(100);
    dialog.show();
}
```

运行效果如图 5.22 所示。该进度条没有任何动态效果。如果需要修改进度条数值,需要启动线程,并借助 Handler 修改取值。真正意义上的文件上传进度条需要与已上传的数据量相匹配,这需要编写一定量的代码才能实现。

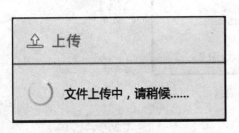

图 5.21 默认进度条对话框

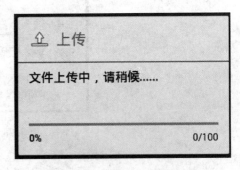

图 5.22 水平进度条对话框

5.4.5 自定义对话框

除了以上几种常见的对话框形式,还可以自定义个性化的对话框。通过自定义对话框的布局,设置对话框中所要表现的外观形式,然后在创建 AlertDialog 时使用 setView 方法来替换 setMessage 或者 setItems 方法,从而实现个性化的自定义对话框。

📖 项目 Proj_5_5 中 MainActivity.java 创建自定义对话框部分的代码

```java
private void dialog() {
    // 获取对话框的布局外观 View
    View loginView = LayoutInflater.from(this).inflate(R.layout.dialog_login, null);
    final EditText etName = (EditText) loginView.findViewById(R.id.login_name);
    final EditText etPwd = (EditText) loginView.findViewById(R.id.login_password);
    new AlertDialog.Builder(this)
        .setTitle("登录")
        .setIcon(android.R.drawable.ic_menu_myplaces)
        .setView(loginView)
        .setPositiveButton("登录", new DialogInterface.OnClickListener(){
            @Override
            public void onClick(DialogInterface dialog, int which) {
                String name = etName.getText().toString();
                String pwd = etPwd.getText().toString();
                Toast.makeText(MainActivity.this, "welcome: " + name + ":" + pwd,
                    Toast.LENGTH_LONG).show();
            }}
        )
    .setNegativeButton("取消",null)
    .create()
    .show();
}
```

运行效果如图 5.23 所示。注意,在使用自定义布局对话框时,不能再使用前几节所介绍的对话框类型,否则会抛出异常。

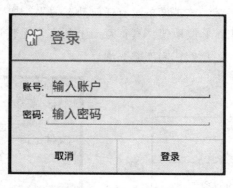

图 5.23 自定义对话框

5.5 习 题

1. 选择题

(1) 在 Android 中使用 Menu 时可能需要重写的方法有(　　)。

  A. onOptionsItemSelected   B. onCreateOptionsMenu

  C. onItemSelected   D. onCreateMenu

（2）创建子菜单的方法是（　　）。

  A. add   B. addSubMenu

  C. createSubMenu   D. createMenu

（3）下列关于如何使用 Notification 的叙述中不正确的是（　　）。

  A. Notification 需要 NotificatinManager 来管理

  B. 使用 NotificationManager 的 notify 方法显示 Notification 消息

  C. 在显示 Notification 时可以设置通知时的默认发声、振动等

  D. Notification 中存在可以清除消息的方法

（4）使进度条变为水平方向的系统样式是（　　）。

  A. @android：style/Widget. ProgressBar. Horizontal

  B. @android：style/ProgressBar. Horizontal

  C. @style/Widget. ProgressBar. Horizontal

  D. @style/ProgressBar. Horizontal

2. 简答题

（1）定义 3 个 Fragment，在主 Activity 的底部添加 3 个 RadioButton 控件，实现单击 RadioButton 时切换主 Activity 中所加载的 3 个 Fragment。

（2）定义一个显示当前所播放歌曲名称的 Notification，并发送该通知。

（3）定义一个城市选择的对话框，并且当用户选择城市后，Toast 弹出用户所选择的结果。

（4）定义一个生日选择对话框。

第6章　二维图像的处理

本章学习目标
- 掌握 Bitmap 与 BitmapFactory 的使用。
- 了解位图放缩技术。
- 掌握 Canvas 与 Paint 的使用。
- 掌握用 Matrix 进行图像变换的方法。
- 掌握 View 与 SurfaceView 的使用。
- 了解线程控制下的动画。

ImageView 是 Android 应用开发中比较常用的图形图像呈现控件,它不但可以直接显示 drawable 文件夹下的资源,还可以显示更广泛的位图。二维图形图像在 Android 应用开发中的运用得非常普遍,了解并掌握二维图像的处理方法,将极大地提升资源使用效率。

View 类是 Android 开发中通用控件的父类,位于 android.view 包中。android.widget 包中的控件几乎都是从 View 类继承的,如 ImageView、ProgressBar、TextView 等。如果基本控件无法满足开发需求,则需要采用自定义类,继承 View 或 SurfaceView,实现个性化操作。

6.1　位图的使用

位图是相对于矢量图而言的,也称为点阵图。位图由像素组成,图像清晰度由单位长度内像素的多少来决定,单位长度内像素越多,分辨率越高,图像的效果越好,图像也会越大。

在 Android 系统中,位图使用 Bitmap 类来表示,它位于 android.graphics 包中,被 final 所修饰,不能被继承。创建 Bitmap 对象可以使用该类的静态方法 createBitmap,也可以借助 BitmapFactory 类来实现。

6.1.1　Bitmap 与 BitmapFactory

Bitmap 是 Android 系统中图像处理的最重要的类之一,它可以获取图像文件信息,如宽高尺寸,可以进行图像剪切、旋转、缩放等操作,还可以将图像保存成指定格式。但是 Bitmap 对象是一种比较消耗内存的数据,尤其是位图比较大时,使用方式稍有不当,就可能引发 OutOfMemoryError 错误。所以在使用 Bitmap 时,应该根据显示控件的大小,利用 BitmapFactory.Options 计算合适的 inSimpleSize(放缩比),对 Bitmap 进行相应的裁剪,以减少 Bitmap 对内存的消耗。

Bitmap 类的常用方法如表 6.1 所示。

表 6.1　Bitmap 常用方法

方　法　名	说　明	返回值类型
createBitmap(Bitmap source, int x, int y, int width, int height)	以 source 为画布资源，x、y 为左上角顶点，width、height 为宽和高，截取新位图	Bitmap
createBitmap(Bitmap source, int x, int y, int width, int height, Matrix m, boolean filter)	在上一方法的基础上进行 Matrix 变换	Bitmap
createBitmap(int width, int height, Bitmap.Config config)	按照指定宽、高和格式创建位图	Bitmap
createScaledBitmap(Bitmap src, int dstWidth, int dstHeight, boolean filter)	将源位图放缩成指定尺寸的位图	Bitmap
compress(Bitmap.CompressFormat format, int quality, OutputStream stream)	按指定格式(JPG、PNG、WEBP)和画质把图片转换为输出流	boolean
getHeight()	获取位图高度	int
getWidth()	获取位图宽度	int
getPixel(int x, int y)	获取指定像素点的颜色	int
isRecycled()	检测位图是否被回收	boolean
recycle()	强制位图立即回收	void

BitmapFactory 是创建 Bitmap 的工具类，能够以文件、字节数组、输入流的形式创建位图对象。BitmapFactory 类提供的都是静态方法，可以直接调用，常用方法如表 6.2 所示。

表 6.2　BitmapFactory 常用方法

方　法　名	说　明	返回值类型
decodeByteArray(byte[] data, int offset, int length, BitmapFactory.Options opts)	从字节数组 offset 位置开始，将 length 长度字节解析为位图，opts 可以为 null	Bitmap
decodeFile(String pathName)	把文件解析为位图	Bitmap
decodeFile(String pathName, BitmapFactory.Options opts)	把文件解析为位图，并指定解析属性	Bitmap
decodeResource(Resources res, int id)	把某个资源文件解析为位图	Bitmap
decodeResource(Resources res, int id, BitmapFactory.Options opts)	把资源文件解析为位图，并设定解析属性	Bitmap
decodeStream(InputStream is)	把输入流解析为位图	Bitmap

在上面的很多方法中，都出现过 BitmapFactory.Options 类型的参数。这是 BitmapFactory 的静态内部类，主要用于设定位图解析参数。BitmapFactory.Options 类包含了很多属性字段，用于标识不同操作，常用属性字段如表 6.3 所示。

表 6.3　BitmapFactory.Options 常用属性字段

属　性	说　明	类　型
inJustDecodeBounds	设置为 true，则不解析图片，不分配内存，只返回图片的宽高信息	boolean
inSampleSize	缩放倍数，当取值大于 1 时，则缩小相应倍数；当取值小于 1 时，则不进行缩小，保持不变。若设为 2，则宽和高都为原来的 1/2，图片大小是原来的 1/4	int
outHeight	位图的高度	int
outWidth	位图的宽度	int

6.1.2 位图的缩略图

由于 Android 设备所限,一个软件中所采用的图片资源不可能全部放入 drawable 文件夹下,这样在读取资源时很容易抛出 OutOfMemoryError(内存溢出)。读取存放在手机中的图片资源时,如果该图片尺寸较大,也会引起 OutOfMemoryError。因此,在解析位图时,将图片进行相应缩放,当位图资源不再使用时,强制资源回收,就可以避免这样的问题。

不加载位图的原有尺寸,而是根据控件的大小呈现图像的缩小尺寸,就是缩略图。设置 BitmapFactory.Options 的属性 inJustDecodeBounds 为 true 后,再解析位图时并不分配存储空间,但可计算出原始图片的宽度和高度,即 outWidth 和 outHeight。有了这两个数值,再与控件的宽高尺寸相除,就可以得到放缩比例,即 inSampleSize 的值。然后,重新设置 inJustDecodeBounds 为 false,inSampleSize 为计算所得比例,重新解析位图文件,即可得原图的缩略图。

下面的代码演示了将 SD 卡上的大尺寸图片解析为控件所指定的尺寸。布局文件中只有一个 ImageView 控件,比较简单,此处省略布局文件代码。位图存放在 SD 卡根目录/Pictures/4.jpg。

📖 Proj06_1 项目 MainActivity.java 文件

```java
public class MainActivity extends Activity {
    ImageView iv1;
    @Override
    protected void onCreate(Bundle savedInstanceState) {
        super.onCreate(savedInstanceState);
        setContentView(R.layout.activity_main);
        iv1 = (ImageView) findViewById(R.id.imageView1);
        initIv();            //加载位图
    }
    //填充图片
    private void initIv() {
        File sdRoot = Environment.getExternalStorageDirectory();
        File imgFile = new File(sdRoot,"/Pictures/4.jpg");
        //获取解析属性对象
        BitmapFactory.Options opt = new BitmapFactory.Options();
        //设置假解析位图
        opt.inJustDecodeBounds = true;
        Bitmap img = BitmapFactory.decodeFile(imgFile.getAbsolutePath(),opt);   ①
        //获取位图尺寸
        int outWidth = opt.outWidth;
        int outHeight = opt.outHeight;
        Log.i("Msg","位图原始尺寸: width = " + outWidth + ",height = " + outHeight);
        //获取控件尺寸
        LayoutParams lp = iv1.getLayoutParams();
        int ivWidth = lp.width;
        Log.i("Msg","控件尺寸: width = " + lp.width + ",height = " + lp.height);
        //计算图片与控件的比例
        int scale   = outWidth/ivWidth;
```

```
        Log.i("Msg", "缩放倍数: " + scale);
        //设置放缩比例
        opt.inSampleSize = scale;
        //设置真解析位图
        opt.inJustDecodeBounds = false;                    ②
        img = BitmapFactory.decodeFile(imgFile.getAbsolutePath(),opt);
        //设置位图
        iv1.setImageBitmap(img);
    }
}
```

上述代码中有两次解析位图的过程，前一次（代码中的①处）只为获取位图尺寸（假解析），因此设定参数 opt.inJustDecodeBounds=true；后一次（代码中的②处）根据缩小比例解析位图（真解析），因此设定参数 opt.inJustDecodeBounds=false。

获取 ImageView 控件尺寸时，需要借助 LayoutParams，而不能直接使用 getWidth 方法。项目运行时，输出的日志如图 6.1 所示。

Level	T	PID	TID	Appl...	Tag	Text
I	C	844	844	ed...	Msg	位图原始尺寸: width=1280,height=800
I	C	844	844	ed...	Msg	控件尺寸: width=360,height=240
I	C	844	844	ed...	Msg	缩放倍数: 3

图 6.1　缩略图输出信息

因为 inSampleSize 属性是整形数据，因此缩放倍数不能为浮点型数据。特别需要注意的是，inSampleSize 的实际取值并非赋值的倍数，而是向下取最接近 2 的指数次方的值。如上述代码中，指定的缩放倍数为 3，而实际值为 2^1。

6.2　使用 View 绘制视图

View 类是 Android 平台中各种控件的父类，是 UI（用户界面）的基础构件。View 相当于屏幕上的一块矩形区域，并负责绘制这个区域和处理事件。View 是所有 widget 类的父类，其子类 ViewGroup 是所有 layout 类的父类。如果需要自定义控件，也需要继承 View，实现 onDraw 方法和事件处理方法。

6.2.1　横竖屏坐标与全屏操作

手机应用程序默认都支持横竖屏切换，但如果不做处理会出现程序运行异常，所以在开发中，多数应用程序都是锁定屏幕方向。设置屏幕方向主要有两种实现方式，通过配置文件确定屏幕方向和通过代码设置屏幕方向。

使用配置文件确定屏幕方向的方式如下，MainActivity 采用横屏设置，由属性 android:screenOrientation="landscape"设定；LitActivity 采用竖屏设置，由属性 android:screenOrientation="portrait"设定。

```xml
<activity
    android:name = "com.freshen.code.MainActivity"
    android:screenOrientation = "landscape"
    android:label = "@string/app_name" />
<activity
    android:name = "com.freshen.code.LitActivity"
    android:screenOrientation = "portrait"
    android:label = "@string/app_name" />
```

使用代码设置屏幕方向的方式如下,为了防止切换后重新启动当前 Activity,需在配置文件中添加 android:configChanges="keyboardHidden|orientation"属性,并在 Activity 中重写 onConfigurationChanged 方法。

```
//设置屏幕为横屏
setRequestedOrientation(ActivityInfo.SCREEN_ORIENTATION_LANDSCAPE);
//设置屏幕为竖屏
setRequestedOrientation(ActivityInfo.SCREEN_ORIENTATION_PORTRAIT);
```

在 Android 系统中,无论横屏、竖屏,屏幕的左上角都是坐标系统的原点(0,0)。水平向右延伸是 X 轴正方向,竖直向下延伸是 Y 轴正方向,如图 6.2 所示。应用程序界面中的控件元素都由坐标来确定,以(0,0)坐标为左上角顶点,(screenWidth, screenHeight)为右下角顶点形成的矩形区域是可视区,该区域的元素可以呈现给用户,其他区域的图片不可见。

为了在屏幕中的合适位置绘制图形,需要使用屏幕的宽和高作为参考,来确定绘制图形的位置。要获得屏幕的宽和高,首先从 Activity 对象中获得 WindowManager 对象,然后从 WindowManager 对象中获得 Display 对象,再从 Display 对象中获得屏幕的宽和高,如下面的代码所示:

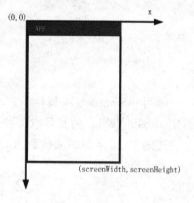

图 6.2　手机屏幕坐标

```
WindowManager wm = getWindowManager();
Display dis = wm.getDefaultDisplay();
//Android 3.2 之前这样做
Log.i("Tag", dis.getWidth() + " " + dis.getHeight());
/* Android 3.2 之后,提倡下面的方法
Point p = new Point();
dis.getSize(p);
Log.i("Tag", p.toString());
*/
```

在很多游戏类 App 开发过程中,都需要在屏幕中绘制游戏元素,确定这些元素的坐标位置。例如,在射击类游戏中需要判断玩家、敌人、子弹等元素的坐标位置。坐标的判断是

对上、下、左、右屏幕边界的判断。如果当前元素的 X 坐标小于零,则当前视图左越界。如果当前元素的 X 坐标加上宽度大于屏幕的宽,则右越界;如果当前元素的 Y 坐标小于零,则当前元素上越界;如果当前元素的 Y 坐标加上高度大于屏幕的高,则下越界。

屏幕中元素位置的移动就是不断改变元素的坐标,然后重新将它们绘制在屏幕上来实现。这种坐标的位置改变和绘制过程是通过一定逻辑来控制的。改变了元素的坐标,再重新绘制,在视觉上就会感觉元素在移动。如果元素水平向左移动,则 X 坐标减小;如果元素水平向右移动,X 坐标增大;如果元素垂直向上移动,Y 坐标减小;如果元素垂直向下移动,Y 坐标增大。

如果要获得最大化的显示区域,则需要将应用程序的标题栏和手机状态栏都隐藏,实现应用程序的全屏显示,这也是很多游戏类应用开发中采用的方式。全屏显示可以通过以下两种方式实现:

(1) 通过配置文件隐藏标题栏和状态栏。

```
android:theme = "@android:style/Theme.Black.NoTitleBar"
android:theme = "@android:style/Theme.Black.NoTitleBar.Fullscreen"
```

隐藏标题栏由上面的第一行代码实现,隐藏状态栏实现全屏由第二行代码实现。

(2) 通过代码隐藏标题栏和状态栏。

```
//隐藏标题栏
requestWindowFeature(Window.FEATURE_NO_TITLE);
//隐藏状态栏
getWindow().setFlags(WindowManager.LayoutParams.FLAG_FULLSCREEN,
        WindowManager.LayoutParams.FLAG_FULLSCREEN);
```

需要注意的是,设置全屏的代码应在 setContentView 方法之前执行,否则会抛出异常。

6.2.2 View 类

View 类位于 android.view 包中,自定义 View 视图时需要继承 View,并提供参数为 Context 类型的构造方法。View 类的常用构造方法如下:

- View(Context context),最简单的构造方法,使用 Context 对象创建视图。在 Activity 中可以借助代码动态创建 View 对象。
- View(Context context,AttributeSet attrs),从 xml 文件创建 View 对象时使用的构造方法。参数 attrs 的作用是接收布局文件中 android:layout_width 和 android:layout_height 参数值。

下面的代码演示如何自定义 View 类,并添加监听器。该 View 类提供了 View(Context context,AttributeSet attrs)类型的构造方法,允许在布局文件中设置宽高尺寸。重写 onDraw 方法,实现界面的绘制功能。

📖 Proj06_2 项目 MyView.java 文件

```java
public class MyView extends View implements OnClickListener {
    Random rand = new Random();
    int color;
    SimpleDateFormat sdf = new SimpleDateFormat("yyyy-MM-dd HH:mm:ss");
    //构造方法,接收 xml 文件的宽高参数
    public MyView(Context context, AttributeSet attrs) {
        super(context, attrs);
        color = Color.rgb(rand.nextInt(255),rand.nextInt(255),rand.nextInt(255));
        setOnClickListener(this);
    }
    //绘制界面方法
    @Override
    protected void onDraw(Canvas canvas) {
        super.onDraw(canvas);
        setBackgroundColor(color);              //设置背景演示
        //绘制文本
        canvas.drawText(sdf.format(new Date()), 80, 20, new Paint());
    }
    //监听方法
    @Override
    public void onClick(View arg0) {
        color = Color.rgb(rand.nextInt(255),rand.nextInt(255),rand.nextInt(255));
        invalidate();                           //更新界面
    }
}
```

View 视图的绘制依赖 onDraw 方法,当该 View 对象显示在屏幕上时,会自动调用该方法。这与 J2SE AWT 和 Swing 编程中的重绘组件类似,很多思想值得借鉴,但是在 Android 中绘制界面的方式与 J2SE 略有不同。

监听器方法比较简单,当 View 对象被单击后,重新创建颜色对象,并通知 UI 线程重新绘制界面。在 View 中更新界面有两个方法:invalidate 和 postInvalidate,前者用于在 UI 线程中的代码更新界面,后者用于在非 UI 线程中的代码更新界面,即在子线程中更新界面。上述代码都是运行在 UI 线程中的,所以使用 invalidate 方法更新。

在布局文件中使用 MyView 控件的方式与普通控件相同,可以设置相应属性。但是,控件的标签需要使用自定义 View 的全路径,具体代码如下。

📖 Proj06_2 项目 activity_main.xml 文件

```xml
…
<edu.freshen.p62.MyView
    android:id = "@+id/myView1"
    android:layout_width = "match_parent"
    android:layout_height = "60dp"
/>
…
```

在 Activity 中引用上述布局文件,不需要添加其他代码,运行效果如图 6.3 所示。MyView 控件自带监听器,被单击时,自动更新控件,更换颜色,显示最新日期和时间。

如果需要自定义更加复杂的 View 控件,需要掌握 Canvas 类和 Paint 类的使用。这样在 onDraw 方法中就可以绘制出更加丰富的图形图像,比如绘制柱形图、折线图等。

图 6.3 自定义 View 运行效果

6.2.3 Canvas 类

Canvas 类位于 android.graphics 包中,它表示一块画布,可以使用该类提供的方法,绘制各种图形图像。一般情况下,Canvas 对象的获取方式有两种:一种是重写 View.onDraw 方法,在该方法中 Canvas 对象会被当做参数传递过来,在该 Canvas 上绘制的图形图像会直接显示在 View 中;另一种是借助位图创建 Canvas,参考代码如下:

```
Bitmap bitmap = Bitmap.createBitmap(100, 100, Bitmap.Config.ARGB_8888);
Canvas canvas = new Canvas(bitmap);
```

Canvas 类提供的方法主要有 3 类应用,分别是绘制图形图像、修改画布状态和绘制剪切区。下面分别介绍 Canvas 在这 3 个方面的应用。

1. 绘制图形图像

在 Android 中绘制图形主要使用 Canvas 提供的各种方法,常用的绘图方法见表 6.4。

表 6.4 Canvas 常用绘制方法

方 法	说 明	返回值类型
drawARGB(int a, int r, int g, int b)	根据 a、r、g、b 绘制颜色,a 是 alpha,表示透明度	void
drawArc(RectF oval, float startAngle, float sweepAngle, boolean useCenter, Paint paint)	绘制弧线	void
drawBitmap(Bitmap bitmap, Matrix matrix, Paint paint)	使用矩阵绘制位图	void
drawBitmap(Bitmap bitmap, float left, float top, Paint paint)	指定坐标绘制位图	void
drawCircle(float cx, float cy, float radius, Paint paint)	指定坐标绘制圆形	void
drawColor(int color)	绘制指定颜色	void
drawLine(float startX, float startY, float stopX, float stopY, Paint paint)	根据两点坐标绘制直线	void
drawLines(float[] pts, Paint paint)	直线的坐标存放在数组中	void
drawOval(RectF oval, Paint paint)	绘制矩形的内切椭圆	void
drawPath(Path path, Paint paint)	绘制路径	void
drawPoint(float x, float y, Paint paint)	绘制一点	void
drawRGB(int r, int g, int b)	绘制 RGB 颜色	void
drawRect(float left, float top, float right, float bottom, Paint paint)	根据坐标绘制矩形	void

续表

方法名	说明	返回值类型
drawRect(RectF rect, Paint paint)	绘制指定圆角矩形	void
drawRect(Rect r, Paint paint)	绘制指定矩形	void
drawRoundRect(RectF rect, float rx, float ry, Paint paint)	绘制圆角矩形	void
drawText(String text, float x, float y, Paint paint)	绘制字符串	void
drawTextOnPath(char[] text, int index, int count, Path path, float hOffset, float vOffset, Paint paint)	沿指定路径绘制字符串	void

项目 Proj06_3 演示如何继承 View 类，使用 Canvas 绘制常见图形。自定义视图类代码如下。

📖 Proj06_3 项目 MyView.java 文件

```java
public class MyView extends View {
    Paint paint;                                    //画笔
    public MyView(Context context) {
        super(context);
        paint = new Paint();                        //创建画笔对象
    }
    //绘图方法
    @Override
    protected void onDraw(Canvas canvas) {
        super.onDraw(canvas);
        //绘制圆弧
        canvas.drawArc(new RectF(0, 0, 100, 100), 0, 120, true, paint);
        paint.setStyle(Style.STROKE);               //设置画笔只划线，不填充
        canvas.drawArc(new RectF(120, 0, 220, 100), 180, 180, false, paint);
        //绘制圆形
        canvas.drawCircle(50, 150, 50, paint);
        //绘制线段
        canvas.drawLine(0, 200, 100, 200, paint);
        //绘制椭圆
        canvas.drawOval(new RectF(0, 200, 100, 250), paint);
        //绘制矩形
        canvas.drawRect(120, 100, 250, 200, paint);
        //绘制圆角矩形
        canvas.drawRoundRect(new RectF(120, 210, 250, 250), 8, 8, paint);
        //绘制字符串
        canvas.drawText("绘制字符串……", 0, 270, paint);
        //绘制位图
        Bitmap bitmap = BitmapFactory.decodeResource(getResources(), R.drawable.icon72);
        canvas.drawBitmap(bitmap, 0, 280, paint);
    }
}
```

绘制圆弧使用 drawArc(RectF oval, float startAngle, float sweepAngle, boolean useCenter, Paint paint)方法。参数 oval 是矩形对象，通过 new RectF(0, 0, 100, 100)创

建，表示以坐标(0,0)和坐标(100,100)所确定的矩形区域。startAngle 是弧开始的度数，sweepAngle 是弧跨过的度数。以 startAngle 为起始度数，在上述矩形内，按照顺时针方向，跨过 sweepAngle 度，就可以形成弧。useCenter 是弧的开始和结束是否与中心位置连接。paint 是画笔，可以决定绘制的图形是否需要填充。paint.setStyle(Style.STROKE)是设置画笔只绘制形状，不填充；如果需要填充，设置 Style.FILL，这是默认值。

绘制圆形使用方法 drawCircle(float cx, float cy, float radius, Paint paint)。参数 cx 和 cy 是圆心坐标，radius 是圆的半径。

绘制线段使用方法 drawLine(float startX, float startY, float stopX, float stopY, Paint paint)。参数 startX 和 startY 是线段的起始坐标，stopX 和 stopY 是线段的结束坐标。

绘制椭圆使用方法 drawOval(RectF oval, Paint paint)。参数 oval 是矩形，它决定了内切椭圆的位置和尺寸。

绘制矩形的方法有多个，可以先创建矩形对象，再将其绘制出来。直接绘制矩形对象使用 drawRect(float left, float top, float right, float bottom, Paint paint)方法。参数 left 和 top 是矩形的左上角顶点坐标，right 和 bottom 是右下角顶点坐标。

绘制圆角矩形使用 drawRoundRect(RectF rect, float rx, float ry, Paint paint)方法。参数 rect 是矩形对象。rx 和 ry 分别指 X 方向和 Y 方向的弧度。

绘制字符串使用 drawText(String text, float x, float y, Paint paint)方法。参数 text 是需要绘制的字符串，x 和 y 是字符串开始绘制的坐标。

绘制位图的方法有多个，简单绘制可以使用 drawBitmap(Bitmap bitmap, float left, float top, Paint paint)方法。bitmap 是需要绘制的位图对象，left 和 top 是位图的左上角顶点坐标，位图的宽高决定了绘制区域的大小。如果绘制的位图需要变换，可以使用 drawBitmap(Bitmap bitmap, Matrix matrix, Paint paint)方法，参数 matrix 在后面会有详细说明。

自定义的 MyView 视图既可以作为一个独立的控件使用，也可以直接设置给 Activity，作为整个应用程序的界面，详细代码如下。

📖 Proj06_3 项目 MainActivity.java 文件

```java
public class MainActivity extends Activity {
    @Override
    protected void onCreate(Bundle savedInstanceState) {
        super.onCreate(savedInstanceState);
        //隐藏标题栏
        requestWindowFeature(Window.FEATURE_NO_TITLE);
        //取消状态栏显示
        getWindow().setFlags(WindowManager.LayoutParams.FLAG_FULLSCREEN,
                WindowManager.LayoutParams.FLAG_FULLSCREEN);
        //使用自定义的 MyView 视图
        setContentView(new MyView(this));
    }
}
```

项目运行效果如图 6.4 所示。

2. 修改画布状态

修改画布状态,其目的是编辑二维图形。常见的二维图形图像变换有平移变换、放缩变换和旋转变换。Canvas 的 API 中对画布状态的修改主要包括平移画布(X 轴和 Y 轴)、放缩画布、旋转画布。这些状态的修改实质上是对画布坐标系的修改。例如,向 X 轴正方向移动画布后,绘制图形图像时就会以当前坐标系为参考,返回原坐标系浏览时,会发现绘制的图形图像被平移了。

Canvas 的坐标系一旦发生变化,此后所有的绘图操作都会参考新的坐标系。如果修改 Canvas 只为某一次绘图,其他绘图还需要参考初始坐标系,则应在修改画布状态前使用 save 方法保存画布状态,绘制完成后,使用 restore 方法恢复画布状态。

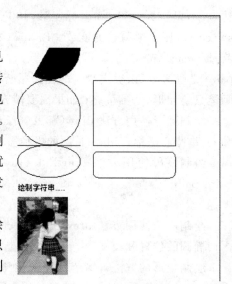

图 6.4　Canvas 绘制图形图像

Canvas 类中关于修改画布状态的方法如表 6.5 所示。

表 6.5　Canvas 中修改画布状态的方法

方法	说明	返回值类型
translate(float dx, float dy)	X 轴移动 dx 大小,Y 轴移动 dy 大小	void
rotate(float degrees)	围绕坐标原点旋转 degrees 度	void
rotate(float degrees, float px, float py)	围绕 px、py 旋转 degrees 度,常用	void
scale(float sx, float sy)	以坐标原点为参考点,X 轴放缩 sx 倍,Y 轴放缩 sy 倍	void
scale(float sx, float sy, float px, float py)	以 px、py 为参考点,X 轴放缩 sx 倍,Y 轴放缩 sy 倍,常用	void
save()	保存画布状态	int
restore()	恢复画布状态	void

项目 Proj06_4 演示了 Canvas 类的平移操作。自定义视图 CanvasView 继承 View,重新绘图方法为 onDraw。在 MainActivity 中创建 CanvasView 对象,并设置为显示界面。

📖 Proj06_4 项目 CanvasView.java 文件

```
public class CanvasView extends View {
    Paint paint;
    Bitmap bitmap;                    //位图
    int bw,bh;                        //位图宽高
    public CanvasView(Context context) {
        super(context);
        paint = new Paint();
        //创建位图
```

```
            bitmap = BitmapFactory.decodeResource(getResources(), R.drawable.qq);
            //获取位图宽高
            bw = bitmap.getWidth();
            bh = bitmap.getHeight();
        }
        //自定义绘图方法
        @Override
        protected void onDraw(Canvas canvas) {
            super.onDraw(canvas);
            //参考初始坐标系绘制位图
            canvas.drawBitmap(bitmap, 0, 0, paint);                            ①
            //保存画布状态
            canvas.save();
            //将画布坐标系向 X 轴正方向平移 bw,向 Y 轴正方向平移 bh
            canvas.translate(bw, bh);
            //参考新坐标系绘制位图
            canvas.drawBitmap(bitmap, 0, 0, paint);                            ②
            //参考新坐标系绘制填充矩形
            canvas.drawRect(0, bh, bw, bh * 2, paint);                         ③
            //恢复画布状态
            canvas.restore();
            //参考初始坐标系绘制无填充矩形
            paint.setStyle(Style.STROKE);
            canvas.drawRect(0, bh, bw, bh * 2, paint);                         ④
        }
}
```

bitmap 是需要绘制的位图,bw 和 bh 是该位图的宽高尺寸。代码①处,参考 Canvas 初始坐标 XOY,在原点位置绘制位图。canvas.translate(bw,bh)代码实现将 Canvas 坐标系向 X 轴正方向平移 bw,向 Y 轴正方向平移 bh。初始坐标系 XOY 变换为 $X'O'Y'$,如图 6.5 所示。

代码②与代码①一样,但绘制出的位图位置已经发生变化,原因就是它是参考了平移之后的 $X'O'Y'$ 坐标系绘制的。读者可以自己对比带有填充效果的矩形和无填充效果的矩形位置,分析一下它们的坐标位置。

相对于平移,旋转和放缩要复杂一些。在对画布进行旋转和放缩操作时,可以设置"参考点",如果不设置,则默认参考坐标原点。以原点为参考点的旋转使用 rotate(float degrees)方法,参数 degrees 是度数,取正值表示顺时针旋转,取负值表示逆时针旋转;自定义参考点的旋转使用 rotate(float degrees, float px, float py)方法,参数 px 和 py 是参考点坐标。

下面的代码演示两种旋转操作,运行效果如图 6.6 所示。

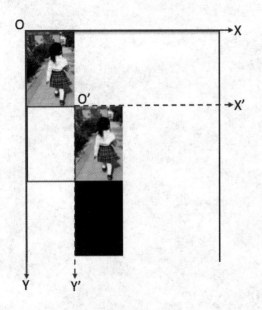
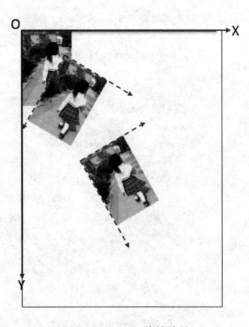

图 6.5 Canvas 平移效果　　　　　　图 6.6 Canvas 旋转效果

📖 Proj06_4 项目 CanvasView.java 文件旋转 Canvas 的相关代码

```
...
//自定义绘图方法
@Override
protected void onDraw(Canvas canvas) {
    super.onDraw(canvas);
    //参考初始坐标系绘制位图
    canvas.drawBitmap(bitmap, 0, 0, paint);
    //第一次旋转,参考点为原点
    canvas.save();
    canvas.rotate(30);                                                          ①
    canvas.drawBitmap(bitmap, bw, 0, paint);
    canvas.restore();
    //第二次旋转,参考点为屏幕中心
    canvas.save();
    canvas.rotate(-30,this.getWidth()/2, this.getHeight()/2);                   ②
    canvas.drawBitmap(bitmap, this.getWidth()/2-bw/2,this.getHeight()/2-bh/2, paint);
    canvas.restore();
}
...
```

第一次旋转是参考原点,顺时针旋转 30°。第二次旋转以屏幕中心为参考点,逆时针旋转 30°。this.getWidth()/2 和 this.getHeight()/2 可以计算出 View 控件的中心坐标。如果需要将位图绘制在屏幕正中心,需要计算位图的左上角顶点坐标,计算过程如同解一道平面几何数学题。

下面的代码演示了 Canvas 放缩操作。以原点为参考点的放缩使用 scale(float sx,

float sy)方法,参数 sx 是 X 轴方向的放缩量,sy 是 Y 轴方向的放缩量。自定义参考点的放缩使用 scale(float sx, float sy, float px, float py)方法,参数 px 和 py 是参考点坐标。

📖 Proj06_4 项目 CanvasView.java 文件,放缩 Canvas 的相关代码

```
…
//自定义绘图方法
@Override
protected void onDraw(Canvas canvas) {
    super.onDraw(canvas);
    //参考初始坐标系绘制位图
    canvas.drawBitmap(bitmap, 0, 0, paint);
    //X轴放大到2倍,Y轴不变
    canvas.save();
    canvas.scale(2, 1);
    canvas.drawBitmap(bitmap, bw, 0, paint);
    canvas.restore();
    //X轴放大到2倍,Y轴缩小到0.5
    canvas.save();
    canvas.scale(2, 0.5f);
    canvas.drawBitmap(bitmap, 0, bh * 2, paint);
    canvas.restore();
}
…
```

坐标系放缩后,会直接影响绘制图形图像时的参考值。放大坐标系,则位图尺寸变大;缩小坐标系,则位图尺寸变小。上述代码的运行效果如图 6.7 所示。

图 6.7 Canvas 放缩效果

Canvas 的放缩方法中,参数 sx 和 sy 取值为 -1,可以实现镜像功能。sx 取值为 -1 时,在 X 轴方向镜像;sy 取值为 -1 时,在 Y 轴方向镜像。下面的代码演示如何实现镜像功能。

📖 Proj06_4 项目 CanvasView.java 文件的镜像命令代码

```
…
//自定义绘图方法
@Override
protected void onDraw(Canvas canvas) {
    super.onDraw(canvas);
    canvas.drawBitmap(bitmap, 0, 0, paint);      ①
    //X轴不变,Y轴镜像
    canvas.save();
    canvas.scale(1, -1);
    canvas.drawBitmap(bitmap, 0, -bh * 2, paint);  ②
    canvas.restore();
```

```
//X轴镜像,Y轴不变
canvas.save();
canvas.scale(-1, 1,bw,0);
canvas.drawBitmap(bitmap,0, 0, paint);
canvas.restore();
}
…
```

canvas.scale(1,-1)没有指定参考点,则以原点为参考点,在Y轴方向镜像。canvas.scale(-1, 1, bw,0)指定了参考点,并在X轴方向镜像。上述代码的运行效果如图6.8所示。

3. 绘制剪切区

创建剪切区使用Canvas类中的clipXXX方法。剪切区创建后,只有处在剪切区中的内容会显示,其他区域的内容被隐藏。

剪切区支持简单图形的剪切,如圆形、矩形,也支持剪切路径(Path)和区域(Region)。剪切Path,需要先创建Path对象。Path可以作为简单的线条路径,也可以是组合图形的路径。给Path添加图形时,使用参数Direction.CW或Direction.CCW分别指明顺时针方向还是逆时针方向绘制。剪切Region相当于多个矩形的组合,代表这些矩形组成的区域,这些矩形通过Region.OP指明运算规则。

图6.8 Canvas镜像效果

Canvas类中常用的绘制剪切区方法如表6.6所示。

表6.6 Canvas绘制剪切区的常用方法

方法	说明	返回值类型
clipPath(Path path, Region.Op op)	剪切路径,根据op参数进行运算	boolean
clipRect(Rect rect, Region.Op op)	剪切矩形,根据op参数进行运算	boolean
clipRect(int left, int top, int right, int bottom)	剪切矩形	boolean
clipRegion(Region region, Region.Op op)	剪切区域,根据op参数进行运算	boolean

Region.Op是多个剪切区的运算方式,可以采用的运算方式如表6.7所示。

表6.7 Region.OP剪切区运算方式

剪切区运算方式	说明
Op.DIFFERENCE	A区域和B区域的差集范围,即A-B,只有在此范围内的绘制内容才会被显示
Op.REVERSE_DIFFERENCE	B区域和A区域的差集范围,即B-A,只有在此范围内的绘制内容才会被显示
Op.INTERSECT	A区域和B区域的交集范围,只有在此范围内的绘制内容才会被显示
Op.REPLACE	B区域将全部进行显示,如果和A有交集,则将覆盖A的交集范围
Op.UNION	A区域和B区域的并集范围,即两者所包括的范围的绘制内容都会被显示
Op.XOR	A区域和B区域的补集范围,只有在此范围内的绘制内容才会被显示

项目 Proj06_5 演示了 Canvas 剪切区的应用，自定义视图类 ClipView 继承 View 类，重写 onDraw 方法。在 MainActivity 中创建 ClipView 对象，并设定为主界面。ClipView 的核心代码如下。

📖 Proj06_5 项目 ClipView.java 文件简单剪切区的相关代码

```
…
@Override
protected void onDraw(Canvas canvas) {
    super.onDraw(canvas);
    canvas.save();
    //创建第一个矩形剪切区
    canvas.clipRect(100, 0, 200, 200);
    //创建第二个矩形剪切区,并设定运算为 UNION
    canvas.clipRect(new Rect(0,50,400,100), Region.Op.UNION);
    //绘制位图
    canvas.drawBitmap(bg, 0, 0, paint);
    canvas.restore();
}
…
```

上述代码中两次创建的矩形剪切区使用 Region. Op. UNION 运算，即求两个区域的并集，形成最终的剪切区。随后绘制的位图 bg 尺寸较大，可以填充满整个屏幕，但只有剪切区内的部分可以显示，如图 6.9 所示。

图 6.9 原图与使用剪切区后的效果

绘制 Path 剪切区，需要先创建路径对象 Path，再借助 Path 类中的 addXXX 方法添加各种形状，最好使用 Canvas 对象的 clipPath 方法。下面的代码演示路径剪切区的使用。

📖 Proj06_5 项目 ClipView.java 文件 Path 剪切区的相关代码

```
…
@Override
protected void onDraw(Canvas canvas) {
    super.onDraw(canvas);
    canvas.save();
    Path path = new Path();                              //创建路径对象
    path.addCircle(50, 50, 50, Direction.CW);            //添加形状
    canvas.clipPath(path);                               //绘制路径
    path.addRect(50, 50, 150,150, Direction.CW);         //添加其他形状
    canvas.clipPath(path, Region.Op.UNION);              //按照"并集"方式绘制路径
    canvas.drawBitmap(bg, 0, 0, paint);                  //绘制位图
    canvas.restore();
}
…
```

路径剪切区是两个形状的"并集"运算，代码运行效果如图 6.10 所示。

绘制 Region 剪切区，先要创建 Region 对象。可以使用无参构造方法或以指定矩形区

域为参数创建 Region 对象。Region 对象可以借助 op 方法与其他区域运算,得到最后的结果。下面的代码演示 Region 剪切区的使用。

📖 Proj06_5 项目 ClipView.java 文件 Region 剪切区的相关代码

```
…
@Override
protected void onDraw(Canvas canvas) {
    super.onDraw(canvas);
    canvas.save();
    Region rg =  new Region(50,0,150,300);              //创建 Region 对象,指定区域
    rg.op(new Rect(0,100,300,200),Region.Op.XOR);       //添加其他区域,并使用补集运算
    canvas.clipRegion(rg);                              //剪切 Region 对象
    canvas.drawBitmap(bg, 0, 0, paint);
    canvas.restore();
}
…
```

Region 对象由两个矩形按照补集运算形成,即两个区域减去交集部分。代码运行效果如图 6.11 所示。

图 6.10 Path 剪切区效果

图 6.11 Region 剪切区效果

6.2.4 Paint 类

Paint 即画笔,在绘图过程中有极其重要的作用。它主要保存了颜色、样式等绘制信息。Paint 对象的参数设置大体上可以分为两类:一类与图形绘制相关,另一类与文本绘制相关。Paint 类中的常用方法如表 6.8 所示。

表 6.8 Paint 类常用方法

方法	说明	返回值类型
getAlpha()	获取画笔的透明度	int
getColor()	获取画笔当前颜色	int
measureText(String text)	测量 text 所占长度	float

续表

方法	说明	返回值类型
setARGB(int a, int r, int g, int b)	设置透明度及颜色,取值为 0～255	void
setAlpha(int a)	设置透明度	void
setAntiAlias(boolean aa)	设置画笔是否具有抗锯齿功能,比较消耗资源	void
setColor(int color)	设置颜色	void
setStrokeWidth(float width)	设置画笔粗细	void
setStyle(Paint.Style style)	设置画笔样式,默认绘制都是填充,Style.STROKE 可以绘制空心图形	void
setTextSize(float textSize)	设置绘制文本时的文字大小	void

项目 Proj06_6 演示了不同笔触的绘图效果。PaintView 类是自定义视图,继承 View。在 MainActivity 中使用 PaintView 作为主界面显示。

📖 Proj06_6 项目 PaintView.java 文件绘图方法的相关代码

```java
…
@Override
protected void onDraw(Canvas canvas) {
    super.onDraw(canvas);
    paint.setColor(Color.RED);              //画笔颜色
    canvas.drawCircle(50, 50, 20, paint);
    canvas.drawText("无抗锯齿", 20, 80, paint);
    //画笔抗锯齿
    paint.setAntiAlias(true);
    canvas.drawCircle(150, 50, 20, paint);
    canvas.drawText("抗锯齿", 140, 80, paint);
    //只绘制线条,不填充
    paint.setStyle(Style.STROKE);
    canvas.drawCircle(50, 150, 20, paint);
    //线条粗细
    paint.setStrokeWidth(7);
    canvas.drawCircle(150, 150, 20, paint);
    //绘制文字
    paint.setStrokeWidth(0.5f);
    canvas.drawText("文字宽度" + paint.measureText("文字宽度"),30,200, paint);
    //文字大小
    paint.setTextSize(22);
    canvas.drawText("文字宽度" + paint.measureText("文字宽度"),120,200, paint);
}
…
```

需要注意的是,画笔的状态设置一次后,会影响其后的所有绘制效果。如果希望画笔状态的修改只影响某一次绘制,则应先保存画笔状态,修改之后再恢复状态。例如在修改画笔颜色时采用如下方法:

```java
int cr = paint.getColor();              //保存画笔原始颜色
paint.setColor(Color.GRAY);             //修改画笔颜色
//执行绘制逻辑
paint.setColor(cr);                     //恢复画笔初始颜色
```

项目 Proj06_6 的运行效果如图 6.12 所示。

6.2.5 使用 View 自定义控件

Android 开发平台提供了大量系统控件，可以比较方便地设计 UI。但是在特定场合下，需要自定义控件，实现个性化 UI。自定义控件的制作可以分为 3 种：自绘控件、组合控件和继承控件。

自绘控件需要继承 View，重写 onDraw 方法，将所展现的内容全部绘制出来。组合控件不需要继承 View，它只是将几个系统原生的控件组合到一起，形成一个独立的控件。继承控件需要继承特定的系统控件，然后在这个控件上增加一些新的功能，形成一个新的自定义的控件。本节主要介绍自绘控件的制作。

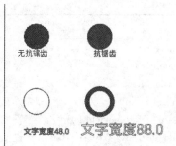

图 6.12 不同笔触的效果

项目 Proj06_7 自定义"柱状图"控件，使用 Activity 传入的数据生成"周销售额柱状图"。该自绘控件比较简单，没有使用 AttributeSet 控制布局参数，因此不适用于在布局文件中引用。创建 ChartView 对象时只能使用 Java 代码。

📖 Proj06_7 项目 ChartView.java 文件

```java
//周销售额柱状图
public class ChartView extends View {
    float data[];                                           //周销售额数据
    Paint paint;
    String week[] = {"一","二","三","四","五","六","日"};
    public ChartView(Context context,float data[]) {
        super(context);
        this.data = data;
        paint = new Paint();
    }
    @Override
    protected void onDraw(Canvas canvas) {
        super.onDraw(canvas);
        int pad = getWidth()/7;                             //计算每组数据宽度
        int w = 10;                                         //每个柱形的宽度
        canvas.save();
        paint.setColor(Color.BLUE);
        canvas.translate(10, 200);                          //平移坐标系
        //绘制 7 组数据对应的柱形图
        for (int i = 0; i < data.length; i++) {
            canvas.drawRect(pad * i, 0,pad * i + w, - data[i], paint);//绘制矩形
            canvas.drawText(week[i], pad * i,25, paint);    //绘制水平坐标的标目
        }
        canvas.drawLine(0, 0,getWidth(), 0, paint);         //绘制水平线
        canvas.restore();
    }
}
```

该柱状图的数据个数是固定的,因此每组数据所占宽度的计算使用 getWidth()/7,即控件宽度除以 7。绘制柱状图时,需要将 7 组数据对应的矩形依次画出。数据的大小即为矩形的高度(实际项目中,矩形高度应通过换算得出相对高度)。

绘制矩形的难点在于计算矩形的坐标值,即左上角顶点坐标和右下角顶点坐标。因为 Canvas 的坐标已经平移过,坐标(pad∗i,0)就是矩形的左上角顶点坐标,坐标(pad∗i+w,−data[i])就是矩形的右下角顶点坐标。因为取了负值:−data[i],矩形被绘制在 Y 轴的负方向。

水平线和每组数据标目的绘制就比较简单了,柱状图的最终运行效果如图 6.13 所示。

MainActivity 使用 activity_main.xml 布局文件,包含一个 TextView 控件和一个 LinearLayout 控件。MainActivity 的代码比较简单,创建柱状图控件对象,并添加到 LinearLayout 中。

📖 Proj06_7 项目 MainActivity.java 文件

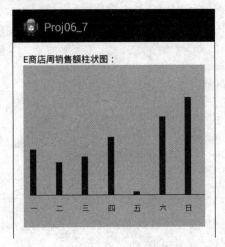

图 6.13 自定义柱状图控件

```
public class MainActivity extends Activity {
    float data[] = {70,50,59,89,5,120,150};
    @Override
    protected void onCreate(Bundle savedInstanceState) {
        super.onCreate(savedInstanceState);
        setContentView(R.layout.activity_main);
        LinearLayout ll = (LinearLayout) findViewById(R.id.chartView);
        ll.addView(new ChartView(this,data));          //添加自定义控件
    }
}
```

6.2.6 Matrix 变换

对二维图形图像的处理,除了上面介绍的方法,还可以借助 Matrix 类。Matrix 是一个 3×3 的矩阵,实现对图形图像的平移、缩放、变形、斜切操作。在 Android 的 API 中对于每一种变换都提供了 3 种操作方式:set(直接设置 Matrix 中的值)、post(相当于矩阵运算中的后乘)、pre(相当于矩阵运算中的前乘)。

与 Canvas 类的放缩和变形类似,Matrix 变换也是默认参考坐标原点(0,0)进行图形图像变换,同时支持自定义参考点。Matrix 的常用方法如表 6.9 所示,其中省略了 preXXX 类方法。

表 6.9 Matrix 常用方法

方 法	说 明	返回值类型
postRotate(float degrees)	设置旋转度数(矩阵后乘)	boolean
postRotate(float degrees, float px, float py)	设置旋转度数,自定义参考点(矩阵后乘)	boolean

方法	说明	返回值类型
postScale(float sx, float sy)	设置放缩量(矩阵后乘)	boolean
postScale(float sx, float sy, float px, float py)	设置放缩量,自定义参考点(矩阵后乘)	boolean
postSkew(float kx, float ky)	设置斜切(矩阵后乘)	boolean
postSkew(float kx, float ky, float px, float py)	设置斜切,自定义参考点(矩阵后乘)	boolean
postTranslate(float dx, float dy)	设置位移(矩阵后乘)	boolean
setRotate(float degrees)	设置旋转度数	void
setRotate(float degrees, float px, float py)	设置旋转度数,自定义参考点	void
setScale(float sx, float sy)	设置放缩量	void
setScale(float sx, float sy, float px, float py)	设置放缩量,自定义参考点	void
setSkew(float kx, float ky)	设置斜切	void
setSkew(float kx, float ky, float px, float py)	设置斜切,自定义参考点	void
setTranslate(float dx, float dy)	设置位移	void

需要注意的是,set 系列方法执行时会直接设置 Matrix 的值。每执行 setXXX 一次,整个 Matrix 矩阵都会重置。post 系列方法是后乘,即当前的矩阵乘以参数给出的矩阵。可以连续多次使用 post 来完成所需的复合变换。pre 系列方法与 post 系列方法的作用类似,只是采用前乘,即参数给出的矩阵乘以当前矩阵。

项目 Proj06_8 演示了如何借助 Matrix 处理位图的变换。自定义视图 MatrixView 继承 View,重写 onDraw 方法,代码如下。

📖 Proj06_8 项目 MatrixView.java 文件

```
… …
@Override
protected void onDraw(Canvas canvas) {
    super.onDraw(canvas);
    int w = bitmap.getWidth();              //位图宽度
    int h = bitmap.getHeight();             //位图高度
    canvas.drawBitmap(bitmap, 0, 0, paint);  //绘制原始位图                     ①
    Matrix matrix = new Matrix();           //创建 Matrix 对象
    matrix.setTranslate(w, 0);              //设置平移
    canvas.drawBitmap(bitmap, matrix, paint); //使用平移后的矩阵,绘制位图        ②
    //重置矩阵
    matrix.reset();
    matrix.setScale(0.5f, 0.5f);            //重置矩阵,缩小到 0.5 倍
    matrix.postTranslate(0,h);              //放缩后的矩阵乘以平移矩阵
    canvas.drawBitmap(bitmap, matrix, paint); //使用缩小、平移后的矩阵绘制位图    ③
    matrix.setScale(1.5f, 1.5f);            //重置矩阵,放大到 1.5 倍
    matrix.postTranslate(w,h);              //放缩后的矩阵乘以平移矩阵
    canvas.drawBitmap(bitmap, matrix, paint); //使用放大、平移后的矩阵绘制位图    ④
    //重置矩阵
    matrix.reset();
    matrix.postTranslate(0,2.5f * h);       //平移矩阵
    matrix.postRotate(30, 0,2.5f * h);      //旋转矩阵 30°,旋转点坐标为(0,2.5 * h)
```

```
        canvas.drawBitmap(bitmap, matrix, paint);    //使用平移、旋转后的矩阵绘制位图    ⑤
        //重置矩阵
        matrix.reset();
        matrix.postTranslate(w,2.5f*h);               //平移矩阵
        matrix.postSkew(0.5f,0.5f, w, 2.5f*h);        //斜切矩阵
        canvas.drawBitmap(bitmap, matrix, paint);    //使用平移、斜切后的矩阵绘制位图    ⑥
    }
    ...
```

在 MainActivity 中创建 MatrixView 对象,并设置为主界面。项目运行效果如图 6.14 所示。

图 6.14　Matrix 变换效果

Matrix 方法中采用的坐标都是相对于 Canvas 对象的原始坐标,不会受到坐标变换的影响,这也是 Matrix 变换与 Canvas 变换最大的区别。

6.3　使用 SurfaceView 绘制视图

SurfaceView 继承了 View,但是它的视图绘制机制不同于 View。在 View 类中,绘图方法 onDraw 会被自动调用,写在该方法中的程序自动执行。SurfaceView 虽然继承了该方法,但 SurfaceView 不会自动调用该方法,使用 SurfaceView 视图时,需要自定义绘图方法,然后主动调用。SurfaceView 是对 View 的扩展,更适合开发 2D 游戏。

6.3.1　SurfaceHolder 介绍

SurfaceView 采用 SurfaceHolder 对象管理视图的创建、更改和销毁。SurfaceHolder 是一个接口,类似于一个 SurfaceView 的监听器,通过 3 个回调方法监听 Surface 的创建、销毁或者改变。

在 SurfaceView 中调用 getHolder 方法,可以获得当前 SurfaceView 对应的 SurfaceHolder。SurfaceHolder 采用 SurfaceHolder.Callback 接口的 3 个方法来实现 SurfaceView 视图的创建、更改和销毁的。因此,自定义 SurfaceView 视图时,需要继承 SurfaceView,并实现 SurfaceHolder.Callback 接口。

表 6.10　SurfaceHolder.Callback 接口的方法

方　　法	说　　明	返回值类型
surfaceChanged（SurfaceHolder holder，int format，int width，int height）	视图发生改变时调用	void
surfaceCreated(SurfaceHolder holder)	视图创建时调用,用于调用绘图方法、启动线程等操作	void
surfaceDestroyed(SurfaceHolder holder)	视图销毁时调用,用于停止线程等操作	void

SurfaceView 和 View 最本质的区别在于,SurfaceView 是在一个新发起的单独线程中重新绘制画面,View 必须在 UI 的主线程中更新画面。在 UI 的主线程中更新画面可能会引发阻塞问题。例如,当更新画面时间过长时,主 UI 线程会被绘图函数阻塞,将无法响应按键、触屏等操作。使用 SurfaceView 则不会阻塞 UI 主线程。

对于 View 与 SurfaceView 的选择,可以参考如下建议:

(1) 处理"被动更新"画面时采用 View 视图,如开发 2D 棋类游戏视图。这类应用画面的更新是依赖于 onTouch 方法的执行,可以直接使用 invalidate 方法。这种情况下,一次触屏和下一次的触屏需要的时间比较长,不会产生阻塞问题。

(2) 处理"主动更新"画面时采用 SurfaceView,如开发 2D 竞技类游戏视图。这类应用的画面需要不断地刷新、重绘,这就需要一个单独的线程不停地执行绘图方法,从而避免阻塞主线程。

项目 Proj06_9 演示了自定义 SurfaceView 视图的使用,MySurfaceView 继承 SurfaceView,实现 android.view.SurfaceHolder.Callback 接口,代码如下。

📖 Proj06_9 项目 MySurfaceView.java 文件

```
public class MySurfaceView extends SurfaceView implements Callback{
    SurfaceHolder holder;                          //处理视图的创建、修改和销毁
    Paint paint;                                   //画笔
    Canvas canvas;                                 //画布
    public MySurfaceView(Context context) {
        super(context);
        paint   = new Paint();
        holder = getHolder();                      //获取 SurfaceHolder 对象
```

```
            holder.addCallback(this);                    //添加回调接口
        }
        //当界面修改时执行
        @Override
        public void surfaceChanged(SurfaceHolder arg0, int arg1, int arg2, int arg3) {
        }
        //当界面创建时执行
        @Override
        public void surfaceCreated(SurfaceHolder arg0) {
            myDraw();                                    //主动调用自定义绘图方法
        }
        //当界面销毁时执行
        @Override
        public void surfaceDestroyed(SurfaceHolder arg0) {
        }
        //自定义绘图方法
        protected void myDraw(){
            canvas = holder.lockCanvas();                //锁定并得到画布对象
            if(canvas!= null){
                canvas.drawColor(Color.WHITE);           //将画笔"涂白"
                canvas.drawCircle(100, 100, 20, paint);  //绘制圆形
                holder.unlockCanvasAndPost(canvas);      //解除锁定画布
            }
        }
    }
```

SurfaceView 视图的使用与 View 视图有很大的差异,主要体现在 SurfaceHolder 类管理视图的变化。SurfaceHolder 使用 surfaceCreated 方法、surfaceChanged 方法和 surfaceDestroyed 方法处理视图的创建、修改和销毁事件。在回调方法 surfaceCreated 中主动调用自定义绘图方法 myDraw,实现自定义绘图。

获取画布对象的方法是 holder.lockCanvas,如果获取画布失败,则返回 null。得到画布之后,绘图逻辑与前面所介绍的基本一致,可以绘制各种图形图像。当绘图结束时,应使用方法 holder.unlockCanvasAndPost(canvas)释放画布。

在 MainActivity 中创建和使用 SurfaceView 对象与使用 View 对象的方法一样,相应的代码就不再给出。

6.3.2 使用子线程绘制视图

SurfaceView 视图主要用于处理需要频繁地重绘的界面。这类视图中的内容即使没有用户的交互也会不停地更新,如游戏视图的开发。因此在 SurfaceView 中启动子线程,在线程中周期性调用自定义的绘图方法,就可以实现绘制动态界面的效果。

在 SurfaceView 中使用子线程,一般是在 surfaceCreated 方法中创建和启动子线程,在 surfaceDestroyed 方法中停止子线程。在项目 Proj06_9 中,新建 ThreadSurfaceView 类,继承 SurfaceView,实现 Callback 接口,使用线程绘制视图,代码如下。

📖 Proj06_9项目 ThreadSurfaceView.java 文件

```java
public class ThreadSurfaceView extends SurfaceView implements Callback{
    SurfaceHolder holder;
    Paint paint;
    Canvas canvas;
    boolean flag = true;                              //线程控制标识
    public ThreadSurfaceView(Context context) {
        super(context);
        paint   = new Paint();
        holder  = getHolder();
        holder.addCallback(this);
    }
    @Override
    public void surfaceChanged(SurfaceHolder arg0, int arg1, int arg2, int arg3) {
    }
    @Override
    public void surfaceCreated(SurfaceHolder arg0) {
        //启动线程,使用匿名内部类创建线程对象
        new Thread(new Runnable(){
            @Override
            public void run() {
                while(flag){
                    myDraw();                         //周期性调用绘图逻辑
                    Log.i("Msg", "绘制线程进行中...");
                    try {
                        Thread.sleep(100);            //线程休眠
                    } catch (InterruptedException e) {
                        e.printStackTrace();
                    }
                }
                Log.i("Msg", "绘制线程结束!");
            }
        }).start();
    }
    @Override
    public void surfaceDestroyed(SurfaceHolder arg0) {
        flag = false;                                 //修改线程控制标识,结束绘图线程
    }
    protected void myDraw(){
        canvas = holder.lockCanvas();
        if(canvas!= null){
            canvas.drawColor(Color.WHITE);
            canvas.drawCircle(100, 100, 20, paint);
        }
        holder.unlockCanvasAndPost(canvas);
    }
}
```

在线程对象中,使用 while 循环周期性地调用 myDraw 方法。循环控制条件是属'

flag，若要该线程结束，只需要将 flag 赋值为 false 即可，这是一种方便控制线程的方法。为避免屏幕绘图过快，在线程中，每调用一次绘图方法，应休眠一段时间。

运行上述代码，线程每隔 100ms 就调用一次绘图方法，但视图中的内容并未发生改变，这是因为 myDraw 方法中的绘制内容是固定的。如果对 myDraw 方法中绘制圆形的代码 canvas.drawCircle(100，100，20，paint)进行修改，把圆心参数用变量（如 px，py）来表示，每次调用绘图方法时都修改圆心参数，就会出现该圆形在视图中"移动"的现象，从而出现动画效果。

6.4 线程控制下的动画效果

Android 中动画的实现方式有很多种，本节要讨论的是在 SurfaceView 中借助子线程实现的动画效果。线程控制下的动画主要分为 3 类，分别是属性动画、帧动画、剪切区动画。属性动画是在绘图时通过修改图形图像属性实现的。帧动画是在绘图时通过按照一定规律依次绘制一组图片实现的。剪切区动画是在绘图时按照一定规律绘制一系列剪切区实现的。

6.4.1 属性动画效果

在 SurfaceView 中实现属性动画，就是修改视图中元素的位置、旋转度数、放缩比例等属性，然后把修改后的图形图像周期性地绘制出来。

通常情况下，视图中需要控制的元素应被封装为一个独立对象，分别对 X 轴和 Y 轴赋予位置属性、旋转属性、放缩属性、速度属性、方向属性等。在绘图线程中，只要修改这些属性，就可以控制视图中元素的动画。

项目 Proj06_10 演示了如何使用线程修改属性，实现属性动画。按照前面所介绍的知识，创建 ThreadSurfaceView 视图，继承 SurfaceView，实现 Callback 接口。在 surfaceCreated 方法中启动线程，调用自定义绘图方法 myDraw。该项目中，被控制的视图元素是一个圆形，它会在视图中以设定的方向和速度不停地移动；当触屏时，圆形直接移动到触屏位置。动画效果比较简单，因此，未将圆形封装为一个类。详细代码如下。

📖 Proj06_10 项目 ThreadSurfaceView.java 文件

```java
public class ThreadSurfaceView extends SurfaceView implements Callback{
    SurfaceHolder holder;
    Paint paint;
    Canvas canvas;
    boolean flag = true;                        //线程控制标识
    int dirx,diry;                              //移动方向,分 X 轴和 Y 轴
    int spx = 5,spy = 8;                        //移动速度,分 X 轴和 Y 轴
    int px,py,pr = 20;                          //圆心坐标和半径
    …
    //重写触屏方法
    @Override
    public boolean onTouchEvent(MotionEvent event) {
        //修改圆形坐标为触屏点坐标
        px = (int) event.getX();
```

```java
            py = (int) event.getY();
            return true;
        }
        @Override
        public void surfaceCreated(SurfaceHolder arg0) {
            //启动线程
            new Thread(new Runnable(){
                @Override
                public void run() {
                    while(flag){
                        move();    //移动圆形
                        myDraw();    //周期性地调用绘图逻辑
                        Log.i("Msg", "绘制线程进行中...");
                        try {
                            Thread.sleep(100);    //线程休眠
                        } catch (InterruptedException e) {
                            e.printStackTrace();
                        }
                    }
                    Log.i("Msg", "绘制线程结束!");
                }
            }).start();
        }
        @Override
        public void surfaceDestroyed(SurfaceHolder arg0) {
            flag = false;    //修改线程控制标识,结束绘图线程
        }
        protected void myDraw(){
            canvas = holder.lockCanvas();
            if(canvas!= null){
                canvas.drawColor(Color.WHITE);
                canvas.drawCircle(px, py, pr, paint);
            }
            holder.unlockCanvasAndPost(canvas);
        }
        protected void move(){
            if(dirx == 0) px += spx;
            else px -= spx;
            if(diry == 0) py += spy;
            else py -= spy;
            //修正方向
            if(px - pr < 0) dirx = 0;
            if(px + pr > getWidth()) dirx = 1;
            if(py - pr < 0) diry = 0;
            if(py + pr > getHeight()) diry = 1;
        }
    }
```

对应需要移动的视图元素,通常需要声明的属性有 X 轴与 Y 轴移动方向、移动速度

位置坐标。上述代码中圆形 X 轴和 Y 轴的移动方向是 dirx 和 diry：当取值为 0 时，表示向正方向移动；当取值为 1 时，表示向负方向移动。移动速度是 spx=5 和 spy=8。位置坐标是 px 和 py，半径是 pr=20。

元素的移动方法 move 主要处理两个逻辑，修改位置坐标和修正移动方向。修改 X 轴位置坐标时，依据 X 轴移动方向；修改 Y 轴位置坐标时，依据 Y 轴移动方向。当元素位置到达屏幕左边时，修改 X 轴移动方向为正；当元素位置到达屏幕右边时，修改 X 轴移动方向为负。当元素位置到达屏幕上边时，修改 Y 轴移动方向为正；当元素位置到达屏幕下边时，修改 Y 轴移动方向为负。上述代码中计算位置时，考虑了半径。

元素在视图中的移动效果是元素在 X 轴和 Y 轴移动的合成，如图 6.15 所示。所谓的"X 轴和 Y 轴速度"实质上是指每次移动时圆形坐标在 X 轴和 Y 轴的变化量。

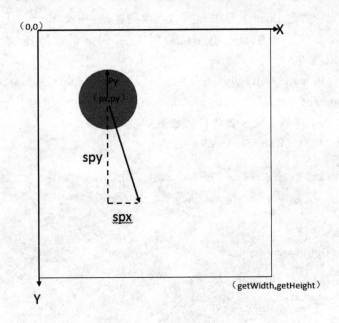

图 6.15 视图元素移动示意图

6.4.2 帧动画效果

帧动画又称逐帧动画，是指把某一个动作分解为一系列位图（通常是 PNG 格式的图片），然后依次"展示"所形成的动画效果。帧动画的实质是将多张有关联的静态图片连续播放，利用人眼的视觉暂留效应形成动画。每一张图片都称为一帧，帧数越多，动画越显流畅；帧数越少，动画越显跳跃。

一般在手机应用程序开发中采用 8 帧图或 4 帧图就可以完成一个动画效果。如果追求较高质量的动画效果，可以采用更多的帧数。图 6.16 所展示的是一个 12 帧图，可以非常详尽地展示动画过程，但帧数较多，也会消耗较多的存储资源和计算资源。

项目 Proj06_11 演示了如何绘制帧动画。将帧图统一命名，如图 6.16 所示，把一个动作的帧图命名为一个序列，编号不能间断，这有利于从代码中加载位图。创建 ThreadSurfaceView，继承 SurfaceView，实现 Callback 接口。

图 6.16 人物移动位图

声明存放帧图的数组 Bitmap frames[]和当前播放帧 currentFrame。由于帧图命名是统一的,它们在 R 文件中所生成的引用值是相邻的,因此借助循环,使用 R.drawable. loading_11+i 读取位图。

绘制位图时使用 canvas.drawBitmap(frames[currentFrame++%frames.length], 0, 0, paint),其中 currentFrame++%frames.length 表示数组中元素的下标,对数组长度"取余"确保不会出现数组越界异常。具体代码如下。

📖 Proj06_11 项目 ThreadSurfaceView.java 文件

```java
public class ThreadSurfaceView extends SurfaceView implements Callback{
    SurfaceHolder holder;
    Paint paint;
    Canvas canvas;
    boolean flag = true;                        //线程控制标识
    Bitmap frames[] = new Bitmap[12];           //存放帧图
    int currentFrame = 0;                       //当前播放帧
    public ThreadSurfaceView(Context context) {
        super(context);
        paint   = new Paint();
        holder = getHolder();
        holder.addCallback(this);
        //循环加载帧图
        for (int i = 0; i < frames.length; i++) {
            frames[i] = BitmapFactory.decodeResource(getResources(), R.drawable.loading_11 + i);
        }
    }
    …                                           //省略 Callback 接口的方法,请参看前面的代码
    protected void myDraw(){
        canvas = holder.lockCanvas();
        if(canvas!= null){
            canvas.drawColor(Color.WHITE);
            canvas.drawBitmap(frames[currentFrame++ % frames.length], 0, 0, paint);
        }
        holder.unlockCanvasAndPost(canvas);
    }
}
```

在 MainActivity 中创建 ThreadSurfaceView 对象,并设置为主界面视图。运行项目,可以看到帧图不停地绘制,出现人物"跑动"的效果。

将本节的帧动画与 6.4.1 节的属性动画相结合,在实现"跑动"的同时移动图片的位置,就可以让人物跑动了。感兴趣的读者可以自己尝试,这是在 Android 中开发 2D 游戏的基础。

6.4.3 剪切区动画效果

剪切区动画可以相对于逐帧动画来分析。在使用逐帧动画时,每一帧都对应一张静态图,因此需要准备多张图片才能实现动画效果,这样会使得动画素材增多,导致 apk 安装文件变大。剪切区动画可以实现将人物的分帧动作都放在一张图上,如图 6.17 所示,然后将需要显示的区域放置在剪切区即可。

项目 Proj06_12 演示了如何借助剪切区实现动画效果。该项目在视图区域中间,创建一个剪切区,依次显示位图中的区域。位图可以被看作是 4 行 4 列的帧图,每一行是一个动作,每个动作由 4 个帧组成。

与剪切区动画相关的 3 组数据是:屏幕宽高 sw 与 sh,位图宽高 w 与 h,每一帧宽高 fw 与 fh。屏幕宽高可以在方法 surfaceCreated 执行后通过 getWidth 和 getHeight 得到。位图宽高在位图创建后得到。帧的宽高由位图宽高计算而得。剪切区与三组数据的关系如图 6.18 所示。

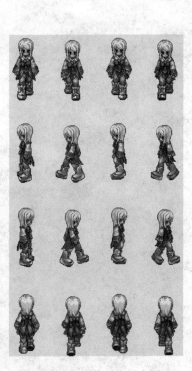

图 6.17 人物行走四向图

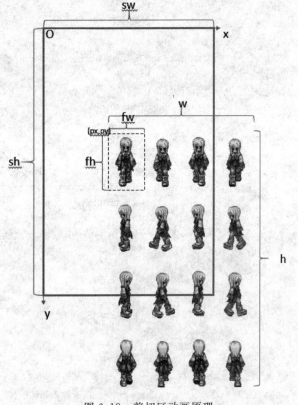

图 6.18 剪切区动画原理

📖 Proj06_12 项目 ThreadSurfaceView.java 文件

```java
public class ThreadSurfaceView extends SurfaceView implements Callback {
    SurfaceHolder holder;
    Canvas canvas;
    Paint paint;
    Bitmap rolePic;                                    //位图
    int frame;                                         //当前绘制帧
    int w, h, px, py;                                  //整幅位图的宽高和位置坐标
    int fw, fh;                                        //每一帧图的宽高
    int sw, sh;                                        //视图区域的宽高
    boolean flag = true;                               //线程控制位
    enum DIR{FRONT,LEFT,RIGHT,BACK};                   //标识人物方向的枚举类型
    DIR dir = DIR.FRONT;                               //人物方向,默认向前
    public ThreadSurfaceView(Context context) {
        super(context);
        paint = new Paint();
        holder = getHolder();
        holder.addCallback(this);
        rolePic = BitmapFactory.decodeResource(getResources(), R.drawable.r0);
        w = rolePic.getWidth();                        //位图宽
        h = rolePic.getHeight();                       //位图高
        fw = w/4;                                      //帧的宽
        fh = h/4;                                      //帧的高
    }
    …
    @Override
    public void surfaceCreated(SurfaceHolder arg0) {
        sw = getWidth();
        sh = getHeight();                              //视图区域宽高
        new Thread(new Runnable(){
            public void run() {
                while(flag){
                    myDraw();                          //调用自定义绘图方法
                    try { Thread.sleep(120);
                    } catch (InterruptedException e) {
                        e.printStackTrace();
                    }
                }
            }
        }).start();
    }
    …
    private void myDraw(){
        canvas = holder.lockCanvas();
        if(canvas!= null){
            canvas.drawColor(Color.WHITE);
            //创建剪切区
            canvas.clipRect(sw/2 - fw/2, sh/2 - fh/2, sw/2 + fw/2, sh/2 + fh/2);
            switch(dir){                               //根据移动方向确定位图坐标 px、py
```

```
                case FRONT:px = sw/2 - fw/2 - frame * fw;py = sh/2 - fh/2 - fh * 0;break;
                case LEFT:px = sw/2 - fw/2 - frame * fw;py = sh/2 - fh/2 - fh * 1;break;
                case RIGHT:px = sw/2 - fw/2 - frame * fw;py = sh/2 - fh/2 - fh * 2;break;
                case BACK:px = sw/2 - fw/2 - frame * fw;py = sh/2 - fh/2 - fh * 3;break;
            }
            frame++;
            if(frame >= 4)frame = 0;
            canvas.drawBitmap(rolePic, px,py, paint);       //绘制位图
        }
        holder.unlockCanvasAndPost(canvas);
    }
    //根据触屏点坐标确定人物方向
    @Override
    public boolean onTouchEvent(MotionEvent event) {
        int x = (int) event.getX();
        int y = (int) event.getY();
        //根据触屏点坐标确定人物方向
        if(x < sw/2 - fw/2 && y > sh/2 - fh/2&&y < sh/2 + fh/2){dir = DIR.LEFT;frame = 0;}
        if(x > sw/2 + fw/2 && y > sh/2 - fh/2&&y < sh/2 + fh/2){dir = DIR.RIGHT;frame = 0;}
        if(y < sh/2 - fh/2 && x > sw/2 - fw/2&&x < sw/2 + fw/2){dir = DIR.BACK;frame = 0;}
        if(y > sh/2 + fh/2 && x > sw/2 - fw/2&&x < sw/2 + fw/2){dir = DIR.FRONT;frame = 0;}
        return true;
    }
}
```

剪切区的位置是确定的,只需要修改位图的左上角顶点坐标 px 和 py,就可以确定显示在剪切区的"帧"。在该项目中(剪切区是固定在视图中间的),剪切区坐标可以通过以下公式计算:

px＝屏幕宽度的一半－帧宽度一半－帧数 x 帧宽度
py＝屏幕高度的一半－帧高度一半－行数 x 帧高度

如上述代码中的计算方式为:px＝sw/2-fw/2-frame * fw,py＝sh/2-fh/2-fh * 0。

人物行进方向通过触屏点修改:单击剪切区左方,修改方向为 DIR.LEFT；单击剪切区右方,修改方向为 DIR.RIGHT；单击剪切区上方,修改方向为 DIR.BACK；单击剪切区下方,修改方向为 DIR.FRONT。

6.5 习　　题

1. 选择题

(1) 使用 BitmapFactory 可以创建位图,其中无法被直接解析为位图的资源是(　　)。

　　A. 存放在 drawable 资源文件夹中的位图

　　B. 存放在 SD 卡中的位图

　　C. 使用 URL 标识的位图

　　D. 输入流中的位图

(2) BitmapFactory.Options 用来控制放缩比例的(　　)属性。
　　A. inJustDecodeBounds　　　　　　B. inSampleSize
　　C. inScaled　　　　　　　　　　　　D. outWidth
(3) 以下关于 Canvas 对象的描述中说法有误的是(　　)。
　　A. Canvas 提供了绘制圆形、矩形等各种图形的方法
　　B. Canvas 提供了直接绘制颜色的方法
　　C. Canvas 提供了多种坐标变化的方法
　　D. Canvas 提供了控制笔触粗细的方法
(4) 以下关于 SurfaceView 的描述中说法正确的是(　　)。
　　A. SurfaceView 中继承了 onDraw 方法,但是不会主动调用
　　B. SurfaceView 直接继承了 View,显示图形图像的原理与 View 一样
　　C. SurfaceView 常用于开发 3D 类应用
　　D. 自定义的 SurfaceView 对象无法作为控件使用
(5) 关于下面代码的功能描述中正确的是(　　)。

```
matrix.reset();
matrix.postRotate(30);
matrix.postTranslate(0,100);
matrix.setScale(1.5f,1.5f);
```

　　A. 将矩阵顺时针旋转 30°,然后平移到(0,100),宽高放大到 1.5 倍
　　B. 将矩阵逆时针旋转 30°,然后平移到(0,100),宽高放大到 1.5 倍
　　C. 将矩阵顺时针旋转 30°,然后平移到(0,100)
　　D. 将矩阵宽高放大到 1.5 倍

2. 简答题

(1) 简述 BitmapFactory 可以使用哪些资源创建位图。
(2) 简要说明在 Canvas 中可以进行哪些变换以及如何实现。
(3) 对比 View 和 SurfaceView,简述二者在处理视图显示时有哪些不同。
(4) 请设计并编写自定义控件,当传入一周销售额后,可以绘制如下所示的折线图。

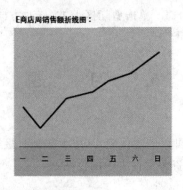

第 7 章　多媒体应用开发

本章学习目标
- 掌握使用 MediaPlayer 创建播放器。
- 掌握使用 MediaPlayer 和广播组件实现播放音乐中的状态进度的控制。
- 掌握使用 SoundPool 的使用方式和不同的创建方式。
- 掌握使用 SoundPool 在手机背景音乐中的应用和多个背景音乐同时运行的效果。
- 了解使用 MediaPlayer 创建模拟简易手机音乐播放器的功能。

Android 在多媒体开发中有很多实用的工具类对象的支持，可以供开发者很容易地集成音频、视频程序到应用程序中。Android 提供了非常丰富的多媒体操作 API，使得实现音频与视频播放操作变得比较简单。Android 支持的音频格式有 mp3、wav 和 3gp 等，支持的视频格式有 mp4 和 3gp 等。Android 系统所支持的全部音频与视频格式可以通过（sdk 路径）/docs/guide/appendix/media-formats.html 查看。

7.1　音频播放

对于手机来说，除了基础的拨打电话、短信等功能外，最重要的一个就是多媒体的应用了。所谓多媒体，简单地说就是音视频，以及文字、图片、动画的总和，通常是指音频和视频应用。Android 音频播放核心是底层 Linux 音频播放内核，提供 Android 音频播放的核心架构系统。

7.1.1　MediaPlayer 对象的创建

Android 系统音频播放主要有两种方式，分别是使用 MediaPlayer 和 SoundPool，这二者都位于 android.media 包下。

MediaPlayer 类位于 android.media 包中，是 Android 系统中重要的多媒体操作类，可以用于播放音频文件，也可以播放视频文件。MediaPlayer 适合播放音质较高、持续时间较长的音频文件；SoundPool 常常用于游戏 App 中音效和背景音乐播放。

Android 音频文件资源可以存放在项目 res/raw（用户独立创建文件夹）下，也可以是位于存储卡或者网络上的某个音频资源。除了在创建 MediaPlayer 对象时设定播放资源，还可以通过 setDataSource 方法设置，这种方法需要先调用 prepare 方法，然后再调用 start 方法播放。常用的设置播放源的方法如下：

- setDataSource(String path),播放指定路径的文件,可用于播放 SD 卡上的音频资源。
- setDataSource(Context context,Uri uri),播放指定 URI 的资源,该 URI 应该是可以被下载的。

MediaPlayer 常用方法如表 7.1 所示。

表 7.1 MediaPlayer 常用方法

方　法	说　　明	返回值类型
create(Context context,Uri uri)	static 方法,利用参数 Uri(包含媒体文件的地址)和上下文对象实例化一个多媒体对象	MediaPlayer
create(Context context,int resid)	static 方法,利用参数 context 和资源 ID 实例化一个多媒体对象	MediaPlayer
create(Context context,Uri uri,SurfaceHolder holder)	static 方法,利用参数 Uri(包含媒体文件的地址)和指定 SurfaceHolder 对象实例化一个多媒体对象	MediaPlayer
getCurrentPosition()	返回当前媒体文件在对应文件列表中的索引	int
getDuration()	返回当前播放媒体文件的播放总时间	int
isLooping()	判断是否循环播放	boolean
isPlaying()	判断是否正在播放	boolean
pause()	暂停播放	void
start()	开始播放	void
stop()	停止播放	void
prepare()	准备同步播放,在 MediaPlayer 生命周期中	void
prepareAsync()	准备异步播放,在 MediaPlayer 生命周期中	void
release()	释放 MediaPlayer 对象资源	void
reset()	重置 MediaPlayer 对象资源	void
seekTo(int msec)	设置当前媒体文件播放的进度位置,参数以毫秒为单位	void
setAudioStreamType(int streamtype)	设置媒体文件流的类型	void
setDataSource(String path)	设置媒体文件的数据来源,参数指定字符串类型的路径	void
setDataSource(FileDescriptor fd,long offset,long length)	设置局限媒体文件数据来源,通过 FileDescriptor 对象,文件从 offset 开始,移动到长度为 length 时结束	void
setDataSource(FileDescriptor fd)	设置媒体文件数据来源,依据 FileDescriptor 参数对象	void
setDataSource(Context context,Uri uri)	设置媒体文件数据来源,依据 Uri 和上下文对象	void
setDisplay(SurfaceHolder sh)	依据 SurfaceHolder 对象来显示媒体文件	void
setLooping(boolean looping)	设置是否循环播放单个媒体文件	void
setOnBufferingUpdateListener(OnBufferingUpdateListener listener)	设置网络媒体文件流的缓冲监听	void
setOnCompletionListener(OnCompletionListener listener)	设置单个网络媒体文件流播放结束时的监听	void

续表

方法	说明	返回值类型
setVolume(float leftVolume, float rightVolume)	根据左右 float 变量设置音量	void
setOnErrorListener(OnErrorListener listener)	设置媒体文件错误信息的监听	void

项目 ch08_ch08_MediaPlayer&SoundPool 演示了 MediaPlayer 的音频播放功能。在 res 根目录下新建一个 raw 文件夹，然后把待播放的 mp3 音乐文件复制到其中，R 文件将自动生成一个 raw 资源的 int 类型变量引用。设计如图 7.1 所示的运行界面，布局文件比较简单，只是 3 个按钮：播放、暂停和停止音乐。

图 7.1　MediaPlayer 播放界面

创建音乐播放对象 MediaPlayer，并设置循环播放当前音乐。

 ch08_ch08_MediaPlayer&SoundPool 项目 MediaPlayerActivity.java 文件

```
…
MediaPlayer mPlayer = MediaPlayer.create(this, R.raw.thesamesong);
mPlayer.setLooping(true);
```

create 是一个静态方法，可以直接调用。它的作用是实例化 MediaPlayer 对象，加载音乐播放文件，并使得音乐播放器处于准备阶段。

 ch08_ch08_MediaPlayer&SoundPool 项目 MediaPlayerActivity.java 文件

```
…
if(!mPlayer.isPlaying()) {
    mPlayer.start();
}
…
```

播放按钮监听器的核心功能，判断播放器是否处于播放状态，如果返回 false 说明是停止状态，则执行 start 方法播放音乐。

 ch08_ch08_MediaPlayer&SoundPool 项目 MediaPlayerActivity.java 文件

```
…
if(!mPlayer.isPlaying()) {
    mPlayer.pause();
}
…
```

暂停按钮监听器的核心功能，判断播放器的状态，如果 mPlayer.isPlaying()==true，则执行 pause 方法暂停音乐。音乐的播放和暂停都是不难理解的，按照同样的逻辑思路，设置音乐停止功能。

  📖 ch08_ch08_MediaPlayer&SoundPool 项目 MediaPlayerActivity.java 文件

```
…
if(!mPlayer.isPlaying())  {
    mPlayer.stop();
}
…
```

  单击停止按钮,音乐播放停止了,可是问题出现了,当再次单击播放按钮后,发现音乐不播放了。其原因是 MediaPlayer 遵循相应的状态转换,各状态之间的跳转需要遵守这个规则。

7.1.2 MediaPlayer 对象的状态转换

  MediaPlayer 的状态控制是通过一个状态机来管理的,如图 7.2 所示。椭圆代表 MediaPlayer 对象可能驻留的状态,箭头表示 MediaPlayer 在各个状态之间转换的控制操作。

  (1) Idle 是空闲状态,是当前播放容器一个初始化的状态。当一个 MediaPlayer 对象被实例化后,或者播放容器又重新调用了 reset 重置方法后会进入该状态。

  (2) End 是结束状态。当 MediaPlayer 退出,释放资源后,就不能再执行其他任何操作,即结束了生命周期。如果媒体文件不再被调用或者操作,那么就应该尽快地进入该状态,从而释放其相关的软硬资源。

  (3) Initialized 是初始化状态。调用 setDataSource 方法,获取相关媒体资源文件,就进入了该状态。

  (4) Prepared 是准备状态。如果程序到这一步没有抛出异常,说明之前的 MediaPlayer 启动操作都是正常的。调用 prepare 或 prepareAsync 方法,可以进入该状态。

  (5) Started 是开始状态。处于 Prepared 状态的 MediaPlayer 对象,调用 start 方法直接进入该状态。此时,可以调用 isPlaying 或者 pause 方法测试 MediaPlayer 是否处于开始状态或暂停状态。

  (6) Paused 是暂停状态。MediaPlayer 只有在播放状态中才可以调用 pause 方法,进入暂停状态。MediaPlayer 暂停后,再次调用 start 则可以继续 MediaPlayer 的播放,转到开始状态,开始和暂停状态可以任意切换。

  (7) Stop 是停止状态。MediaPlayer 停止状态不能直接返回开始状态和暂停状态,如果想重新播放,需要通过 prepare 方法回到准备状态,再重新开始才可以。

  (8) PlaybackCompleted 是回放已完成状态。这是 MediaPlayer 的一个监听器可以处理的状态。当 MediaPlayer 正常播放完毕后,若没有设置播放完后的操作,就会触发 OnCompletionListener 监听器。此时 MediaPlayer 没有受影响,可以调用 start 方法重新播放,或调用 stop 方法停止 MediaPlayer,也可以调用 seekTo 方法来重新定位播放位置。

  (9) Error 是出错状态。由于某种原因导致 MediaPlayer 出现错误,会触发 OnErrorListener.onError 事件,进入 Error 状态。在 Error 状态下,调用 reset 方法可以恢复,使得 MediaPlayer 能重新返回到空闲状态继续运行。

  MediaPlayer 提供了很多监听器,方便对播放器的工作状态进行监听,通过这些监听器

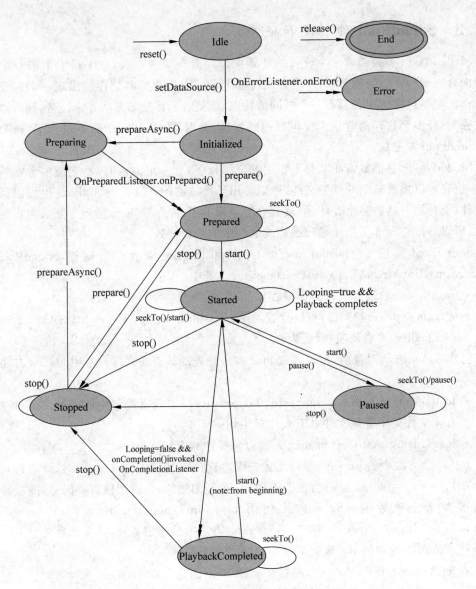

图 7.2 MediaPlayer 状态转换图

对播放过程中事件进行处理。常用的监听器如下：

- setOnCompletionListener(MediaPlayer. OnCompletionListener listener)，当播放完成时触发监听器。
- setOnErrorListener(MediaPlayer. OnErrorListener listener)，监听错误发生。
- setOnPreparedListener(MediaPlayer. OnPreparedListener listener)，监听播放资源准备就绪。
- setOnSeekCompleteListener(MediaPlayer. OnSeekCompleteListener listener)，监听 seek 方法执行。
- setOnVideoSizeChangedListener (MediaPlayer. OnVideoSizeChangedListener listener)，监听视频变化。

7.1.3 SoundPool 的创建和使用

MediaPlayer 适合播放一个较长的音频文件。但在很多时候，在手机上单击一个按钮时会配合出现一个短暂的音乐，或者在跳转一个新的手机页面时会伴随出现一个声音来完成操作。类似这样的声音都有一个共同的特点，就是音乐播放的时间非常短暂，而且连续性非常强，另外多个这样的背景声音可以同时发出，此时 MediaPlayer 就不适合了，这需要使用 SoundPool 来完成。

MediaPlayer 在播放音乐文件时占用的资源较多，有一定的延时，不支持多个音频同时播放，但是可以播放较大的音频文件，所以常用于播放器的开发。SoundPool 也可以播放音频文件，支持同时播放多个音频，但最大只能申请 1MB 内存空间，仅能用于播放很短的声音片段，所以 SoundPool 常用于播放游戏或软件中的声音特效。

SoundPool 类位于 android.media 包中，可以通过构造方法创建 SoundPool(int maxStreams, int streamType, int srcQuality)。其中：

- maxStreams，最大支持的音频数量。
- streamType，音频类型，可以设置为 AudioManager.STREAM_MUSIC。
- srcQuality，声音品质，默认为 0。

创建 SoundPool 对象后，可以通过 load 方法预先加载音频文件。常用的 load 方法描述如下：

- load(Context context, int resId, int priority)，加载资源包中的音频文件，resId 是 raw 文件夹中音频文件对于 R 中的引用。
- load(String path, int priority)，通过路径加载音频文件。

上述两个方法中的 priority 参数是优先级，当前没有启用(adk4.2 文档这样解释)，建议设置为 1，便于向后兼容。load 方法会返回 int 类型的数值，该返回值需要保存，播放音频文件时需要指定该数值。SoundPool 使用 play(int soundID, float leftVolume, float rightVolume, int priority, int loop, float rate)方法播放指定音频，其参数含义如下：

- soundID，音频的 ID，该值是在 load 方法执行时的返回值。
- leftVolume、rightVolume，左右音量，取值范围是 0.0~1.0。
- priority，优先级，数值越大优先级越高，最低是 0。
- loop，是否循环，0 为不循环，-1 为循环。
- rate，播放速率，取值范围是 0.5~2.0，正常播放速率为 1。

📖 ch08_ch08_MediaPlayer&SoundPool 项目 MediaPlayerActivity.java 文件

```
SoundPool soundPool = new SoundPool(4,AudioManager.STREAM_MUSIC,100);
Map< Integer, Integer > soundPoolMap = new HashMap< Integer, Integer >();
soundPoolMap.put(1,soundPool.load(this,R.raw.thesamesong,1));
soundPoolMap.put(2,soundPool.load(this, R.raw.dingdong, 2));
...
```

初始化 SoundPool 对象构造方法中的 3 个参数分别是：int maxStreams 为 4，表示支持同时播放声音的数量；streamType 为 AudioManager.STREAM_MUSIC,表示声音的类型；srcQuality 为 100，表示声音的品质。

📖 ch08_ch08_MediaPlayer&SoundPool 项目 MainActivity.java 文件

```java
public class MainActivity extends Activity implements OnClickListener {
    SoundPool sp;
    Map<String,Integer> sm = new HashMap<String,Integer>();   //保存load后的返回值   ①
    Button b1,b2,b3;
    @Override
    protected void onCreate(Bundle savedInstanceState) {
        super.onCreate(savedInstanceState);
        setContentView(R.layout.activity_main);
        b1 = (Button) findViewById(R.id.button1);
        b2 = (Button) findViewById(R.id.button2);
        b3 = (Button) findViewById(R.id.button3);
        b1.setOnClickListener(this);
        b2.setOnClickListener(this);
        b3.setOnClickListener(this);
        //创建SoundPool
        sp = new SoundPool(5,AudioManager.STREAM_MUSIC,5);
        //加载音效文件
        sm.put("1",sp.load(this, R.raw.dummy_break_05, 1));                        ①
        sm.put("2",sp.load(this, R.raw.footstep_water_01, 1));
        sm.put("3",sp.load(this, R.raw.woman_scream, 1));
    }
    @Override
    public void onClick(View arg0) {
        //更换按钮播放不同音效
        switch(arg0.getId()){
            case R.id.button1:     sp.play(sm.get("1"), 1.0f, 1.0f, 1, 0, 1.0f);    ②
                break;
            case R.id.button2:     sp.play(sm.get("2"), 1.0f, 1.0f, 1, 0, 1.0f);
                break;
            case R.id.button3:     sp.play(sm.get("3"), 1.0f, 1.0f, 1, 0, 1.0f);
                break;
        }
    }
}
```

在使用 SoundPool 时注意，代码①处加载音频文件时一定要保存返回值；代码②处播放时指定相应的音频 ID，该 ID 不是资源文件在 R 中生成的 ID，而是 load 时的返回值。

对于 MediaPlayer 和 SoundPool 的使用场景应该注意，MediaPlayer 支持播放较大的音频文件，占用资源较大，常用于开发播放器。SoundPool 支持同时播放多个音频文件，但申请的内存空间有限制，可用于播放声音片段。SoundPool 播放的音频文件建议使用 ogg 格式。

7.2 视频播放

在 Android 视频播放中,VideoView 控件可以使用本地存储卡或在线网络资源,可以计算视频显示的尺寸,可以方便地使用各种显示的选项,如缩放、着色等。

7.2.1 VideoView 播放本地资源

VideoView 是 Android 提供的系统控件,用于播放视频。VideoView 类位于 android. widget 包中,使用方法与 MediaPlayer 较为类似。VideoView 提供以下两个方法用于加载视频文件:

- setVideoPath(String path),指定文件路径加载视频,用于播放 sdcard 中的视频。
- setVideoURI(Uri uri),指定 URI 加载视频,用于播放网络中的视频。

Android 播放视频资源基本上可以分为网络和本地两种方式。项目 ch08_VideoView 演示了如何播放本地资源。该项目布局比较简单,主要涉及 3 个 ImageButton 控件,代码如下。

📖 ch08_VideoView 项目 activity_main.xml 文件

```xml
<VideoView
    android:id="@+id/videoView1"
    android:layout_width="match_parent"
    android:layout_height="match_parent" />
<LinearLayout
    android:layout_width="fill_parent"
    android:layout_height="wrap_content"
    android:layout_alignParentBottom="true"
    android:background="#22cccccc"
    android:gravity="center_horizontal"
    android:orientation="horizontal" >
    <ImageButton
        android:id="@+id/ib_play"
        android:layout_width="50dp"
        android:layout_height="50dp"
        android:layout_marginRight="8dp"
        android:scaleType="centerCrop"
        android:src="@drawable/play" />
    <ImageButton
        android:id="@+id/ib_pause"
        android:layout_width="50dp"
        android:layout_height="50dp"
        android:scaleType="centerCrop"
        android:src="@drawable/stop" />
    <ImageButton
        android:id="@+id/ib_stop"
        android:layout_width="50dp"
        android:layout_height="50dp"
        android:scaleType="centerCrop"
```

```
        android:src = "@drawable/stop" />
</LinearLayout>
```

3个图片按钮从左向右分别是播放、暂停和停止,如图 7.3 所示。在测试项目时,如果使用 SDK 模拟器,需要在模拟器对应 sdcard 中先添加对应的视频文件。本项目中视频文件路径是/mnt/sdcard/beargolf.3gp,在模拟器中播放视频,画质不应该太高,否则会不流畅,建议使用 Android 真机测试。

图 7.3　VideoView 控件的界面效果

加载视频资源代码如下,读取 SD 卡指定位置的视频资源,把路径信息设置给 VideoView 控件即可。测试项目时,需要申请读写 SD 卡权限。

📖 ch08_VideoView 项目 VedioActivity.java 文件

```
//创建 File 文件对象
File file = new File("/mnt/sdcard/beargolf.3gp");
if(file.exists()){
    vv.setVideoPath("/mnt/sdcard/beargolf.3gp");
}else{
    Toast.makeText(getApplicationContext(),"没有对应的视频文件",Toast.LENGTH_LONG).show();
}
…
if(v.getId() == R.id.ib_play){
    vv.start();
}else if(v.getId() == R.id.ib_pause){
    vv.pause();
}else if(v.getId() == R.id.ib_stop){
    vv.stopPlayback();
    //和 MediaPlayer 一样,重新加载视频文件
    vv.setVideoPath("/mnt/sdcard/beargolf.3gp");
}
…
```

以上代码的逻辑不难理解。这里需要注意两点：一是加载视频文件路径 vv.setVideoPath("/mnt/sdcard/beargolf.3gp")；二是停止后需要重新加载视频文件，才能继续播放，否则依旧无法再次播放，这可以参考 7.1.2 节的内容。

7.2.2 MediaController

VideoView 控件的控制操作也可以由 MediaController 完成，该类位于 android.widget 包中。MediaController 带有播放、暂停、上一个、下一个等按钮，使用这些按钮就可以完成对 VideoView 的控制，不过在使用之前需要将 VideoView 和 MediaController 建立关联。

MediaContraller 对象将创建后，默认放在应用程序 XML 布局文件之上，即该对象的视图将"悬浮"在 VideoView 之上。MediaContraller 常用方法如表 7.2 所示。

表 7.2 MediaController 常用方法

方　法	说　明	返回值类型
onFinishInflate()	加载 XML 文件中所有的子视图后调用的方法	void
setAnchorView(View view)	设置 MediaController 加载到参数对应的 view 视图对象上	void
setMediaPlayer(MediaPlayerControl player)	设置 player 对应的媒体播放器	void
show()	显示 MediaController 面板，默认显示 3s 后自动隐藏	void
show(int timeout)	设置 MediaController 在显示 timeout 毫秒后自动隐藏	void
isShowing()	判断 MediaController 是否已显示	boolean
hide()	隐藏 MediaController	void
onTouchEvent(MotionEvent event)	判断 event 触摸事件	boolean
onTrackballEvent(MotionEvent ev)	处理触摸轨迹球的运动事件	boolean
setEnabled(boolean enabled)	设置 MediaController 是否可用	void

MediaController 与 VideoView 结合之后，运行效果如图 7.4 所示。布局文件非常简单，只需要一个 VideoView 控件即可。MediaController 与 VideoView 建立关联的代码如下。

图 7.4 MediaController 对象的界面效果

📖 ch08_MediaController 项目 MainActivity.java 文件

```
MediaController mc = new MediaController(this);
File file = new File("/mnt/sdcard/beargolf.3gp");
if(file.exists()){
    //设置视频播放路径
    VideoView vv.setVideoPath("/mnt/sdcard/beargolf.3gp");
    //播放器和控制建立连接
    vv.setMediaController(mc);
    mc.setMediaPlayer(vv);
    //设置请求焦点
    vv.requestFocus();
}
```

MediaController 默认有 3 个按钮,对应 4 个功能：前进、后退、播放和暂停,进度条显示当前播放进度。对于简单的音视频播放,借助 MediaContrraller 是不错的选择。但如果需要个性化的操作,还需要自定义控制键。

7.2.3 播放网络资源

加载网络资源需要借助 URI 类,这涉及一些网络编程的知识,具体可以参考第 10 章的相关内容。URI 是统一资源标示符（Uniform Resource Identifier）,可以对任何（包括本地和互联网）资源通过特定的协议进行交互操作。

📖 加载网络视频资源的代码

```
//获取网络发布视频文件地址信息作为 Uri parse 方法的一个参数
Uri uri = Uri.parse("http://192.168.0.1/web/video?soundName=video.3gp");
VideoView video = (VideoView)this.findViewById(R.id.video_view);
//设置对应的 MediaController
video.setMediaController(new MediaController(this));
//把获取的 Uri 资源对象赋值给 VideoView 对象
video.setVideoURI(uri);
videoView.requestFocus();
```

代码中访问的地址 http://192.168.0.1/web/video?soundName=video.3gp 是发布在本机 Tomcat 服务器中的一个视频资源。关于服务器程序的开发知识,可以参考其他相关资料。

7.3 MediaRecorder

MediaRecorder 位于 android.media 包中,用于录制声音和视频。MediaRecorder 控制需要按照合法的状态转换进行,这与 MediaPlayer 类似。MediaRecorder 的状态转换如图 7.5 所示。

MediaRecorder 对象创建之后,可以调用 setAudioSource 和 setVideoSource 方法设置音频和视频的来源。确定了录制来源,就可以根据应用需求确定输出文件的格式。只有确

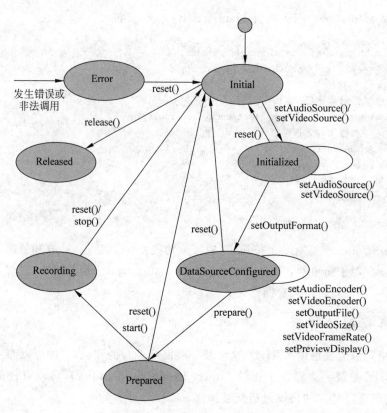

图 7.5 MediaRecorder 状态图

定了音频的输出格式,才能设定编码格式,包括音频和视频的编码格式、视频的帧速率、视频大小等参数。确定输出格式和设置编码格式不可以颠倒顺序,否则会抛出异常。如果需要重新设定输出格式,需要执行 reset 方法回到初始状态。设定完编码格式之后,调用 prepare、start 就可以开始录制了。

MediaRecorder 类常用方法如表 7.3 所示。

表 7.3 MediaRecorder 常用方法

方 法	说 明
setAudioSource(int audio_source)	设置音频来源,其值可以调用内部类 MediaRecorder.AudioSource 的静态值,MIC 表示来源于手机麦克
setOutputFormat(int output_format)	设置输出格式,其值可以调用内部类 MediaRecorder.OutputFormat 的静态值:AMR_NB、AMR_WB、DEFAULT、MPEG_4、RAW_AMR、THREE_GPP
setAudioEncoder(int audio_encoder)	设置音频的编码格式,值为内部类 MediaRecorder.AudioEncoder 的静态值:AAC、AMR_NB、AMR_WB、DEFAULT
setAudioEncodingBitRate(int bitRate)	设置音频的编码率,可以不设置,采用默认值
setAudioSamplingRate(int sRate)	设置音频采样率,可以不设置,采用默认值
setOutputFile(String path)	录制文件的保存位置
prepare()	准备录制
start()	开始录制

续表

方法	说明
stop()	停止录制
release()	释放资源
setVideoSource(int video_source)	设置视频来源,其值采用内部类 MediaRecorder.VideoSource 的静态值,CAMERA 表示来源于摄像头
setVideoEncoder(int video_encoder)	设置视频的编码格式,值为内部类 MediaRecorder.VideoEncoder 的静态值:DEFAULT、H263、H264、MPEG_4_SP
setVideoEncodingBitRate(int bitRate)	设置视频编码的比特率
setVideoSize(int width, int height)	设置视频的宽高尺寸(必须设置)
setVideoFrameRate(int rate)	设置帧速率,设置完视频来源后,必须设置该方法
setPreviewDisplay(Surface sv)	设置预览视频的界面

使用 MediaRecorder 录像时,可以通过调用 setPreviewDisplay(SurfaceView sv)方法设置预览视图,这需要传入 SurfaceView 类型的参数。

要借助手机的硬件功能,比如摄像头、麦克、存储介质,需要申请相应的权限,如下所示:

```
<!-- 授予该程序录制声音的权限 -->
<uses-permission android:name="android.permission.RECORD_AUDIO"/>
<!-- 授予使用读取 SD 卡权限 -->
<uses-permission android:name="android.permission.WRITE_EXTERNAL_STORAGE"/>
```

7.3.1 录制音频

Android 录制声音的过程相对固定,按照如下代码中的 6 个步骤进行即可。

📖 ch08_VoiceRecord 项目 RecorderActivity.java

```java
//录音
public void recorde(){
    //录音文件需要放在 sdcard 上
    //MEDIA_MOUNTED 状态为 sdcard 正常使用状态
    if(!Environment.getExternalStorageState().equals(Environment.MEDIA_MOUNTED)){
        Toast.makeText(this, "SD卡无法正常使用,录音结束!", Toast.LENGTH_LONG).show();
        return;
    }
    try {
        //sd 卡根路径
        File sdroot = Environment.getExternalStorageDirectory();
        //保存录音文件
        File file = new File(sdroot,System.currentTimeMillis() + ".amr");
        //1. 创建录音器
        mrecorder = new MediaRecorder();
        //2. 设置音频来源
        mrecorder.setAudioSource(MediaRecorder.AudioSource.MIC);
        //3. 设置输出格式
        mrecorder.setOutputFormat(MediaRecorder.OutputFormat.THREE_GPP);
```

```java
        //4. 设置编码格式
        mrecorder.setAudioEncoder(MediaRecorder.AudioEncoder.AMR_NB);
        //5. 设置输出音频路径
        mrecorder.setOutputFile(file.getAbsolutePath());
        //6. 准备和开始
        mrecorder.prepare();
        mrecorder.start();
    } catch (IllegalStateException e) {
        // TODO Auto-generated catch block
        e.printStackTrace();
    } catch (IOException e) {
        // TODO Auto-generated catch block
        e.printStackTrace();
    }
}
```

项目运行后,单击录制便可在 SD 卡上创建录音文件。对于 SD 卡的读写将在第 9 章详细介绍。录音结束后需要释放 MediaRecorder 所占用的资源。注意开启相应的权限:

- < uses-permission android:name = "android.permission.RECORD_AUDIO"/>,录制声音权限。
- < uses-permission android:name = "android.permission.WRITE_EXTERNAL_STORAGE"/>,写 SD 卡权限。

7.3.2 同时录制音视频

MediaRecorder 用于录制视频的过程与录制声音基本一致,不同于单独录制音频的是在相应步骤要设置视频的录制格式、编码等。录制音视频需要申请如下权限:

- < uses-permission android:name = "android.permission.RECORD_AUDIO"/>,启用录制声音权限。
- < uses-permission android:name = "android.permission.CAMERA"/>,打开摄像机权限。
- < uses-permission android:name = "android.permission.WRITE_EXTERNAL_STORAGE"/>,启用写入 SD 卡权限。

录制音视频的代码如下:

📖 ch08_VoiceRecord 项目 RecorderActivity.java

```java
//录制视频
public void recorde(){
    //录制的视音频文件需要放置在 sdcard 上
    //MEDIA_MOUNTED 状态为 sdcard 正常使用状态
    if(!Environment.getExternalStorageState().equals(Environment.MEDIA_MOUNTED)){
        Toast.makeText(this, "SD 卡无法正常使用,录制结束!",
                                    Toast.LENGTH_LONG).show();
        return;
    }
```

```java
try {
    //sd卡根路径
    File sdroot = Environment.getExternalStorageDirectory();
    //保存录制的视音频文件
    File file = new File(sdroot,System.currentTimeMillis() + ".3gp");
    //1. 创建 MediaRecorder
    mrecorder = new MediaRecorder();
    //2. 设置音频、视频来源
    mrecorder.setVideoSource(MediaRecorder.VideoSource.CAMERA);
    mrecorder.setAudioSource(MediaRecorder.AudioSource.MIC);
    //3. 设置输出格式
    mrecorder.setOutputFormat(MediaRecorder.OutputFormat.THREE_GPP);
    //4. 设置音频、视频编码格式
    mrecorder.setVideoEncoder(MediaRecorder.VideoEncoder.H264);
    mrecorder.setVideoSize(320, 240);
    //mrecorder.setVideoFrameRate(15);
    mrecorder.setAudioEncoder(MediaRecorder.AudioEncoder.DEFAULT);
    //5. 设置输出路径
    mrecorder.setOutputFile(file.getAbsolutePath());
    //6. 视频预览设置
    mrecorder.setPreviewDisplay(sv.getHolder().getSurface());
    //7. 准备和录制
    mrecorder.prepare();
    mrecorder.start();
} catch (IllegalStateException e) {
    e.printStackTrace();
} catch (IOException e) {
    e.printStackTrace();
}
```

该项目布局文件比较简单，含有 3 个控件，两个按钮用于录制和停止；一个 SurfaceView 控件用于预览录制的视频，代码如下。该项目在虚拟机上运行没有效果，原因是虚拟机没有摄像头硬件支持。在真机上的运行效果如图 7.6 所示。

📖 ch08_VoiceRecord 项目 main.xml 布局文件

```xml
<RelativeLayout xmlns:android = "http://schemas.android.com/apk/res/android"
    xmlns:tools = "http://schemas.android.com/tools"
    android:layout_width = "match_parent"
    android:layout_height = "match_parent"
    tools:context = ".MainActivity" >
    <SurfaceView
        android:id = "@ + id/sv"
        android:layout_width = "match_parent"
        android:layout_height = "match_parent" />
    <LinearLayout
        android:layout_width = "fill_parent"
        android:layout_height = "wrap_content"
```

```
            android:orientation = "horizontal"
            android:gravity = "center_horizontal"
            android:layout_alignParentBottom = "true">
        <Button
            android:id = "@ + id/ib_recorde"
            android:layout_width = "wrap_content"
            android:layout_height = "wrap_content"
            android:text = "录制"/>
        <Button
            android:id = "@ + id/ib_stop"
            android:layout_width = "wrap_content"
            android:layout_height = "wrap_content"
            android:text = "停止"/>
    </LinearLayout>
</RelativeLayout>
```

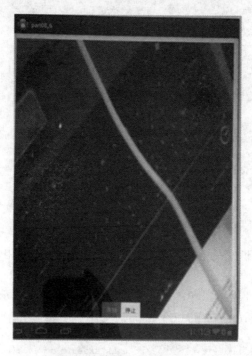

图 7.6 真机视频录制效果

使用 MediaRecorder 记录声音、视频时需要注意，MediaRecorder 方法的执行要遵循图 7.5 所示的先后顺序，比如在执行 setOutputFormat 方法之后再执行 setAudioEncoder 等方法。另外，录制声音和视频需要麦克和摄像头，不要忘记开启相应权限。

7.4 使用 Camera 拍照

Android 调用 Camera，一是拍照，二是摄像。Android 提供了功能强大的 API，可以非常方便地在 Android 系统上进行 Camera 应用的开发。通常调用摄像头有两个方式，一是借助 Intent 和 MediaStore 调用系统 Camera App 程序，实现拍照和摄像功能；二是根据 Camera API 自定义 Camera 程序，并可以处理各种拍照特效，如曝光、饱和度等。要自定义 Camera 程序，需要对 Camera API 有充分的了解。

7.4.1 启动相机与拍照

带有摄像头的 Android 设备通常会提供一个 Camera 应用程序，实现拍照功能。该应用程序通常是设备生产厂商实现的，不同设备的 Camera 应用程序功能不同。有的 Camera 应用集成了各种照片处理效果，如变色、曝光，有的 Camera 应用可以添加水印。系统集成 Camera 应用程序虽然在功能上不同，但都可以通过 IntentFilter 响应其他应用的"调用"。开发人员可以借助 Camera 应用获取照片，而不需要自己实现 Camera 功能。

使用隐式 Intent，设定 action 属性为 android. provider. MediaStore. MediaStore 类的常量 ACTION_IMAGE_CAPTURE，就可以打开 Camera 应用。ACTION_IMAGE_CAPTURE 常量对应的字符串是 android. media. action. IMAGE_CAPTURE。在开发过程中，一般选择使用 MediaStore 类的常量，应避免直接使用字符串，以防止 Android 版本升级带来的变化。实现代码如下：

```
Intent intent = new Intent(MediaStore.ACTION_IMAGE_CAPTURE);
startActivity(intent);
```

项目 Proj07_6 演示如何通过系统集成的 Camera 应用拍照。布局文件中包含 Button 按钮和 ImageView 控件，布局效果可以参考图 7.7，分别实现打开 Camera 应用和现实照片的功能，代码如下：

📖 Proj07_6 项目 activity_main. xml 布局文件

```xml
<RelativeLayout xmlns:android = "http://schemas.android.com/apk/res/android"
    xmlns:tools = "http://schemas.android.com/tools"
    android:layout_width = "match_parent"
    android:layout_height = "match_parent"
    tools:context = ".MainActivity" >
    <Button
        android:id = "@ + id/button1"
        android:layout_width = "wrap_content"
        android:layout_height = "wrap_content"
        android:layout_alignParentBottom = "true"
        android:layout_centerHorizontal = "true"
        android:text = "拍照" />
    <ImageView
        android:id = "@ + id/imageView1"
        android:layout_width = "match_parent"
        android:layout_height = "match_parent"
        android:layout_centerInParent = "true"/>
</RelativeLayout>
```

MainActivity 是主界面代码，在按钮监听器中打开 Camera 应用。当 Camera 应用完成时，需要返回照片数据，因此需要使用 startActivityForResult 方法启动。

📖 Proj07_6 项目 MainActivity.java 文件

```java
public class MainActivity extends Activity implements OnClickListener {
    private ImageView iv;
    @Override
    protected void onCreate(Bundle savedInstanceState) {
        super.onCreate(savedInstanceState);
        setContentView(R.layout.activity_main);
        Button bt = (Button) findViewById(R.id.button1);
        iv = (ImageView) findViewById(R.id.imageView1);
        bt.setOnClickListener(this);
    }
    @Override
    public void onClick(View arg0) {
        Intent intent = new Intent(MediaStore.ACTION_IMAGE_CAPTURE);
        //启动 Camera 应用并等待返回数据
        startActivityForResult(intent, 0x100);
    }
}
```

startActivityForResult(intent，0x100)方法中的参数 0x100 是自定义的标识，用于区分从什么位置启动了 Camera 应用。在真机上运行上述项目，初始化界面如图 7.7 所示，打开 Camera 应用的界面如图 7.8 所示。该 Camera 应用的界面是华为手机中集成的效果，其他品牌手机的 Camera 应用会不同。当 Camera 应用打开后，用户直接操作的界面就不再是本项目中的 Activity 了，所使用的拍照功能也是 Camera 应用自己实现的。我们需要做的就是等用户拍照结束后获取照片数据。

图 7.7 拍照应用初始界面

图 7.8 系统集成的 Camera 应用

7.4.2 获取相机返回数据

在启动 Camera 应用程序时使用 startActivityForResult 方法,而不是 startActivity,目的是为了获取 Camera 的返回数据。当 Camera 应用结束后,会将照片数据存入 Intent 中返回给"调用者",获取返回数据的方式是使用 onActivityResult 方法响应,处理接收的数据,详细代码如下。

📖 Proj07_6 项目 MainActivity.java 文件

```
@Override
protected void onActivityResult(int requestCode, int resultCode, Intent data) {
    super.onActivityResult(requestCode, resultCode, data);
    //判定返回是否正确,是否是本 Activity 调用引用的
    if(){
        Bundle bd = data.getExtras();
        Bitmap bitmap = (Bitmap) bd.get("data");
        iv.setImageBitmap(bitmap);
    }
}
```

onActivityResult 方法与 startActivityForResult 方法相关,当启动的 Activity 正常结束后,onAcitityResult 方法会自动调用。该方法中有 3 个参数,分别表示请求代码、结果代码和返回 Intent。requestCode 表示发起请求时传入的标志,在该项目中是 startActivityForResult(intent,0x100)中的 0x100。resultCode 表示 Activity 的返回状态,如果正常结束并返回数据,则取值为 RESULT_OK 常量,值为 -1。data 是 Intent 类型,封装了 Activity 返回时的数据。

Camera 应用结束后,将照片数据封装在 Bundle 对象中,键为 data。在 onActivityResult 方法中,首先要获取 Bundle 对象,再使用 Bundle.get("data")方法获取 Bitmap 类型的数据,设置给 ImageView 控件,就可以显示拍照所得照片,如图 7.9 所示。

需要注意的是,Camera 返回的照片尺寸很小(不同设备的表现不同),所以在 ImageView 中显示时出现了模糊现象。这并不意味着 Camera 的拍照像素低,这是因为返回的照片数据是经过计算变换后的。Camera 应用不会把全尺寸的照片数据返回给调用者,因为那样会消耗大量内存,而移动设备的内存都有限制,所以 Camera 应用默认只返回一幅很小的缩略图。

图 7.9 Camera 应用返回的照片

7.4.3 获取原尺寸照片

如果要获得更大尺寸的照片,在启动 Camera 应用时需要传入一个参数,设置照片的存放位置。参数名是 MediaStore.EXTRA_OUTPUT,值是 URI 类型。

项目 Proj07_7 演示了如何获取全尺寸照片,布局文件与 7.4.1 节一致,此处不再给出,主要的逻辑代码如下。imageFile 是 File 类型的,用于存放照片。该项目中照片的存放位置是 SD 卡根目录,名称是 capture.jpg。获取全尺寸照片,在启动 Camera 应用时需要设置参数 intent.putExtra(MediaStore.EXTRA_OUTPUT,imageUri),其中 imageUri 是存放文件的 URI 格式。

把数据写入 SD 卡需要获取外部存储权限,本项目需要在配置文件 AndroidManifest.xml 中添加如下代码:

```xml
<uses-permission android:name="android.permission.WRITE_EXTERNAL_STORAGE"/>
```

📖 Proj07_7 项目 MainActivity.java 文件

```java
public class MainActivity extends Activity implements OnClickListener {
    private ImageView iv;
    File imageFile;          //照片文件
    @Override
    protected void onCreate(Bundle savedInstanceState) {
        super.onCreate(savedInstanceState);
        setContentView(R.layout.activity_main);
        Button bt = (Button) findViewById(R.id.button1);
        iv = (ImageView) findViewById(R.id.imageView1);
        bt.setOnClickListener(this);
    }
    @Override
    public void onClick(View arg0) {
        //获取 SD 卡根目录
        File sdCardRoot = Environment.getExternalStorageDirectory();
        //创建文件
        imageFile = new File(sdCardRoot,"capture.jpg");
        Log.i("Msg", "文件暂存 " + imageFile.getAbsolutePath());
        //转换文件格式
        Uri imageUri = Uri.fromFile(imageFile);
        //准备开启 Camera 应用
        Intent intent = new Intent(MediaStore.ACTION_IMAGE_CAPTURE);
        //传入存放参数
        intent.putExtra(MediaStore.EXTRA_OUTPUT, imageUri);
        startActivityForResult(intent, 0x100);
    }
    @Override
    protected void onActivityResult(int requestCode, int resultCode, Intent data) {
        super.onActivityResult(requestCode, resultCode, data);
        //Camera 返回时处理后续逻辑
```

```
        if(resultCode == RESULT_OK && requestCode == 0x100){
            Log.i("Msg", "imageFile size:" + imageFile.length() + "byte");
            Bitmap bm = BitmapFactory.decodeFile(imageFile.getAbsolutePath());
            Log.i("Msg", "bitmap Size" + bm.getByteCount());
            iv.setImageBitmap(bm);
        }
    }
}
```

Camera 拍照时会使用传入的 Uri 位置存放照片,该照片数据是全尺寸文件,文件大小与摄像头分辨率有关。本项目测试手机是华为的低端手机,摄像头分辨率是 1920×2650,照片文件大小是 1 579 803B,约 1.5MB。

BitmapFactory.decodeFile 方法可以将图片文件转为位图对象,转换后的 Bitmap 对象大小为 19 660 800B,约 18.75MB,保存了照片的所有信息,显示在 ImageView 控件时,可以保持比较清晰的分辨率,如图 7.10 所示。对比图 7.10 和图 7.9,分辨率完全不同。

将全尺寸照片全部读入内存,虽然不一定会造成内存不足,但肯定会加速手机内存的消耗,因此这种做法存在较大的弊端。要使照片以适当的清晰度显示,又要节约内存使用,在显示照片时就要进行缩放。

图 7.10 全尺寸照片显示

7.4.4 照片缩略图

分辨率高的摄像头所拍摄的照片会更加清晰，照片尺寸也更大，但在手机中显示时，通常不需要分辨率如此高的图片，因此需要对图形做放缩处理。BitmapFactory 类在解析位图时可以设定 BitmapFactory.Options 类，用于控制位图的解析参数。能够控制缩放比例的参数是 inSampleSize，如果设置 inSampleSize=6，则 BitmapFactory 解析位图时会产生一幅是原始尺寸 1/6 的图像。

设置 inSampleSize 参数时，通常需要设定 inJustDecodeBounds 参数取值为 true，该参数将设置 BitmapFactory 在解析位图时不需要加载位图的数据，而只返回尺寸数据。获取这些尺寸时，只需要使用 BitmapFactory.Options.outWidth 和 BitmapFactory.Options.outHeight 变量。

修改项目 Proj07_7，显示位图的功能由方法 showImage 实现，代码如下。

📖 Proj07_7 项目 MainActivity.java 文件

```
...
protected void onActivityResult(int requestCode, int resultCode, Intent data) {
    super.onActivityResult(requestCode, resultCode, data);
    Log.i("Msg", "onRes rc = " + resultCode + "   R_OK = " + RESULT_OK);
    if(resultCode == RESULT_OK && requestCode == 0x100){
        showImage();
    }
}
//按比例显示位图
private void showImage(){
    //获取屏幕尺寸
    Display disp = getWindowManager().getDefaultDisplay();
    int screenWidth = disp.getWidth();
    int screenHeight = disp.getHeight();
    //设置 BitmapFactory 解析位图的方式
    BitmapFactory.Options ops = new BitmapFactory.Options();
    ops.inJustDecodeBounds = true;         //只加载位图尺寸,不加载位图数据
    Bitmap bt1 = BitmapFactory.decodeFile(imageFile.getAbsolutePath(),ops);
    //位图尺寸
    int imageWidth = ops.outWidth;
    int imageHeight = ops.outHeight;
    //根据屏幕尺寸,计算宽高的放缩比例
    int widthRatio = (int) Math.ceil((float)imageWidth/screenWidth);
    int heightRatio = (int) Math.ceil((float)imageHeight/screenHeight);
    Log.i("Msg", "宽度比例: " + widthRatio + ",高度比例: " + heightRatio);
    //为保证图像的放缩不失真,宽高放缩比例取较大者
    if(widthRatio>1 || heightRatio>1){
        if(widthRatio >= heightRatio)
            ops.inSampleSize = widthRatio;
        else
            ops.inSampleSize = heightRatio;
    }
```

```
    //重新设定加载模式
    ops.inJustDecodeBounds = false;
    bt1 = BitmapFactory.decodeFile(imageFile.getAbsolutePath(),ops);
    Log.i("Msg", "bitmap size = " + bt1.getByteCount());
    iv.setImageBitmap(bt1);
}
```

上述代码的放缩比例是通过位图宽高和屏幕宽高计算出来的，高度比例和宽度比例哪一个作为最终放缩比例，取决于它们取值的大小。在宽度比例和高度比例中取值较大者作为 BitmapFactory.Options.inSampleSize 的值。

在该项目中最终计算宽度比例为 4，高度比例为 3，因此以宽度比例为放缩比例。经放缩后，BitmapFactory 加载的位图大小为 1 228 800B，约 1.17MB，所占空间是原始尺寸的 1/16。该项目运行效果如图 7.11 所示，对比图 7.10，清晰度没有很大变化。

图 7.11　放缩后的图像效果

7.5 习　　题

1. 选择题

(1) 关于 Android 视频播放的说法错误的是(　　)。

　　A. 可以使用 SurfaceView 组件来播放视频

　　B. VideoView 组件可以控制播放位置和播放图像的大小

　　C. 可以使用 VideoView 组件来播放视频

　　D. VideoView 播放的格式只能是 3gp

(2) MediaPlayer 中 isLooping 方法是(　　)。

　　A. 初始化播放状态　　　　　　　　B. 暂停播放

　　C. 判断是否循环播放　　　　　　　D. 抛出异常

(3) 关于 MediaPlayer 停止状态的说法正确的是(　　)。

　　A. 停止的时候如果想重新播放，需要通过 prepare 方法回到先前的 Prepared 状态再重新开始才可以

　　B. 和暂停状态可以同时搭配使用

　　C. 停止状态很容易抛出异常，使播放的音乐停止

　　D. 停止的时候需要重新初始化后才可以播放音乐

(4) 关于 SoundPool 的说法正确的是(　　)。

　　A. 可以支持 mp3 等多个音乐格式播放

　　B. 适合各种形式的背景音乐播放

　　C. 支持多个音乐文件同时播放，那就说明它支持多线程模式下同时播放

　　D. 和 MediaPlayer 不同，是由独立的播放框架支持的

2. 简答题

(1) 设置音频文件"上一首""下一首"的功能，也就是读取 sdcard 音乐文件中列表的下一项。

(2) 音乐播放进度条显示，根据音乐播放的时间长短来设置音乐文件进度条的显示和滚动的频率。

(3) 单击音乐列表中的某一首歌曲，进行自定义歌曲播放。

(4) 显示对应音视频播放的时间动态的进度。

(5) 控制视频文件的播放、暂停和快进。

第 8 章　Service 与 BroadcastReceiver

本章学习目标
- 掌握 Service 的创建与配置。
- 掌握 Activity 与 Service 的通信。
- 掌握 BroadcastReceiver 的应用。
- 了解 Android 短信收发过程。

Service 可以看作是没有交互、不可见的 Activity，用于在后台执行某些操作，为其他对象提供接口，以便于在应用程序中调用。Service 组件不是独立的线程，它运行在启动它的主线程中。如果 Service 中存在比较耗时的操作（如联网），会影响到主线程，甚至阻塞主线程。Service 作为没有界面的长生命周期的组件，常常被应用于媒体播放，检测 SD 卡上的文件变化，记录用户地理位置等信息的变化。

BroadcastReceiver 是 Android 系统中的广播接收器，实现广播接收功能，既可以接收其他应用程序中的广播，也可以接收 Android 系统发出的广播。

8.1　创建并配置 Service

Service 作为一种服务组件，不被用户所见。它用于处理一些不干扰用户交互的后台操作，如更新应用、播放音乐等。Service 可以通过 Intent 来启动（Start），也可以绑定（Bind）到宿主对象（启动 Service 的对象）。Service 位于 android.app 包中，间接继承 Context，与 Activity 一样，可以调用 Context 中的方法。

8.1.1　自定义 Service

Service 是四大组件之一，与 Activity 一样，需要在 AndroidManifest.xml 中配置。Service 类位于 android.app 包中，继承 android.content.ContextWrapper 类。Service 是抽象类，必须被继承才能使用。Service 的应用场景分为本地服务（local service）和远程服务（remote service）。本地服务运行在主进程中，可以访问同一个应用程序中的其他资源，不需要应用 AIDL（Android Interface Definition Language，Android 内部进程通信接口描述语言，用于实现进程通信），启动或绑定 Service 组件相对简单。远程服务运行在独立的进程中，因此，不受其他进程影响，有利于为多个进程提供服务，具有较高的灵活性。远程服务会占用一定资源,需要借助 AIDL 实现进程间通信,使用这种组件稍微麻烦一些。本节先介绍本地服务的应用。

Service 组件既不会开启新的进程,也不会开启新的线程,它运行在应用程序的主线程中。Service 中实现的逻辑代码不能阻塞整个应用程序的运行,否则会引起应用程序抛出 ANR(Application Not Responding)异常,即应用程序不响应用户操作。

Service 组件常被用于实现以下两种功能:

(1) 使用 startService 方法启动 Service 组件,运行在系统的后台,在不需要用户交互的前提下,实现某些功能。

(2) 使用 bindService 方法启动 Service 组件,启动者与服务组件之间建立"绑定关系",应用程序可以与 Service 组件交互。

下面的代码通过继承 Service 类实现自定义服务组件,并重写了 Service 组件的生命周期方法。

📖 Proj08_1 项目 MyService.java 文件

```java
public class MyService extends Service {
    @Override
    public IBinder onBind(Intent arg0) {
        Log.i("Msg", "onBind run!");
        return null;
    }
    @Override
    public void onCreate() {
        super.onCreate();
        Log.i("Msg", "onCreat run!");
    }
    @Override
    public void onDestroy() {
        super.onDestroy();
        Log.i("Msg", "onDestroy run!");
    }
    @Override
    public int onStartCommand(Intent intent, int flags, int startId) {
        Log.i("Msg", "onStartCommand run!");
        return super.onStartCommand(intent, flags, startId);
    }
    @Override
    public boolean onUnbind(Intent intent) {
        Log.i("Msg", "onUnbind run!");
        return super.onUnbind(intent);
    }
}
```

Service 组件也具有自己的生命周期,并且会根据不同的启动模式执行不同的生命周期,这与 Activity 组件非常相似。Service 类中的常用方法如表 8.1 所示。

表 8.1　Service 常用方法说明

方　法	说　　明	返回值类型
onBind(Intent intent)	抽象方法,绑定模式时执行该方法,通过 IBinder 对象访问 Service	abstract IBinder
onCreate()	Service 组件创建时执行	void
onDestroy()	Service 组件销毁时执行	void
onStartCommand(Intent intent, int flags, int startId)	开始模式时执行该方法。每次执行 startService,该方法都会被执行	int
onUnbind(Intent intent)	接触绑定时执行	boolean
stopSelf()	停止 Service 组件	void

　　Service 作为系统组件,需要在配置文件中设置才可以使用,这一点与 Activity 的要求一致。Service 的配置标签是< service ></service >,嵌套在< application >标签内部。android:name 属性设置 Service 的全路径信息。< service >标签内部可以配置< intent-filter >标签,用于响应隐式 Intent。下面的代码演示了 MyService 组件的配置信息。

　📖 Proj08_1 项目 AndroidManifest.xml 文件

```
…
<!-- 配置 Service -->
<service android:name = "edu.freshen.service.MyService">
    <intent-filter>
        <action android:name = "edu.freshen.service.MyService.ACTION"/>
    </intent-filter>
</service>
…
```

8.1.2　Service 的生命周期

　　启动 Service 组件有两种方式,分别是执行 Context 中的 startService 方法或 bindService 方法。启动方式决定了 Service 组件的生命周期和它所执行的生命周期方法。Service 的生命周期方法比 Activity 的生命周期要简单,由于 Service 是运行在后台,用户无法感知它的存在,所以掌握 Service 的生命周期是比较重要的。

　　1. 执行 startService 启动 Service 组件

　　在一个应用程序中的任何位置,当需要启动 Service 时,都可以通过执行 startService 方法实现。系统会自动回调 Service 组件的 onCreate 方法,创建 Service 对象,随后执行 onStartCommand 方法。如果需要启动的 Service 组件已经被创建,并处于运行状态,那么系统将不再创建新的 Service 组件对象,直接调用 onStartCommand 方法响应启动操作。

　　这种方式启动的 Service 组件会一直运行在后台,与"启动者"没有关联。Service 组件的运行状态不受"启动者"的影响,即使"启动者"被销毁,Service 组件还会继续运行,直到调用 Context 中的 stopService 方法,或者执行 stopSelf 方法。

　　2. 执行 bindService 启动 Service 组件

　　执行 bindService 方法,可以启动一个 Service 组件。如果该 Service 组件对象不存在,则首先创建它,然后再与之"绑定",如果目标 Service 组件存在,则不再创建新的对象,直接与之绑定。这种启动模式会把 Service 组件与"启动者"绑定,Service 返回 IBinder 对象。启动者借助 ServiceConnection 对象,实现与 Service 组件的交互。在这种模式下,Service 组件与"启动者"建立关联,如果"启动者"被销毁,Service 组件也将会结束。

需要特别注意的是,Service 组件的这两种启动方式并不是相互独立的,允许出现交叉调用的情况。即允许先以 startService 方法启动 Service 组件,然后再使用 bindService 方法与之绑定。如果出现这种情况,需要按照启动的顺序依次调用与启动对应的结束方式。比如,先调用 startService,然后再调用 bindService,启动并绑定 Service。此时,"启动者"既可以保持和 Service 交互,又使得 Service 不会随着"启动者"的退出而退出。当不需要绑定时,先调用 unbindService 方法,此时 Service 会执行 onUnbind,但不会把这个 Service 销毁,再调用 stopService 时,Service 才会销毁。

Service 组件的两个生命周期方法如同 8.1 所示。当执行 startService 方法时,实现左侧的生命周期过程;当执行 bindService 方法时,实现右侧的生命周期过程。

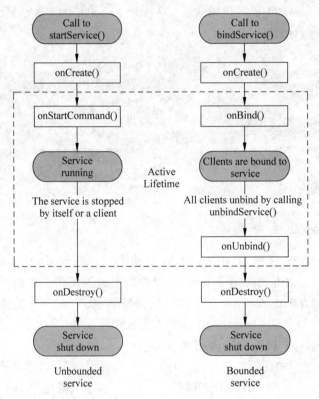

图 8.1 Service 生命周期

8.2 Service 的启动模式

启动 Service 组件和停止 Service 组件的方法都位于 Context 类,在 Activity 中可以直接使用。方法的功能与参数意义如下:

- startService(Intent service),以 Start 模式启动 Service 组件,参数 service 是目标

组件。
- stopService(Intent service),停止 Start 模式启动的 Service 组件,参数 Service 是目标组件。
- bindService (Intent service, ServiceConnection conn, int flags),以 Bind 模式启动 Service 组件,参数 service 是目标组件;conn 是与目标组件链接的对象,不可以是 null;flags 是绑定模式,可以取值为 Context. BIND_AUTO_CREATE、Context. BIND_NOT_FOREGROUND 等。前者表示当收到绑定请求时,如果服务尚未创建,则即刻创建。若系统内存不足,需要先销毁优先级低的组件来释放内存,并且只有当绑定该服务的"启动者"被销毁时,服务才可被销毁。后者表示绑定该服务的"启动者"不具有前台优先级,仅在后台运行。
- unbindService(ServiceConnection conn),解除绑定模式的 Service 组件,conn 是绑定时的链接对象。

8.2.1 startService

项目 Proj08_1 使用 Service 组件实现音乐播放器,Activity 组件作为 Service 的"启动者",负责音乐播放的控制操作。

Activity 界面比较简单,详细内容可以查看项目 Proj08_1 的布局文件 activity_main.xml,运行效果如图 8.2 所示。Activity 通过 5 个按钮实现播放、暂停、停止、上一曲、下一曲功能,它们的监听器代码如下。

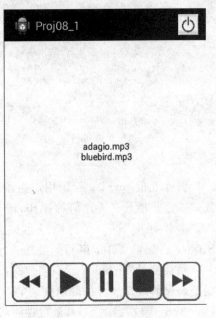

图 8.2 播放器控制界面

📖 Proj08_1 项目 MainActivity.java 文件

```java
//播放器控制监听器方法
@Override
public void onClick(View v) {
    Intent intent = new Intent("edu.freshen.service.SongService.ACTION");
    switch(v.getId()){
    case R.id.op1:                          //上一曲按钮单击
        curIndex = curIndex - 1 < 0?0:curIndex - 1;
        curFlag = 0;
        break;
    case R.id.op2:                          //播放按钮单击
        if(curFlag == 2){
            curFlag = 1;
            op3.setEnabled(true);
            op2.setEnabled(false);
        }else{
            curFlag = 0;
```

```
            }
            break;
        case R.id.op3:                    //暂停按钮单击
            if(curFlag == 1){
                curFlag = 2;
                op2.setEnabled(true);
                op3.setEnabled(false);
            }
            break;
        case R.id.op4:                    //停止按钮单击
            curFlag = 3;
            break;
        case R.id.op5:                    //下一曲按钮单击
            curIndex = curIndex + 1 >= songs.length - 1?songs.length - 1:curIndex + 1;
            curFlag = 0;
    }
    intent.putExtra("opFloag",curFlag);
    intent.putExtra("currSongPath", songs[curIndex].getAbsolutePath());
    startService(intent);
}
```

属性变量 curFlag 标识用户操作，取值含义为：0——播放新曲，1——继承播放，2——暂停播放，3——停止播放。无论何种操作，最后都会执行 startService 方法，启动指定的 Service 组件。本程序中 Service 的响应事件是 edu.freshen.service.SongService.ACTION，这与 Service 在配置文件中声明的信息一致。

属性变量 curIndex 标识当前需要播放的曲目在歌曲列表中的下标，上一曲、下一曲功能是通过修改该属性的取值实现的。

Activity 在启动 Service 组件时将获取"启动者"的"操作意图"，即需要播放歌曲的路径和控制信息，封装在 Intent 中，传递给 Service。Service 组件在 onStartCommand 方法中获取这些信息，根据控制执行不同的逻辑。

Activity 处于开启状态时（显示在顶层），通过 5 个操作按钮向 Service 传递信息；Activity 不在顶层显示时，Service 将继续执行，不会受到 Activity 的影响。Activity 通过 ActionBar 右侧的退出系统按钮执行 stopService 方法，并退出系统。

Service 组件执行 onCreate→onStartCommand→onDestroy 生命周期方法，与绑定模式相关的方法将不会执行。在 onCreate 方法中创建播放器对象。在 onStartCommand 方法中获取"启动者"的意图，并分配逻辑。在 onDestroy 方法中释放资源。

📖 Proj08_1 项目 SongService.java 文件

```java
public class SongService extends Service {
    MediaPlayer mp;                      //播放器
    String currSongPath;                 //需要播放歌曲的路径
    int opFlag;                          //操作标识,0——播放新曲,1——继承播放,2——暂停播放,3——停止播放
    @Override
    public IBinder onBind(Intent arg0) {
        return null;
```

```java
    }
    @Override
    public void onCreate() {
        super.onCreate();
        //创建播放器
        mp = new MediaPlayer();
    }
    @Override
    public void onDestroy() {
        super.onDestroy();
        //释放资源
        mp.release();
        mp = null;
    }
    @Override
    public int onStartCommand(Intent intent, int flags, int startId) {
        //获取启动者的"操作意图"
        currSongPath = intent.getStringExtra("currSongPath");
        opFlag = intent.getIntExtra("opFloag", -1);
        switch(opFlag){
        case 0:play();break;
        case 1:conti();break;
        case 2:pause();break;
        case 3:stop();break;
        }
        return super.onStartCommand(intent, flags, startId);
    }
    //继续播放
    private void conti() {
        mp.start();
    }
    //停止播放
    private void stop() {
        mp.stop();
    }
    //暂停播放
    private void pause() {
        mp.pause();
    }
    //播放新曲
    private void play() {
        try {
            mp.reset();
            mp.setDataSource(currSongPath);
            mp.prepare();
            mp.start();
        } catch (IllegalArgumentException e) {
            e.printStackTrace();
        } catch (SecurityException e) {
```

```
                    e.printStackTrace();
            } catch (IllegalStateException e) {
                    e.printStackTrace();
            } catch (IOException e) {
                    e.printStackTrace();
            }
        }
        @Override
        public boolean onUnbind(Intent intent) {
            return super.onUnbind(intent);
        }
    }
```

需要注意的是,使用 Start 模式启动 Service 组件,应执行 stopService 方法停止 Service 组件,否则该组件不会自动停止,会持续消耗移动设备的运算能力和电量。

8.2.2 bindService

项目 Proj08_2 使用 Service 组件,以绑定模式实现彩票号码随机生成的功能。Activity 作为 Service 的"启动者",并与 Service 组件绑定,通过滚动号码和确定号码按钮选出一组彩票号码。

Activity 布局界面比较简单,详细内容可以查看项目 Proj08_2,布局文件 activity_main.xml,运行效果如图 8.3 所示。

Activity 启动后使用下面的代码与 Service 组件绑定,Intent 中设定的 Action,即 edu.freshen.service.LotteryService.ACTION,与 Service 组件在配置文件中声明的信息一致。bindService 方法有 3 个参数,具体参数含义可以查看 8.2 节开始的说明。

图 8.3 彩票机运行界面

```
Intent intent = new Intent("edu.freshen.service.LotteryService.ACTION");
bindService(intent, conn, Service.BIND_AUTO_CREATE);
```

conn 是 ServiceConnection 对象,充当"启动者"与 Service 组件之间的连接桥,可以获取 Service 组件的 onBind 方法返回的 IBinder 类型对象,从而可以访问 Service 中的方法。ServiceConnection 是一个接口,它的匿名对象创建代码如下。

📖 Proj08_2 项目 MainActivity.java 文件

```
public class MainActivity extends Activity implements OnClickListener {
    ImageButton bt1,bt2;              //声明控件的引用
    TextView tv;
    LotteryBinder lb;                 //Ibinder 接口的实现对象
    boolean ispick;                   //线程控制参数
```

```java
//修改主UI
Handler handler = new Handler(){
    public void handleMessage(android.os.Message msg) {
        tv.setText(lb.productTicket());
    };
};                          //注意,此处有分号
//与Service组件绑定的连接桥
ServiceConnection conn = new ServiceConnection() {
    @Override
    public void onServiceDisconnected(ComponentName arg0) {
    }
    @Override
    public void onServiceConnected(ComponentName arg0, IBinder arg1) {
        lb = (LotteryBinder) arg1;
        Log.i("Msg", "Service绑定Ok");
    }
};                          //注意,此处有分号
@Override
protected void onCreate(Bundle savedInstanceState) {
    super.onCreate(savedInstanceState);
    setContentView(R.layout.activity_main);
    bt1 = (ImageButton) findViewById(R.id.imageButton1);
    bt2 = (ImageButton) findViewById(R.id.imageButton2);
    tv = (TextView) findViewById(R.id.textView1);
    bt1.setOnClickListener(this);
    bt2.setOnClickListener(this);
    //绑定Service
    Intent intent = new Intent("edu.freshen.service.LotteryService.ACTION");
    bindService(intent, conn, Service.BIND_AUTO_CREATE);
}
@Override
protected void onDestroy() {
    super.onDestroy();
    unbindService(conn);    //解除绑定Service组件
}
@Override
public void onClick(View arg0) {
    switch(arg0.getId()){
        case R.id.imageButton1:
            roundTicket();    break;
        case R.id.imageButton2:
            pickTicket();     break;
    }
}
//选定彩票
private void pickTicket() {
    ispick = false;
}
//彩票号码滚动
```

```java
    private void roundTicket() {
        ispick = true;
        new Thread(new Runnable(){
            @Override
            public void run() {
                while(ispick){
                    handler.sendEmptyMessage(0x100);
                    try {
                        Thread.sleep(200);
                    } catch (InterruptedException e) {
                        e.printStackTrace();
                    }
                }
            }
        }).start();
    }
}
```

Activity 中子线程修改 UI 时需要借助 Handler 对象。执行 Service 组件返回的 IBinder 对象中的方法 lb.productTicket 产生一组随机彩票号码。

IBinder 接口的对象允许跨进程访问。一般情况下，普通对象只能在当前进程中访问，如果希望对象能被其他进程访问，那就必须实现 IBinder 接口（或者使用 AIDL）。IBinder 接口可以指向本地对象，也可以指向远程对象，调用者不需要关心指向的对象是本地的还是远程的。

Binder 类是 IBinder 接口的一个实现类，它们都位于 android.os 包中。LotteryService 组件中的 LotteryBinder 是通过继承 Binder 类实现的，它所提供的功能比较简单——产生一组随机数序列，具体代码如下。

📖 Proj08_2 项目 LotteryService.java 文件

```java
public class LotteryService extends Service {
    //自定义 Binder 类
    class LotteryBinder extends Binder{
        public String productTicket(){
            StringBuffer sb = new StringBuffer();
            Random r = new Random();
            //产生 7 个[1,35]区间的数字
            for (int i = 0; i < 7; i++) {
                sb.append((r.nextInt(34) + 1) + "  ");
            }
            return sb.toString();
        }
    }
    @Override
    public IBinder onBind(Intent arg0) {
        return new LotteryBinder();
    }
    @Override
```

```java
    public void onCreate() {
        super.onCreate();
        Log.i("Msg", "LotteryService onCreate!");
    }
    @Override
    public void onDestroy() {
        super.onDestroy();
        Log.i("Msg", "LotteryService onDestroy!");
    }
    @Override
    public boolean onUnbind(Intent intent) {
        Log.i("Msg", "LotteryService onUnbind!");
        return super.onUnbind(intent);
    }
}
```

当 Service 组件被"启动者"绑定后，onBind 方法得到执行，随即返回 LotteryBinder 对象。"启动者"通过 ServiceConnection 对象获取该 LotteryBinder 对象。

需要注意的是，处于绑定模式的 Service 组件，如果"启动者"没有执行 unbindService 方法解除绑定，退出时将造成 ServiceConnectionLeaked 错误。在上面的代码中，MainActivity 通过在 onDestroy 方法中执行 unbindService(conn) 解除绑定，此时 LotteryService 组件也将自动"终结"。

8.3 远程 Service

远程 Service 之所以有远程的特征，是因为这种 Service 组件运行在独立的进程中，它与应用程序所运行的进程不是同一个。因此，从应用程序的视角来看，这种 Service 组件是外部的、远程的。

一般情况下，一个进程不能访问另一个进程的内存空间。如果在特定应用场景下确实需要实现这种功能，就需要借助 RPC（Remote Procedure Call，远程进程调用）来完成进程之间的通信。Andorid 系统采用了一种轻量级 RPC 通信实现方式，即定义 AIDL。

AIDL 是 Android 系统的一种接口描述语言，运行时会被编译成一段代码，借助接口中定义的方法，就可以达到两个进程内部通信的目的。

使用 AIDL 定义远程 Service 可以按照以下两步进行：

(1) 在项目中创建 AIDL 文件，这是一种后缀名为 .aidl 的普通文件。在该文件中定义接口，声明方法，与在 Java 文件中的操作一样。不同的是，AIDL 中所用到的类型（除基本数据、String、List、Map 和 CharSequence）即使在同一个包中，都需要明确引入。

(2) 新建 Service 组件，使用内部类实现 AIDL 中定义接口的相关方法，在 onBind 生命周期方法中返回该内部类的对象。

创建完 AIDL 文件后，编译器会在项目的 gen 包中自动生成一个与 AIDL 文件同名的接口，并为该接口创建内部抽象类 Stub，该类继承了 Binder 类。Service 组件中创建的实现 IBinder 接口的内部类，其选项继承 Stub 类即可（在 8.2 节中是继承 Binder 类的）。

项目 Proj08_3 演示了如何创建远程 Service,返回时间戳字符串。创建包 edu.freshen.aidl,并创建文件 IDateStamp.aidl,文件的代码如下。

📖 Proj08_3 项目 IDateStamp.aidl 文件

```
package edu.freshen.aidl;
interface IDateStamp{
    String getSerDate();
}
```

远程 Service 组件与本地 Service 组件一样,都需要继承 Service。不同之处是,远程 Service 返回 IBinder 接口的实现对象与本地 Service 不同。MyBinder 继承 IDateStamp.Stub,并实现 AIDL 文件中定义的接口。具体代码如下。

📖 Proj08_3 项目 RemoteService.java 文件

```
public class RemoteService extends Service {
    //自定义 IBinder 内部类
    class MyBinder extends IDateStamp.Stub{
        SimpleDateFormat sdf = new SimpleDateFormat("yyyy-MM-dd HH:mm:ss");
        @Override
        public String getSerDate() {
            String dateStamp = sdf.format(new Date());
            Log.i("Msg", "RemoteService 产生时间" + dateStamp);
            return dateStamp;
        }
    }
    @Override
    public IBinder onBind(Intent arg0) {
        return new MyBinder();
    }
    ...
}
```

方法 getSerDate 是接口文件 AIDL 中定义的方法,用于返回一个时间戳字符串。

访问远程 Service 的方式一般有两种:一是"启动者"(或称客户端)与远程 Service 组件在同一个应用中,同属于一个项目;二是"启动者"与远程 Service 组件不在同一个应用中,分属于两个项目。

1. 在同一个项目中访问远程 Service

远程 Service 的配置信息与本地 Service 不同,需要在 service 标签中增加属性 android:process=":remote",指明该 Service 启动后会使用另外的进程。":"之后的变量名一般设置为 remote,也可以自定义,与":"之前的包名组合在一起,作为 Service 组件运行的应用程序名。项目 Proj08_3 中的 Service 配置信息如下:

```
<service android:name="edu.freshen.rs.RemoteService"
    android:process=":remote">
    <intent-filter>
```

```xml
        <action android:name = "edu.freshen.rs.RemoteService.ACTION"/>
    </intent-filter>
</service>
```

与绑定本地 Service 组件不同,绑定远程 Service 组件时,不再使用 IBinder 接口,而是使用 AIDL 文件定义的接口类型 IdateStamp,如下面代码中的①所示。当 ServiceConnection 对象的 onServiceConnected 方法执行时,将 IBinder 类型转换为 IDateStamp 类型,如下面代码中的②所示。获取接口实现对象后,执行 AIDL 中声明的方法,如下面代码中的③所示,即可以访问远程 Service 组件中的内容。

📖 Proj08_3 项目 MainActivity.java 文件

```java
public class MainActivity extends Activity {
    TextView tv;
    IDateStamp binder;                                    //①AIDL 中定义的接口
    ServiceConnection sc = new ServiceConnection() {
        @Override
        public void onServiceDisconnected(ComponentName arg0) {
        }
        @Override
        public void onServiceConnected(ComponentName arg0, IBinder arg1) {
            binder = IDateStamp.Stub.asInterface(arg1);   //②转换为接口对象
            Log.i("Msg", "Service Connected!");
        }
    };
    @Override
    protected void onCreate(Bundle savedInstanceState) {
        super.onCreate(savedInstanceState);
        setContentView(R.layout.activity_main);
        tv = (TextView) findViewById(R.id.textView1);
        Intent intent = new Intent("edu.freshen.rs.RemoteService.ACTION");
        bindService(intent, sc, Service.BIND_AUTO_CREATE);
    }
    public void getDateStamp(View v){
        try {
            String dt = binder.getSerDate();              //③调用 AIDL 接口中的方法
            Log.i("Msg", "Activity 获取时间戳: " + dt);
            tv.setText("远程时间: " + dt);
        } catch (RemoteException e) {
            e.printStackTrace();
        }
    }
    @Override
    protected void onDestroy() {
        super.onDestroy();
        unbindService(sc);                                //解除绑定
    }
}
```

启动项目中 Activity,执行绑定 RemoteService,访问其中的内部类,获取时间戳。该应用在 Log 日志的输出信息如图 8.4 所示。PID 列是进程编号,RemoteService 分配的进程编号是 17260,Activity 分配的进程编号是 17247,二者运行在不同的进程中。进程编号的分配有系统决定,每次运行都会发生变化。

```
L.. Ti... PID    TID    Application          Tag  Text
D  0... 17247   17247  edu.freshen.rs       g... Emulator without GPU emulation detected.
I  0... 17260   17260  edu.freshen.rs:remote Msg RemoteService create!
I  0... 17247   17247  edu.freshen.rs       Msg  Service Connected!
I  0... 17260   17272  edu.freshen.rs:remote Msg RemoteService产生时间2016-04-08 02:46:35
I  0... 17247   17247  edu.freshen.rs       Msg  Activity获取时间戳: 2016-04-08 02:46:35
```

图 8.4　同项目中访问远程 Service 输出日志

在同一个项目中访问远程 Service 的应用比较常见。下面的代码是使用百度定位服务时,需要在 AndroidManifest.xml 中配置的信息。android:process=":remote"标明该定位服务将运行在独立的进程中。

```
<service android:name="com.baidu.location.f"
    android:enabled="true"  android:process=":remote">
</service>
```

2. 在不同的项目中访问远程 Service

新建项目 Proj08_4 的包名为 edu.freshen.client(远程 Service 所在项目的包名为 edu.freshen.rs),并创建与远程 Service 所在项目包结构一致的 IDateStamp.aidl 文件。在 Activity 中绑定并访问远程 Service 组件 edu.freshen.rs.RemoteService.ACTION,这是 Proj08_3 中的 Service。

Activity 执行绑定与访问 Service 组件的方法与在同一个项目中执行绑定和访问一样,可以参考上面的代码。执行该项目,Log 日志的输出如图 8.5 所示。远程 Service 输入信息与 Activity 输出信息在 PID(进程编号)、TID(线程编号)和 Application(应用程序名)序列都不同。

```
L.. Ti... PID    TID    Application           Tag  Text
I  0... 21605   21605  edu.freshen.client    Msg  Proj08_4 MainActivity Connected!
I  0... 18166   18180  edu.freshen.rs:dateSer Msg RemoteService产生时间2016-04-08 03:44:22
I  0... 21605   21605  edu.freshen.client    Msg  Activity获取时间戳: 2016-04-08 03:44:22
I  0... 18166   18179  edu.freshen.rs:dateSer Msg RemoteService产生时间2016-04-08 03:45:00
I  0... 21605   21605  edu.freshen.client    Msg  Activity获取时间戳: 2016-04-08 03:45:00
```

图 8.5　不同项目中访问远程 Service 的输出日志

8.4　BroadcastReceiver

广播机制是 Android 系统中非常重要的通信机制,它可以实现在应用程序内部或应用程序之间传递消息的作用。发出广播(或称广播)和接收广播是两个不同的动作,发出广播就如同无线电台,接收广播就如同调频收音机。

8.4.1 发出广播与接收广播

广播是 Android 系统级的事件。当移动设备状态发生变化时,如开机、时区改变、电池电量降低、网络已连接等,都会以广播的形式向外通知。如果某个应用程序对某个广播"感兴趣",则只需要注册可以接收该广播的接收器(BroadcastReceiver)就可以响应该事件。

在应用程序开发过程中,多数情形是需要注册接收器,响应广播,从而启动某个功能。如果有需要,应用程序也可以主动发出广播。Android 系统主动发出的广播称为系统广播,应用程序发出的广播称为自定义广播。一个应用程序发出的广播数量没有限制,接收广播的数量也没有限制。

所谓广播,实质上是一个 Intent 对象。通过设置它的 Action、Category 等信息,以有序或无序的方式发送到系统中。而所谓接收广播,实质上就是接收 Intent 对象。

发送广播需要用到的方法位于 Content 类中,具体功能描述如表 8.2 所示。

表 8.2 发送广播的方法

方法	说明	返回值类型
sendBroadcast(Intent intent, String receiverPermission)	发送广播,接收该广播时需要相应权限	void
sendBroadcast(Intent intent)	发送广播,不设权限	void
sendOrderedBroadcast(Intent intent, String receiverPermission)	发送有序广播,接收该广播时需要相应权限	void
sendStickyBroadcast(Intent intent)	发送粘性广播,API 17 取消了该方法	void
sendBroadcastAsUser(Intent intent, UserHandle user)	发送广播,带有用户组权限,API 17 新增	void

接收广播需要自定义广播接收器类,继承 BroadcastReceiver 类。该类位于 android.content 包中,是 Android 四大组件之一。

与 Activity 和 Service 组件不同,BroadcastReceiver 的生命周期很短。当系统或其他程序发送广播时,Android 系统检查所有已安装的应用程序,筛查配置文件有没有匹配的 action。如果存在对应的广播接收器,并且有权接收,那么就创建 BroadcastReceiver 对象,然后执行 onReceiver 方法。方法执行完成时 BroadcastReceiver 对象被销毁,所以说 BroadcastReceiver 的生命周期是非常短的。

由于 BroadcastReceiver 生命周期很短,因此,在 onReveiver 方法中不能执行比较耗时的操作(一般应不超过 10s),否则会弹出 ANR(应用程序不响应)对话框。如果需要完成耗时操作,可以启动 Service 组件,让 Service 执行业务逻辑。此外,在 onReceiver 方法中也不能开启子线程,因为当父线程被杀死后,它的子线程也会被杀死,所以,这是很不安全的做法。

下面的代码演示了如何自定义 BroadcastReceiver。

📖 Proj08_5 项目 MyBroadcastReceiver.java 文件

```
public class MyBroadcastReceiver extends BroadcastReceiver {
    @Override
```

```
        public void onReceive(Context arg0, Intent arg1) {
            Log.i("Msg", "Broadcast Receive!");
            String msg = arg1.getStringExtra("Msg");
            Log.i("Msg", "收到广播中的数据: " + msg);   }
    }
```

定义广播接收器后,需要进行注册才能接收广播。Android 支持静态注册和动态注册两种方式,前者将接收器的信息配置在 AndroidManifest.xml 文件中,后者使用代码进行。

8.4.2 广播的分类与权限

Android 系统中的广播可以分为普通广播(normal broadcasts)和有序广播(ordered broadcasts),它们的发送方式和特性都有较大区别。普通广播使用 sendBroadcast 方法发送,有序广播使用 sendOrderedBroadcast 方法发送。普通广播有时也被称为一般广播。

普通广播对于接收器来说是无序的、没有优先级的,每个接收器都无须等待即可以接收到广播,接收器之间相互没有影响,因此广播效率较高。这种广播无法被终止,即无法阻止其他接收器的接收动作。

有序广播对于接收器来说是有序的、有优先级区分的。广播首先发送到优先级高的接收器,然后再传播到优先级低的接收器,并且,优先级高的接收器可以选择终止这个广播。广播接收器的优先级在注册时设置,使用 intent-filter 类的属性 android:priority,取值范围是-1000~1000,数值越大,优先级越高。终止广播使用 BroadcastReceiver 类的 abortBroadcast 方法。普通广播与有序广播的工作方式如图 8.6 所示。

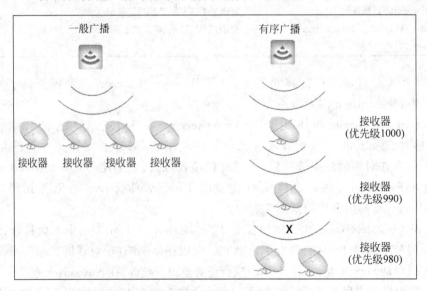

图 8.6 普通广播与有序广播

广播的权限管理主要为了解决如下两个问题:
- 何种接收器有权接收广播,这通过配置 uses-permission 实现。
- 何种广播有权向接收器发送广播,这通过设置 receiver 的 android:permission 属性实现。

无论普通广播还是有序广播,在发出时都可以设定接收权限。拥有接收权限的接收器可以响应广播,没有权限的接收器不响应。使用带有参数 receiverPermission 的 sendBroadcast 方法或 sendOrderedBroadcast 方法,可以发出带有权限的广播。而所谓的"权限",其实质就是一个字符串,接收器需要明确配置 permission 属性,拥有该权限即可接收广播。

8.4.3 注册广播接收器

Android 系统中注册 BroadcastReceiver 有两种方式,分别是配置文件静态注册和代码动态注册。在 Android 3.1 版本之前,静态注册的广播接收器即使 App 已经退出,当有相应的广播发出时,接收器依然可以接收到,在 Android 3.1 版本之后,这种功能被修改了。不过对于自定义的广播,可以通过设置 Intent 的 flag 属性值 FLAG_INCLUDE_STOPPED_PACKAGES 启动已经停止的 App。系统广播是无法实现这种功能的,如果确实需要实现类似功能,可以借助 Service 组件。

1. 使用配置文件静态注册接收器接收普通广播

BroadcastReceiver 组件使用 receiver 标签配置,写在 application 内部。receiver 标签的具体配置信息如下:

```
< receiver android:enabled = ["true" | "false"]
android:exported = ["true" | "false"]
android:icon = "drawable resource"
android:name = "string"
android:permission = "string"
android:process = "string" >
…
</receiver >
```

- android:enabled 属性说明 BroadcastReceiver 是否可用。
- android:exported 属性说明 BroadcastReceiver 能否接收其他 App 发出的广播。
- android:name 属性说明 BroadcastReceiver 类名。
- android:permission 属性说明具有相应权限的广播才能被 BroadcastReceiver 所接收。
- android:process 属性说明 BbroadcastReceiver 运行所处的进程,默认为 App 的进程。

下面的代码演示如何发出普通广播、静态注册和接收广播数据。详细代码请参考项目 Proj08_5。MainActivity 的布局文件比较简单,放置按钮,并设置 onClick 属性,执行下面的方法即可。发送广播时,设定 Intent 动作是 edu.freshen.broadcast.ACTION(见 1 处),这相当于说明该广播发出的"频段"。

📖 Proj08_5 项目 MainActivity.java 文件

```java
//发出普通广播的按钮单击
public void sendNormalBr(View v){
```

```
        Intent broadIntent = new Intent("edu.freshen.broadcast.ACTION");     ①
        broadIntent.putExtra("Msg","没有权限的普通广播");
        sendBroadcast(broadIntent);
    }
```

MyBroadcastReceiver 配置信息如下。receiver 标签的 name 属性是必须的,其他几个属性可以不设置。内部标签 intent-filter 用于设置广播接收器能够响应的广播动作(见 2 处),相当于说明广播接收器接收的"频段"。

📖 Proj08_5 项目 AndroidManifest.xml 文件

```
<receiver android:name = "edu.freshen.broadcast.MyBroadcastReceiver">
    <intent-filter>
        <action android:name = "edu.freshen.broadcast.ACTION"/>              ②
    </intent-filter>
</receiver>
```

发出广播后,广播接收器 onReceive 方法执行,输出如图 8.7 所示的信息。

Level	PID	TID	Application	Tag	Text
I	772	772	edu.freshen.broadcast	Msg	Broadcast Receive!
I	772	772	edu.freshen.broadcast	Msg	收到广播中的数据:没有权限的普通广播

图 8.7 普通广播接收器的输出信息

2. 使用配置文件静态注册接收器接收带有权限的广播

发出带有权限的广播应使用带有参数 receiverPermission 的方法。"权限"的实质就是一个字符串,如下面代码③处所示,该字符串由程序员定义(不可重复),并在配置文件中声明。

📖 Proj08_5 项目 MainActivity.java 文件

```
//发出带有权限广播的按钮单击
public void sendPermissionBr(View v){
    Intent broadIntent = new Intent("edu.freshen.broadcast.ACTION");
    broadIntent.putExtra("Msg","没有权限的普通广播");
    sendBroadcast(broadIntent, "edu.freshen.broadcast.PERMISSION");          ③
}
```

在配置文件中声明自定义权限使用 permission 标签,如下面代码 4 处所示。配置应用程序具有接收权限,使用 uses-permission 标签,如下面代码 5 处所示。

📖 Proj08_5 项目 AndroidManifest.xml 文件

```
<permission android:name = "edu.freshen.broadcast.PERMISSION"></permission>  ④
<uses-permission android:name = "edu.freshen.broadcast.PERMISSION"/>         ⑤
...
```

```
    <receiver android:name = "edu.freshen.broadcast.MyBroadcastReceiver">
        <intent-filter>
            <action android:name = "edu.freshen.broadcast.ACTION"/>
        </intent-filter>
    </receiver>
...
```

3. 代码动态注册接收器

使用代码注册广播接收器需要使用 Context 中的 registerReceiver(BroadcastReceiver receiver, IntentFilter filter)方法。参数 receiver 是需要注册的广播接收器,参数 filter 用于选择相匹配的广播接收器。

在项目 Proj08_6 中定义了 3 个广播接收器:MyReceiver1、MyReceiver2 和 MyReceiver3。MainActivity 主界面上有 3 个按钮,分别实现发送广播、注册接收器和解除接收器的功能,详细代码可以查看 activity_main.xml 文件。MainActivity 的代码如下。

📖 Proj08_6 项目 MainActivity.java 文件

```java
public class MainActivity extends Activity implements OnClickListener {
    Button bt1,bt2,bt3;
    BroadcastReceiver br1,br2,br3;           //广播接收器
    @Override
    protected void onCreate(Bundle savedInstanceState) {
        super.onCreate(savedInstanceState);
        setContentView(R.layout.activity_main);
        bt1 = (Button) findViewById(R.id.button1);
        bt2 = (Button) findViewById(R.id.button2);
        bt3 = (Button) findViewById(R.id.button3);
        bt1.setOnClickListener(this);
        bt2.setOnClickListener(this);
        bt3.setOnClickListener(this);
        //创建广播接收器对象
        br1 = new MyReceiver1();
        br2 = new MyReceiver2();
        br3 = new MyReceiver3();
    }
    //按钮监听器
    @Override
    public void onClick(View arg0) {
        switch(arg0.getId()){
            case R.id.button1:
                sendBr();    break;
            case R.id.button2:
                regBr();     break;
            case R.id.button3:
                unRegBr();   break;
        }
    }
```

```java
    //解除注册的广播接收器
    private void unRegBr() {
        unregisterReceiver(br1);
        unregisterReceiver(br2);
        unregisterReceiver(br3);
    }
    //注册广播接收器
    private void regBr() {
        IntentFilter filter = new IntentFilter();
        filter.addAction("edu.freshen.p85.Broadcast.ACTION");      ②
        registerReceiver(br1, filter);
        registerReceiver(br2, filter);
        registerReceiver(br3, filter);
    }
    //发送普通广播
    private void sendBr() {
        Intent intent = new Intent("edu.freshen.p85.Broadcast.ACTION");   ①
        intent.putExtra("msg", "0000");    //给广播增加数据
        sendBroadcast(intent);
    }
}
```

由于广播接收器需要使用代码注册,因此广播接收器的对象必须先创建出来。注册广播接收器时,IntentFilter 的动作必须与发出广播时的动作一致,如代码①②处所示。上面的代码中,3 个广播接收器使用了同一个 IntentFilter 对象。

解除注册的广播接收器使用 Context 中的 unregisterReceiver(BroadcastReceiver receiver)方法,参数 receiver 是待解除的接收器对象。需要特别注意的是,如果广播接收器没有解除注册,Activity 退出时会抛出 android.app.IntentReceiverLeaked 错误。因此建议将解除广播接收器的代码写在 Activity 的生命周期方法 onStop 或 onDestroy 中,将注册广播接收器的代码写在 onStart 或 onCreate 中。

在该项目中,广播接收器的代码比较简单,使用日志输出了收到的广播信息,相应的输出信息如图 8.8 所示。

L...	Ti...	PID	TID	Application	Tag	Text
I	0...	771	771	edu.freshen.p85	Msg	MyReceiver1 onReceiver run!
I	0...	771	771	edu.freshen.p85	Msg	收到广播中的数据: 0000
I	0...	771	771	edu.freshen.p85	Msg	MyReceiver2 onReceiver run!
I	0...	771	771	edu.freshen.p85	Msg	收到广播中的数据: 0000
I	0...	771	771	edu.freshen.p85	Msg	MyReceiver3 onReceiver run!
I	0...	771	771	edu.freshen.p85	Msg	收到广播中的数据: 0000

图 8.8 动态注册接收器的输出信息

4. 配置有序广播接收器

接收带有优先级顺序的有序广播,既可以使用静态配置,也可以使用动态注册。在项目 Proj08_6 的 MainActivity 文件中增加可以发送有序广播的方法,代码如下。

📖 Proj08_6 项目 MainActivity.java 文件,发送有序广播

```java
//发送有序广播
private void sendOrderBr() {
    Intent intent = new Intent("edu.freshen.p85.Broadcast.ACTION");
    intent.putExtra("msg", "0000");
    sendOrderedBroadcast(intent,"edu.freshen.p85.Broadcast.PERMISSION");
}
```

有序广播都是带有权限的,需要在配置文件中声明权限信息。

注册有序广播接收器的方法与注册普通广播接收器一样,不同之处是,可以使用 setPriority 方法设置该广播接收器的优先级。优先级数值范围是 −1000~1000,数值越大优先级越高。具体代码如下。

📖 Proj08_6 项目 MainActivity.java 文件

```java
//注册带有优先级的广播接收器
private void regPropertyBr() {
    //广播接收器 1,优先级 1000
    IntentFilter filter1 = new IntentFilter();
    filter1.addAction("edu.freshen.p85.Broadcast.ACTION");
    filter1.setPriority(1000);
    registerReceiver(br1, filter1);
    //广播接收器 2,优先级 900
    IntentFilter filter2 = new IntentFilter();
    filter2.addAction("edu.freshen.p85.Broadcast.ACTION");
    filter2.setPriority(900);
    registerReceiver(br2, filter2);
    //广播接收器 3,优先级 800
    IntentFilter filter3 = new IntentFilter();
    filter3.addAction("edu.freshen.p85.Broadcast.ACTION");
    filter3.setPriority(800);
    registerReceiver(br3, filter3);
}
```

有序广播会按照接收器的优先级顺序,逐个传递。优先级高的广播接收器在收到广播后,可以选择是否终止此次广播(这是各种广播拦截器实现的原理),也可以在广播中追加数据,继续向下传递。

下面是 MyReceiver1 代码,使用参数 arg1 获取广播中的数据,创建 Bundle 对象,追加数据信息。使用 BroadcastReceiver 类中的方法 setResultExtras(Bundle extras)将新创建的 Bundle 对象更新到广播中,该方法仅对有序广播有效。

📖 Proj08_6 项目 MyReceiver1.java 文件

```java
public class MyReceiver1 extends BroadcastReceiver {
    @Override
    public void onReceive(Context arg0, Intent arg1) {
        Log.i("Msg", "MyReceiver1 onReceiver run!");
```

```java
        String msg = arg1.getStringExtra("msg");
        Log.i("Msg", "收到广播中的数据:" + msg);
        Bundle bundle = new Bundle();
        bundle.putString("msg", msg + " 1111");
        setResultExtras(bundle);
    }
}
```

在该项目中 MyReceiver2 的优先级低于 MyReceiver1,它将在 MyReceiver1 处理完广播之后再执行。MyReceiver2 接收器使用 BroadcastReceiver 类中的方法 getResultExtras (boolean makeMap)获取广播中数据信息。参数 makeMap 取值为 true,表示如果广播中的数据信息是 null,则创建一个对象并返回;取值为 false,表示不创建对象,此时该方法有可能会返回 null。

终止广播使用 BroadcastReceiver 类中的 abortBroadcast 方法,该方法只能终止有序广播。

📖 Proj08_6 项目 MyReceiver2.java 文件

```java
public class MyReceiver2 extends BroadcastReceiver {
    @Override
    public void onReceive(Context arg0, Intent arg1) {
        Log.i("Msg", "MyReceiver2 onReceiver run!");
        Bundle bundle = getResultExtras(true);
        String msg = bundle.getString("msg");
        Log.i("Msg", "收到广播中的数据:" + msg);
        bundle.putString("msg", msg + " 2222");
        setResultExtras(bundle);
        //abortBroadcast();          //终止广播
    }
}
```

广播接收器 MyReceiver3 的代码如下。它的优先级最低,如果前面的接收器终止了广播,它将不再执行。

📖 Proj08_6 项目 MyReceiver3.java 文件

```java
public class MyReceiver3 extends BroadcastReceiver {
    @Override
    public void onReceive(Context arg0, Intent arg1) {
        Log.i("Msg", "MyReceiver3 onReceiver run!");
        Bundle bundle = getResultExtras(true);
        String msg = bundle.getString("msg");
        Log.i("Msg", "收到广播中的数据:" + msg);
    }
}
```

有序广播接收器运行输出信息如图 8.9 所示,优先级低的接收器可以接收到优先级高的接收器传递过来的数据信息。如果将 MyReceiver2.java 中的终止广播语句启用,则

MyReceiver3 将不再输出。

L...	Ti...	PID	TID	Application	Tag	Text
I	0...	1201	1201	edu.freshen.p85	Msg	MyReceiver1 onReceiver run!
I	0...	1201	1201	edu.freshen.p85	Msg	收到广播中的数据: 0000
I	0...	1201	1201	edu.freshen.p85	Msg	MyReceiver2 onReceiver run!
I	0...	1201	1201	edu.freshen.p85	Msg	收到广播中的数据: 0000 1111
I	0...	1201	1201	edu.freshen.p85	Msg	MyReceiver3 onReceiver run!
I	0...	1201	1201	edu.freshen.p85	Msg	收到广播中的数据: 0000 1111 2222

图 8.9 有序广播接收器的输出信息

8.4.4 接收系统广播

Android 系统中内置了很多广播，只要涉及手机的基本操作，都会发出相应广播，这些广播不同于上面所讲的自定义广播。它们是当特定事件发生时由系统内部自动发出的，因此被看作是系统广播。

系统广播主要是反映移动设备的状态发生变化或者设备的某个功能发生变化，如开机、网络状态改变、拍照、屏幕关闭与开启、电量不足等。每个系统广播都具有特定的 intent-filter，包含了对应的 Action。系统广播发出后，将被相应的 BroadcastReceiver 接收。如果需要在应用程序中响应这些广播，则应开发对应的广播接收器，声明必要权限，指明响应的 Action，就可以收到广播。

Android 系统中常用的广播事件如表 8.3 所示。

表 8.3 常用系统广播事件

广播事件	说明
ACTION_BATTERY_CHANGED	充电状态，或者电池的电量发生变化时发送广播，只能通过动态注册接收
ACTION_BOOT_COMPLETED	系统启动完成后发送广播，该广播仅发送一次
ACTION_PACKAGE_ADDED	成功安装 APK 之后发送广播
ACTION_PACKAGE_REMOVED	成功删除某个 APK 之后发送广播
ACTION_POWER_CONNECTED	插上外部电源时发送广播
ACTION_POWER_DISCONNECTED	断开外部电源连接时发送广播
ACTION_REBOOT	重启设备时发送广播
ACTION_SHUTDOWN	关闭系统时发送广播
ACTION_SCREEN_OFF	屏幕被关闭之后发送广播
ACTION_SCREEN_ON	屏幕被打开之后发送广播
ACTION_TIME_CHANGED	时间被设置时发送广播
ACTION_TIME_TICK	当前时间改变时发送广播；每分钟都发送，只能通过动态注册接收
ACTION_BATTERY_CHANGED	充电状态或者电池的电量发生变化时发送广播，只能通过动态注册接收
Telephony.SMS_RECEIVED	短信到达广播，使用 pdus 作为 key 即可从 Intent 中获取短信内容

项目 Proj08_7 新建广播接收器 SysBroadcastReceiver，接收屏幕点亮和关闭广播事件，

并将事件输出到日志。接收器代码如下。

📖 Proj08_7 项目 SysBroadcastReceiver.java 文件

```java
public class SysBroadcastReceiver extends BroadcastReceiver {
    @Override
    public void onReceive(Context arg0, Intent arg1) {
        Log.i("Msg", "系统广播到达!" + arg1.getAction());
    }
}
```

SysBroadcastReceiver 广播接收器采用静态注册的方式,在 AndroidManifest.xml 配置文件中增加如下信息。

📖 Proj08_7 项目 AndroidManifest.xml 文件

```xml
<service android:name="edu.freshen.p87.SysBroadcastReceiver">
    <intent-filter>
        <action android:name="android.intent.action.ACTION_SCREEN_ON"/>
        <action android:name="android.intent.action.ACTION_SCREEN_OFF"/>
    </intent-filter>
</service>
```

运行该项目,当设备启动,屏幕点亮时会收到"屏幕打开"广播;当设备待机,屏幕关闭时会收到"屏幕关闭"广播。日志输出信息如图 8.10 所示。

Level	PID	TID	Application	Tag	Text
I	780	780	edu.freshen.p87	Msg	系统广播到达! android.intent.action.SCREEN_ON
I	780	780	edu.freshen.p87	Msg	系统广播到达! android.intent.action.SCREEN_OFF
I	780	780	edu.freshen.p87	Msg	系统广播到达! android.intent.action.SCREEN_ON

图 8.10 接收屏幕打开与关闭广播的日志输出信息

8.5 实现短信拦截

当 Android 系统收到短信时,会以广播的形式向外通知。该广播属于有序广播,可以被优先级高的接收器拦截。广播的 action 名称为 Android.provier.Telephony.SMS_RECEIVED,并且在 Intent 中封装了短信内容。使用参数值 pdus 即可从 Intent 中获取短信内容。pdus 是一个 Object 类型的数组,每一个 Object 对象都是一个 byte[]字节数组,即每一元素都对应一条短信内容。

本节通过项目 Proj08_8 演示如何在应用程序中实现响应短信广播、接收广播数据、根据"来电黑名单"拦截短信、显示短信内容等功能。该项目旨在进一步介绍 Android 系统广播机制,请勿用于非法功能的实现。随着 Android 系统版本的升级,短信拦截与读取的方式有可能会随之改变,具体信息请参阅最新的 Android 开发文档。

SmsReceiver 是自定义广播接收器,list 属性用于模拟黑名单列表,列表元素在构造方法中初始化。参考代码如下。

📖 **Proj08_8 项目 SmsReceiver.java 文件**

```java
public class SmsReceiver extends BroadcastReceiver {
    //格式化来短信时间
    private SimpleDateFormat format = new SimpleDateFormat("yyyy-MM-dd HH:mm:ss");
    //模拟电话黑名单
    List<String> list = new ArrayList<String>();
    public SmsReceiver(){
        list.add("16012345678");
        list.add("16098985566");
    }
    //该方法中的代码执行时间不应超过10s
    @Override
    public void onReceive(Context arg0, Intent arg1) {
        //读取广播中的数据
        Object[] pduses = (Object[])arg1.getExtras().get("pdus");
        //解析每一个元素的数据
        for(Object pdus: pduses){
            byte[] msgArr = (byte[])pdus;
            //创建短信对象
            SmsMessage sms = SmsMessage.createFromPdu(msgArr);
            //获取短信详情
            String phone = sms.getOriginatingAddress();        //发送短信的手机号码
            String content = sms.getMessageBody();              //短信内容
            Date date = new Date(sms.getTimestampMillis());
            String dt = format.format(date);                    //得到发送时间
            Log.i("Msg", phone + ": " + content + "\n" + dt);
            if(list.contains(phone)){
                abortBroadcast();                                //停止广播
                Log.i("Msg", "黑名单短信,已被拦截");
            }
        }
    }
}
```

收到短信广播后,短信数据封装在 Intent 中,以 pdus 为 key,从 Bundle 中直接读取即可。由于一条短信的字符长度有限制(一般不超 70 个汉字或 160 个英文字母),所以一个短信可能会分成多条,因此从 Bundle 中获取的是一个 Object 对象数组。

短信内容以字节数组的形式存放,可以直接将 Object 数组的元素转换为字节数组。SmsMessage 类位于 android.telephony 包中,对应一条短信对象。使用 SmsMessage 类的静态方法 createFromPdu（byte[] pdu）可以用字节数组创建短信对象,然后使用 SmsMessage 类的方法读取短信的详细信息。

该项目中的短信拦截功能比较简单,只要发送短信的电话号码在黑名单中,就调用 abortBroadcast 方法终止广播。需要注意的是,如需终止此次广播,那么 SmsReceiver 接收器的优先级必须足够高。

SmsReceiver 接收器的配置信息如下,属性 android:priority="1000"设置优先级为最大值。短信接收广播事件是 android.provider.Telephony.SMS_RECEIVED。

📖 Proj08_8 项目 AndroidManifest.xml 文件

```
< receiver android:name = "edu.freshen.p88.SmsReceiver">
    < intent - filter android:priority = "1000">
        < action android:name = "android.provider.Telephony.SMS_RECEIVED"/>
    </intent - filter >
</receiver >
```

接收短信需要声明以下权限，否则在应用程序中收不到短信。

```
< uses - permission android:name = "android.permission.RECEIVE_SMS"/>
```

运行项目后，在 DDMS 视图下（Window 菜单→Open Perspective→DDMS）模拟向"虚拟机"发送短信，如图 8.11 所示。

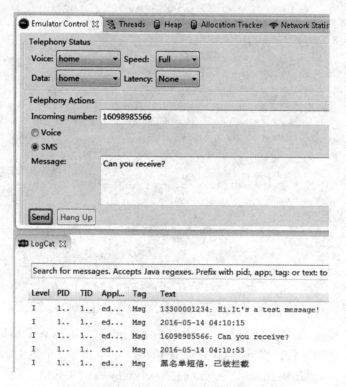

图 8.11　模拟短信发送与日志输出

切换到 Emulator Control 面板，在此可以模拟电话呼叫、短信发送、GPS 定位模拟。Incoming Number 是指发送短信的电话号码，Voice 是电话呼叫模拟，SMS 是短信模拟，Message 是短信内容。如果发送中文，可以使用真机进行测试，否则还要处理汉字转码的问题。

模拟发送的短信都会被短信广播接收器收到，具体信息都输出在日志列表中。如果是黑名单中的电话号码，将提示"已被拦截"，此时手机的状态通知栏中将不再显示有短信到达，否则会显示有短信到达，如图 8.12 所示。

图 8.12 短信通知

8.6 习　　题

1. 选择题

(1) 使用 startService 方法启动的 Service 组件，Service 不会执行的方法是(　　)。

　　A. onCreate()

　　B. onBind(Intent intent)

　　C. onStartCommand(Intent intent, int flags, int startId)

　　D. onDestroy()

(2) 以下关于 Service 组件的描述中正确的是(　　)。

　　A. 使用 startService 启动 Service，必须以 stopService 停止

　　B. 使用 startService 启动 Service，必须以 endService 停止

　　C. 使用 bindService 启动 Service，必须以 stopService 停止

　　D. 使用 bindService 启动 Service，必须以 endService 停止

(3) 以 bindService 方法启动 Service 组件，以下关于启动者与 Service 之间关系的说法中正确的是(　　)。

　　A. 启动者的生命周期与 Service 的生命周期无关

　　B. 启动者销毁时，必须解除绑定 Service，否则抛出异常

　　C. Service 销毁时，会导致启动者也被销毁

　　D. Service 只能与一个启动者绑定

(4) 以下关于远程 Service 的说法中正确的是(　　)。

　　A. 远程 Service 不支持 bindService 启动模式

　　B. 远程 Service 运行在服务器中，与本地应用程序无关

　　C. 远程 Service 运行在独立的进程中，与本地应用程序无关

　　D. 远程 Service 借助 AIDL 实现线程间通信

(5) 以下关于 Android 广播的说法中有误的是(　　)。
　　A. 普通广播无法被终止
　　B. 有序广播按照接收器的优先级顺序将广播发送到接收器
　　C. 普通广播和有序广播都可以使用静态注册
　　D. 动态注册指的是将广播接收器配置在 AndroidManifest.xml 文件中

2. 简答题

(1) 简述 Service 的生命周期有哪两种，分别对应哪些生命周期方法。
(2) 简要说明 startService 与 bindService 两种方式对 Service 组件的影响有哪些不同。
(3) Android 中的广播有哪些分类？使用什么方法发送这些广播？
(4) 对比普通广播和有序广播，请简要说明二者的区别有哪些。

第 9 章　数据存储与 ContentProvider

本章学习目标
- 掌握 SharedPreferences 接口的使用。
- 掌握读写 SD 卡中文件的方式。
- 掌握 SQLiteDatabase 增删改查操作。
- 掌握 ContentProvider 类的使用。
- 了解访问手机联系人的基本操作。

Android 应用程序能以文件或数据库的形式存储数据，前者可以是普通文件，也可以是 XML 文件，后者是 SQLite 数据库，这是一种 Android 系统内置的数据库，没有后台服务，是真正的轻量级数据库，整个数据库对应一个文件。Android 读写 XML 文件需要借助 SharedPreferences，读写普通文件需要借助 J2SE 中 I/O 流的技术，读写数据库需要借助 SQLiteDatabase 和 SQLiteOpenHelper。如果文件或数据库不是位于手机内部存储，而是在 SD 卡中，还需要了解 Android 如何控制 SD 卡的访问。

如果需要访问其他应用程序中的数据，或者需要向其他应用程序提供数据访问接口，都需要借助 ContentProvider 类。它是对不同应用程序中数据访问的统一封装，屏蔽不同应用程序中数据格式的差异，采用统一的 API 执行读写操作。ContentProvider 也是 Android 应用开发中的四大组件之一。

9.1　以文件形式存储数据

文件通常分为文本文件和二进制文件，读写这些文件可以直接使用 I/O 技术。Android 平台将 J2SE 平台中有关文件操作的 API 直接移植过来，并增加了一些专门的 I/O API，使其可以更好地操作文件读写。

9.1.1　读写 XML 文件

Android 平台专门用于读写 XML 文件的是 SharedPreferences 类，它按照键-值对的格式存储简单数据。SharedPreferences 是一个接口，位于 android.content 包中。它的实现对象可以通过 Context 对象的 getSharedPreferences(String name, int mode)方法获取，参数 name 是 XML 文件的名称，mode 的取值如下：
- Context.MODE_PRIVATE，指定数据只能由本应用程序读写，这是比较常用的一个取值。

- Context.MODE_WORLD_READABLE，指定数据可以被其他应用程序读取。
- Context.MODE_WORLD_WRITEABLE，指定数据可以被其他应用程序读写。
- Context.MODE_MULTI_PROCESS，多线程访问标志。SDK 2.3 之后如果要进行多线程访问，需要明确设定该参数。

Activity 类是 Context 类的间接子类，因此在 Activity 中可以直接使用 getSharedPreferences 方法。SharedPreferences 接口中的常用方法如表 9.1 所示。

表 9.1 SharedPreferences 常用方法

方法	说明	返回值类型
contains(String key)	判断是否包含特定 key 的数据	boolean
edit()	获取内部 Editor 对象，用于写入数据	Editor
getAll()	获取所有的键-值对	Map<String,？>
getString(String key, String defValue)	获取 key 所对应的数据，如果 key 不存在，则返回 defValue	String

SharedPreferences 读取数据的方法还有 getBoolean、getInt 等，分别对应读取不同类型的数据，使用方式与 getString 类似。SharedPreferences 无法直接保存数据，必须借助它的内部类 Editor。Editor 接口常用的方法如表 9.2 所示。

表 9.2 Editor 接口常用方法

方法	说明	返回值类型
clear()	清空所有数据	Editor
commit()	提交修改，保存数据之后必须执行该方法	boolean
putString(String key, String value)	保存 String 类型的数据，key 为键，value 为需要保存的值	Editor
remove(String key)	移除 key 所对应的数据	Editor

Editor 保存数据的方法还有 putBoolean、putFloat、putInt 等，分别对应不同类型的基本数据，使用方法与 putString 类似。

下面的代码演示了 SharedPreferences 读写 XML 文件，布局文件 activity_main.xml 比较简单，包括两个 EditText 和两个 Button 控件。

📖 Proj09_1 项目 MainActivity.java 文件

```java
public class MainActivity extends Activity {
    private EditText et1,et2;
    private Button bt1,bt2;
    @Override
    protected void onCreate(Bundle savedInstanceState) {
        super.onCreate(savedInstanceState);
        setContentView(R.layout.activity_main);
        et1 = (EditText) findViewById(R.id.editText1);
        et2 = (EditText) findViewById(R.id.editText2);
        bt1 = (Button) findViewById(R.id.button1);
        bt2 = (Button) findViewById(R.id.button2);
        //button1 添加监听器,保存数据
```

```java
        bt1.setOnClickListener(new OnClickListener(){
            @Override
            public void onClick(View arg0) {
                //调用保存数据
                saveData(et1.getText().toString(),et2.getText().toString());
            }
        });
        //button2 添加监听器,读取数据
        bt2.setOnClickListener(new OnClickListener(){
            @Override
            public void onClick(View arg0) {
                readData();
            }
        });
    }
    //使用 SharedPreferences 保存数据
    public void saveData(String d1,String d2){
        SharedPreferences sp = getSharedPreferences("temp", Context.MODE_PRIVATE);
        Editor editor = sp.edit();                //获取保存数据的编辑器
        //保存数据
        editor.putString("name", d1);
        editor.putString("pwd", d2);
        editor.commit();                          //提交保存
    }
    //使用 SharedPreferences 读取数据
    public void readData(){
        SharedPreferences sp = getSharedPreferences("temp", Context.MODE_PRIVATE);
        //读取数据
        String name = sp.getString("name", "为找到需要的数据");
        String pwd = sp.getString("pwd", "为找到需要的数据");
        //显示读取到的数据
        et1.setText(name);
        et2.setText(pwd);
    }
}
```

运行上述代码,会在手机内部存储中创建 temp.xml 文件。选择 Window 菜单→Open Perspective→DDMS 视图,选择 File Explore 选项卡,会列出手机模拟器中的文件。展开 data 文件夹中的 data 文件夹,找到本应用程序包名所对应的文件夹,如图 9.1 所示。展开 share_prefs 就会 SharedPreferences 保存文件的位置。单击右上角的图标 ,从 DDMS 视图切换为 Java 视图。

图 9.1 SharedPreferences 保存文件路径

选择 temp.xml 文件,单击右上角的图标 ■ ■ ,选择 pull a file from the device,可以把文件从手机模拟器中复制到计算机中。打开该文件,其内容如下。

📖 Proj09_1 项目 temp.xml 文件

```xml
<?xml version = '1.0' encoding = 'utf-8' standalone = 'yes' ?>
<map>
    <string name = "pwd">root</string>
    <string name = "name">admin</string>
</map>
```

从文件内容可以看出,SharedPreferences 以 XML 格式组织数据,根元素为 map,元素标签是数据类型,如 string、int、float 等,元素属性 name 的取值是数据对应的 key,元素体是数据。

9.1.2 读写普通文件

Android 系统读写 XML 文件使用 SharedPreferences 类,其他格式文件的读写,包括文本文件和二进制文件,都需要借助 JDK I/O 流 API 实现。这些文件既可以位于手机内部存储中,也可以位于手机外部存储(即 SD 卡)中。读写手机内部存储中的文件时,可以通过 Context 提供的两个方法获取文件输入输出流。

- FileInputStream openFileInput(String name),获取输入文件流。name 是文件名,不能含有路径分隔符。
- FileOutputStream openFileOutput(String name, int mode),获取输出文件流。name 是文件名,不能含有路径分隔符。mode 是占用文件的模式,与 SharedPreferences 中的模式类似,但可以使用 MODE_APPEND 表示追加内容。

下面的程序演示上述两个方法的使用,获取文件输入输出流,向手机内部存储中保存登录日期。布局文件 activity_main.xml 包含两个 Button 控件和一个 TextView 控件。

📖 Proj09_2 项目 MainActivity.java 文件

```java
public class MainActivity extends Activity {
    private Button bt1,bt2;
    private TextView tv;
    @Override
    protected void onCreate(Bundle savedInstanceState) {
        super.onCreate(savedInstanceState);
        setContentView(R.layout.activity_main);
        bt1 = (Button) findViewById(R.id.button1);
        bt2 = (Button) findViewById(R.id.button2);
        tv = (TextView) findViewById(R.id.textView1);
        //添加监听器
        bt1.setOnClickListener(new OnClickListener(){
            @Override
            public void onClick(View arg0) {
                FileOutputStream fos = null;
                PrintWriter pw = null;
```

```java
            try {
                //打开文件输出流
                fos = openFileOutput("temp.sav", Context.MODE_PRIVATE);
                //字节输出流转换为字符输出流
                pw = new PrintWriter(fos);
                SimpleDateFormat sdf = new SimpleDateFormat("yyyy - MM - dd HH:mm:ss");
                pw.println(sdf.format(new Date()));
                pw.flush();
            } catch (FileNotFoundException e) {
                e.printStackTrace();
            }finally{   pw.close();}
        }
    });
    //添加监听器
    bt2.setOnClickListener(new OnClickListener(){
        @Override
        public void onClick(View arg0) {
            FileInputStream fis = null;
            BufferedReader br = null;
            try {
                fis = openFileInput("temp.sav");
                //字节输入流转换为字符输入流
                br = new BufferedReader(new InputStreamReader(fis));
                String txt = br.readLine();
                tv.setText(txt);
            } catch (FileNotFoundException e) {
                e.printStackTrace();
            } catch (IOException e) {
                e.printStackTrace();
            }finally{
                try {   if(br!= null)br.close(); } catch (IOException e) {
                    e.printStackTrace();
                }
            }
        }
    });
  }
}
```

运行上述程序,会在程序文件夹下创建 temp.sav 文件。在 DDMS 视图下,可以查看到该文件的详细路径,如图 9.2 所示。

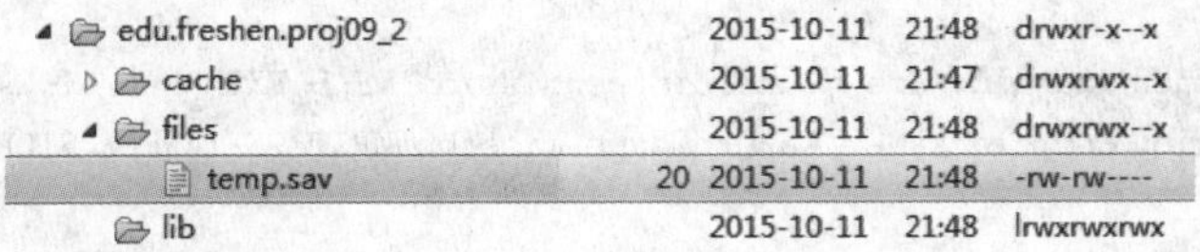

图 9.2 内部存储文件的保存路径

与 SharedPreferences 保存文件的路径不同，temp.sav 存放在 files 文件夹中，路径信息是固定的，默认为 /data/data/包名/files/文件名。

9.1.3 读写 SD 中的文件

Android 系统支持设备插入 SD 卡（或称外部存储媒介、外部储存设备），用以扩展手机的存储空间，但 SD 卡对于设备的运行并不是必需的。因此在读写 SD 卡时，需要先判断设备中是否已插入 SD 卡，并且该 SD 卡处于可用状态（如插入 SD 卡，但未格式化，则处于不可用状态）。

不同设备对 SD 卡的管理方式不同，其路径信息也不统一。Android 系统使用 Environment 类（位于 android.os 包中）获取 SD 卡状态信息和路径信息，从而屏蔽不同设备的差异。读写 SD 的程序可以按照下面的步骤进行。

(1) 使用 Environment.getExternalStorageState 方法，获取 SD 卡状态信息，返回值为状态字符串。返回字符串取值及意义如表 9.3 所示。

表 9.3　SD 卡状态字符串

状态字符串	类型	说明
MEDIA_BAD_REMOVAL	String	存储媒介已被移除
MEDIA_CHECKING	String	存储媒介正在检查
MEDIA_MOUNTED	String	存储媒介已插入，处于可用状态
MEDIA_MOUNTED_READ_ONLY	String	存储媒介已插入，处于只读模式
MEDIA_NOFS	String	存储媒介文件系统无法识别
MEDIA_REMOVED	String	存储媒介已移除
MEDIA_SHARED	String	存储媒介处于共享模式，比如 USB 模式
MEDIA_UNKNOWN	String	无法识别存储媒介
MEDIA_UNMOUNTABLE	String	存储媒介无法安装挂载
MEDIA_UNMOUNTED	String	没有存储媒介

(2) 当 SD 卡的状态处于 MEDIA_MOUNTED 时，可以对其进行读写操作。通过 Environment.getExternalStorageDirectory 方法获取 SD 卡根路径，返回值是 File。

(3) 当成功获取 SD 卡根路径 File 后，读写文件的方式与 J2SE 中 I/O 流操作文件一致，可以使用字节流和字符流。

(4) 运行程序前，需要在 AndroidManifest.xml 文件申请对应权限。具备写权限时，自动具备读权限。SD 卡读写权限如下：

```
<uses-permission android:name="android.permission.READ_EXTERNAL_STORAGE"/>
<uses-permission android:name="android.permission.WRITE_EXTERNAL_STORAGE"/>
```

下面的程序演示了 SD 读写操作，实现手机截屏，保存到 SD 卡，并将截屏图片显示在 ImageView 控件中。布局文件 activity_main.xml 包含两个 Button 控件，用于截屏和显示，一个 ImageView 控件，用于显示图片。

📖 Proj09_3 项目 MainActivity.java 文件

```java
public class MainActivity extends Activity {
    private Button bt1,bt2;
    private ImageView iv;
    private File root;                          //SD卡的根路径
    @Override
    protected void onCreate(Bundle savedInstanceState) {
        super.onCreate(savedInstanceState);
        setContentView(R.layout.activity_main);
        bt1 = (Button) findViewById(R.id.button1);
        bt2 = (Button) findViewById(R.id.button2);
        iv = (ImageView) findViewById(R.id.imageView1);
        BtListener bl = new BtListener();
        //添加监听器
        bt1.setOnClickListener(bl);
        bt2.setOnClickListener(bl);
        Log.i("Msg", Environment.getExternalStorageState());
        //获取SD卡的根路径
          if(Environment.getExternalStorageState().equalsIgnoreCase(Environment.MEDIA_MOUNTED)){
            root = Environment.getExternalStorageDirectory();
        }
    }
    //内部类,实现按钮的监听
    class BtListener implements OnClickListener{
        @Override
        public void onClick(View vw) {
            //判定是否得到SD卡根路径
            if(root == null){
                Toast.makeText(MainActivity.this, "SD卡不可用", Toast.LENGTH_LONG).show();
                return;
            }
            switch(vw.getId()){
            case R.id.button1:                  //截屏操作
                screenShot();break;
            case R.id.button2:                  //读取截屏图片
                loadScreen();
            }
        }
    }
    //截屏方法
    public void screenShot(){
        View view = bt1.getRootView();          //获取按钮控件所在根视图
        view.setDrawingCacheEnabled(true);      //缓存视图
        view.buildDrawingCache();
        Bitmap bitmap = view.getDrawingCache(); //从缓存绘制位图
        FileOutputStream fos = null;
        try {                                   //建立文件输出流
```

```
            fos = new FileOutputStream(root.getAbsoluteFile() + "/screen.png");
            bitmap.compress(Bitmap.CompressFormat.PNG, 100,fos);
            Toast.makeText(this, "截屏已保存", Toast.LENGTH_SHORT).show();
        } catch (FileNotFoundException e) {
            e.printStackTrace();
        }
    }
    //加载截屏图片
    public void loadScreen(){
        String filePath = root.getAbsoluteFile() + "/screen.png";
        Bitmap bitmap = BitmapFactory.decodeFile(filePath);
        iv.setImageBitmap(bitmap);
    }
}
```

运行上述程序前,需要在配置文件 AndroidManifest.xml,声明读写 SD 卡的权限。运行该程序的手机或模拟器必须具备 SD 卡,并处于可用状态。执行截屏后,生成的位图保存在 SD 卡根路径下,文件名是 screen.png,如图 9.3 所示。

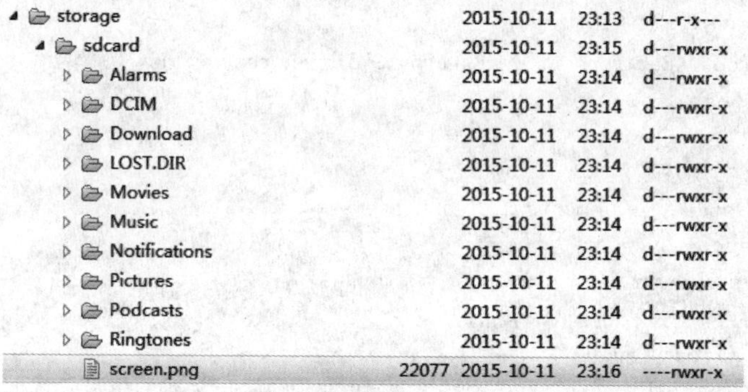

图 9.3 截屏保存路径位图

9.2 以数据库形式存储数据

使用文件系统存储图片和结果较为简单的数据比较方便,但无法保存数据间的关系。使用数据库保存结构化数据,既可以存储数据,还可以维持数据间的关系。Android 系统采用的数据是 SQLite,这是一种嵌入式的关系型数据库。它即不同于 Oracle、MySQL、SQL Server 等数据库,需要启动服务;又可以像它们一样支持 SQL 语句。SQLite 实际上是一个文件,后缀名通常为.db、.db2 或.db3 等,它占用的资源非常少,适合移动设备暂存数据。Android 系统提供了丰富 API,可以比较方便地操作 SQLite 数据库。

9.2.1 SQLiteDatabase 介绍

SQLiteDatabase 位于 android.database.sqlite 包,被 final 修饰,不可以被继承。它本身就表示一个 SQLite 数据库,同时拥有创建表、插入、查询、更新、删除等操作。获取

SQLiteDatabase 对象有两种方式：

（1）执行 Context 对象中的方法，返回值都是 SQLiteDatabase。
- openOrCreateDatabase（String name，int mode，SQLiteDatabase.CursorFactory factory）
- openOrCreateDatabase（String name，int mode，SQLiteDatabase.CursorFactory factory，DatabaseErrorHandler errorHandler）

（2）执行 SQLiteDatabase 类中的静态方法，返回值都是 SQLiteDatabase。
- openDatabase（String path，SQLiteDatabase.CursorFactory factory，int flags，DatabaseErrorHandler errorHandler）
- openDatabase（String path，SQLiteDatabase.CursorFactory factory，int flags）
- openOrCreateDatabase（String path，SQLiteDatabase.CursorFactory factory，DatabaseErrorHandler errorHandler）
- openOrCreateDatabase（String path，SQLiteDatabase.CursorFactory factory）
- openOrCreateDatabase（File file，SQLiteDatabase.CursorFactory factory）

SQLiteDatabase 常用方法如表 9.4 所示。

表 9.4 SQLiteDatabase 常用方法

方 法 名	说 明	返回值类型
beginTransaction()	开始自动控制事务	void
setTransactionSuccessful()	自动事务成功完成	void
endTransaction()	结束自动控制事务	void
delete（String table，String whereClause，String[] whereArgs）	相当于 delete 语句，由参数组合实现删除记录操作	int
insert（String table，String nullColumnHack，ContentValues values）	相当于 insert 语句，由参数组合实现插入记录操作	long
update（String table，ContentValues values，String whereClause，String[] whereArgs）	相当于 update 语句，由参数组合实现更新记录操作	int
execSQL（String sql）	执行 SQL 语句，该 SQL 不能有返回数据	void
query（String table，String[] columns，String selection，String[] selectionArgs，String groupBy，String having，String orderBy，String limit）	相当于 select 语句，参数含义如下： table 指待操作的表； columns 指待选列； selection 指 where 字句； selectionArgs 指 where 字句参数； groupBy 指分组； having 指分组限制条件； orderBy 指排序条件； limit 指限制查询记录分页	Cursor

Android 系统请求数据库的方法与 JDBC 操作数据访问类似，只是将原生态的 SQL 语句拆分成对应的方法，语句中的表、字段、条件等都以参数的方式传入。在 Android 开发文档中建议使用上述方式实现增删改查操作。

数据库操作方法中还涉及 ContentValues 和 Cursor。前者类似一个 Map，可以直接创建,字段名为"键",字段对应的数据为"值"。后者是查询数据时产生的一个结果集,如同 JDBC 中的 ResultSet,可以通过游标上下移动读取记录。

9.2.2 执行增删改操作

通常对数据库执行的操作有 4 种,分别是向数据库某表中插入记录、删除记录、更新记录和查询符合条件的记录,即增、删、改、查操作。与操作对应的 SQL 语句分别是 insert、delete、update 和 select。4 种操作中比较复杂的是"查"操作,尤其是需要根据条件进行查询时,该部分知识在 9.2.3 节会专门介绍。

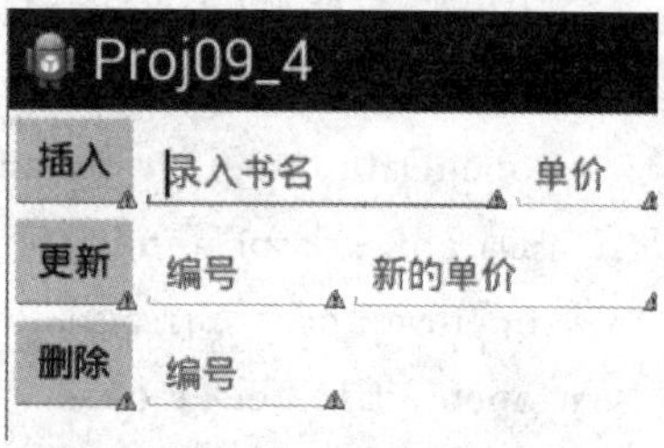

下面的项目演示了 SQLiteDatabase 数据库增、删、改 3 种操作,项目采用布局文件 activity_main.xml,设计样式如图 9.4 所示。

图 9.4 数据库基本操作 UI 设计

📖 Proj09_4 项目 MainActivity.java 文件

```java
public class MainActivity extends Activity {
    //布局文件中的控件
    private Button bt1,bt2,bt3;
    private EditText et1,et2,et3,et4,et5;
    //数据库引用
    SQLiteDatabase db;
    @Override
    protected void onCreate(Bundle savedInstanceState) {
        super.onCreate(savedInstanceState);
        setContentView(R.layout.activity_main);
        bt1 = (Button) findViewById(R.id.button1);
        bt2 = (Button) findViewById(R.id.button2);
        bt3 = (Button) findViewById(R.id.button3);
        et1 = (EditText) findViewById(R.id.editText1);
        et2 = (EditText) findViewById(R.id.editText2);
        et3 = (EditText) findViewById(R.id.editText3);
        et4 = (EditText) findViewById(R.id.editText4);
        et5 = (EditText) findViewById(R.id.editText5);
        //给按钮添加监听器
        BtListener bl = new BtListener();
        bt1.setOnClickListener(bl);
        bt2.setOnClickListener(bl);
        bt3.setOnClickListener(bl);
        //初始化数据库
        initDB();
    }
    //创建数据库并创建表
    private void initDB() {
        //创建数据库
        db = openOrCreateDatabase("bk.db", Context.MODE_PRIVATE, null);
        //准备创建数据库的 SQL 语句
```

```java
        String sql = "create table tb_bookInfo(_id integer primary key autoincrement," +
                "bookName varchar,bookPrice float)";
        //执行 SQL,创建数据表,同一个数据库中禁止出现同名表
        db.execSQL(sql);
        Log.i("Msg", "数据库已经创建完成");
    }
    //插入操作
    public long insertBook(){
        //创建 ContentValues 对象,并封装需要插入的数据
        ContentValues cv = new ContentValues();
        cv.put("bookName", et1.getText().toString());
        cv.put("bookPrice", et2.getText().toString());
        //执行插入,并得到 id
        long id = db.insert("tb_bookInfo", null, cv);
        Log.i("Msg", "插入数据完成   id = " + id);
        return id;
    }
    //更新操作
    public int updateBook(){
        //创建 ContentValues 对象,并封装需要更新的数据
        ContentValues cv = new ContentValues();
        cv.put("bookPrice", et4.getText().toString());
        //执行插入,返回影响的记录数
        int n = db.update("tb_bookInfo", cv, "_id = ?", new String[]{et3.getText().toString()});
        Log.i("Msg", "更新数据完成   n = " + n);
        return n;
    }
    //删除操作
    public int deleteBook(){
        int n = db.delete("tb_bookInfo", "_id = ?",  new String[]{et5.getText().toString()});
        Log.i("Msg", "删除数据完成   n = " + n);
        return n;
    }
    //按钮监听器类
    class BtListener implements OnClickListener{
        @Override
        public void onClick(View arg0) {
            switch(arg0.getId()){
                case R.id.button1:insertBook();break;
                case R.id.button2:updateBook();break;
                case R.id.button3:deleteBook();break;
            }
        }
    }
}
```

初始化数据库要完成两项工作：创建数据库文件和创建数据库表。这两项工作只需要在用户第一次运行该程序时执行，如果数据库已经存在，再创建同名表，会抛出 SQLiteException。所以上面的程序第一次可以正常运行,第二次就无法运行了。重新运行

项目不会删除已经创建的文件,除非卸载软件。为了解决这个问题,可以在创建数据库之前进行逻辑判定,或者将数据库的创建放入 try…catch。这两种方案虽然能保证程序正常运行,但都不是很优雅。在 9.2.4 节会介绍 SQLiteOpenHelper 类,它将很好地解决这个问题。

上述代码创建数据库时调用的 Context 中的 openOrCreateDatabase 方法,它会在手机内部存储中创建数据库,如图 9.5 所示。如果需要在 SD 卡中创建数据库,可以使用 SQLiteDatabase 静态方法 openDatabase,采用如下代码:

```
File root = Environment.getExternalStorageDirectory();
File dbFile = new File(root,"/bk.db");
try {
    dbFile.createNewFile();           //创建数据库文件
    db = SQLiteDatabase.openDatabase(dbFile.getAbsolutePath(),
                       null,SQLiteDatabase.OPEN_READWRITE);
    //准备创建数据库的 SQL 语句
    String sql = "create table tb_bookInfo(_id integer primary key autoincrement," +
            "bookName varchar,bookPrice float)";
    //执行 SQL,创建数据表,同一个数据库中禁止出现同名表
    db.execSQL(sql);
    Log.i("Msg", "数据库已经创建完成");
} catch (IOException e) {
    e.printStackTrace();
}
```

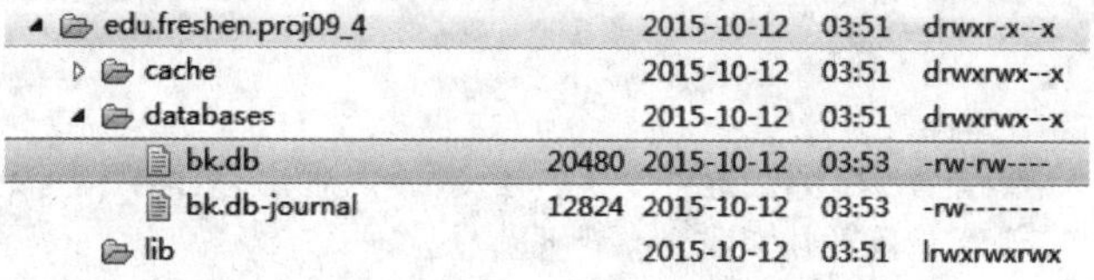

图 9.5　SQLiteDatabase 文件路径

插入和更新操作需要创建 ContentValues 对象,封装数据时 key 必须与数据库表中的字段一致。插入方法的返回值表示该记录对应的行号,更新和删除操作的返回值表示影响的记录条数。

插入操作的 SQL 语句通常是"insert into [tableName](字段1,字段2,…)values(值1,值2,…)"。更新操作的 SQL 语句通常是"update [tableName] set 字段1=值1,字段2=值2,…where 条件字段=值"。删除操作的 SQL 语句通常是"delete from [tableName] where 条件字段=值"。将这些语句拆分,字段、取值和条件就是操作方法对应的参数。

9.2.3　Cursor 与查询操作

SQLiteDatabase 执行查询操作时调用 query 方法,其返回值是 Cursor(游标)对象。Cursor 是一个接口,位于 android.database 包中。Cursor 对象包含查询的结果集,可以使用指针上下移动访问每一条记录。Cursor 的常用方法如表 9.5 所示。

表 9.5 Cursor 常用方法

方法	说明	返回值类型
close()	关闭结果集,内部数据将被清空	void
getBlob(int columnIndex)	获取指定列的数据,列下标从 0 开始计数	byte[]
getColumnCount()	获取列数	int
getColumnIndex(String columnName)	获取指定字段名对应列的下标	int
getColumnName(int columnIndex)	获取指定下标的列的字段名	String
getColumnNames()	获取所有字段名	String[]
getCount()	获取结果集中的行数	int
getXxx(int columnIndex)	获取指定列的数据,类型有 int、float 等	Xxx
getString(int columnIndex)	获取指定列的数据,返回字符串类型	String
isAfterLast()	判定当前游标是否位于最后一行之后	boolean
isBeforeFirst()	判定当前游标是否位于第一行之前	boolean
isFirst()	判定游标是否指向第一行	boolean
isLast()	判定游标是否指向最后一行	boolean
move(int offset)	从当前位置移动游标	boolean
moveToFirst()	游标移动到第一行	boolean
moveToLast()	游标移动到最后一行	boolean
moveToNext()	游标向后移动一次	boolean
moveToPrevious()	游标向前移动一次	boolean

在项目 Proj09_4 中添加查询功能代码。首先,在布局文件中添加查询按钮和显示结果的 TextView 控件,如图 9.6 所示。其次,在 MainActivity.java 中初始化控件,添加监听器并实现查询操作,代码如下。

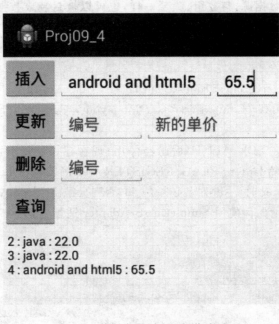

图 9.6 数据库增删改查操作演示

📖 Proj09_4 项目 MainActivity.java 文件

```java
//查询操作
public void queryBook(){
    Cursor c = db.query("tb_bookInfo", new String[]{"_id","bookName","bookPrice"},
                                            null, null, null, null, null);
    if(c == null){
        Toast.makeText(this, "查询结果错误!", Toast.LENGTH_SHORT).show();
        return;
    }
    StringBuffer bf = new StringBuffer();
    //遍历结果集
    while(c.moveToNext()){
        bf.append( c.getInt(c.getColumnIndex("_id")) + " : ");
        bf.append( c.getString(c.getColumnIndex("bookName")) + " : ");
        bf.append( c.getFloat(c.getColumnIndex("bookPrice")) + "\n");
    }
    tv.setText(bf.toString());          //显示查询结果
}
```

查询操作的 SQL 语句通常是"select 字段 1,字段 2,… from [tableName] where 条件字段=值 group by 字段 having 条件 order by 字段[AES/DESC] limit 分页",将该语句中的字段、取值和条件拆分,就是 query 方法中对应的参数。如果查询条件不需要设置,则参数赋值为 null。

在读取 Cursor 中的记录时,因为没有根据字段名直接读取对应数据的方法,只能先获取字段对应的列下标,然后再读取对应的数据,这一点与 JDBC 中的 ResultSet 不同。

Cursor 作为一种数据源,与数组、集合一样,也可以被封装成数据适配器,直接显示在 ListView 或 GridView 等控件中。CursorAdapter 位于 android.widget 包,继承了 BaseAdapter,是一个抽象类。比较常用的是它的间接子类 SimpleCursorAdapter,构造方法如下:

SimpleCursorAdapter(Context context, int layout, Cursor c, String[] from, int[] to)

SimpleCursorAdapter(Context context, int layout, Cursor c, String[] from, int[] to, int flags)

上面的构造方法在 SDK API 11 后已经停用,现在建议使用下面的构造方法。参数 flags 的取值为 CursorAdapter.FLAG_REGISTER_CONTENT_OBSERVER。

使用 CursorAdapter 时,主键列名必须是 _id,否则会抛出 IllegalArgumentException 异常。下面的代码演示了如何使用 SimpleCursorAdapter 适配器,列表布局文件是 listview_item.xml。

```java
//查询操作,将查询结果显示在 ListView 中
public void queryBookToListView(){
    Cursor c = db.query("tb_bookInfo", new String[]{"_id","bookName","bookPrice"},
                                            null, null, null, null, null);
    if(c == null){
        Toast.makeText(this, "查询结果错误!", Toast.LENGTH_SHORT).show();
```

```
        return;
    }
    //创建 CursorAdapter 适配器
    SimpleCursorAdapter ca = new SimpleCursorAdapter(this,R.layout.listview_item,c,
            new String[]{"_id","bookName","bookPrice"},
            new int[]{R.id.listview_item_id,R.id.listview_item_bname,
                                        R.id.listview_item_bprice},
            CursorAdapter.FLAG_REGISTER_CONTENT_OBSERVER);
    lv.setAdapter(ca);
}
```

9.2.4 SQLiteOpenHelper 的使用

SQLiteOpenHelper 位于 android.database.sqlite 包中，是一个抽象类，必须被继承才能使用，是创建 SQLiteDatabase 和管理 SQLiteDatabase 版本的帮助类。使用它可以避免前面章节所遇到的重复创建数据库和数据表的问题。

继承 SQLiteOpenHelper，至少需要重写如下两个构造方法之一：

SQLiteOpenHelper(Context context, String name, SQLiteDatabase.CursorFactory factory, int version)

SQLiteOpenHelper(Context context, String name, SQLiteDatabase.CursorFactory factory, int version, DatabaseErrorHandler errorHandler)

参数 context 是上下文环境，传入 Activity 即可。name 是数据库的名称。factory 用于创建 Cursor，如果输入 null，则使用默认配置。version 是数据库版本，若版本不同，则执行更新操作。errorHandler 是数据库出现错误时的处理。

SQLiteOpenHelper 常用方法如表 9.6 所示，其中抽象方法必须被实现。

表 9.6 SQLiteOpenHelper 常用方法

方法名	说明	返回值类型
onCreate(SQLiteDatabase db)	创建数据库时调用，是抽象方法，一般将创建表等初始化操作在该方法中执行	abstract void
onUpgrade(SQLiteDatabase db, int oldVersion, int newVersion)	版本更新时调用，是抽象方法	abstract void
getReadableDatabase()	创建或打开一个只读数据库	SQLiteDatabase
getWritableDatabase()	创建或打开一个读写数据库	SQLiteDatabase
close()	关闭打开的数据库	void

下面的代码演示了 SQLiteOpenHelper 的使用，与 Proj09_4 项目的功能一样，只是数据库的创建由 SQLiteOpenHelper 实现。

📖 Proj09_5 项目 MyDbHelper.java 文件

```
public class MyDbHelper extends SQLiteOpenHelper {
    public MyDbHelper(Context context, String name, CursorFactory factory, int version) {
        super(context, name, factory, version);
    }
```

```
@Override
public void onCreate(SQLiteDatabase db) {
    //准备创建数据库的 SQL 语句
    String sql = "create table tb_bookInfo(_id integer primary key autoincrement," +
            "bookName varchar,bookPrice float)";
    db.execSQL(sql);
}
@Override
public void onUpgrade(SQLiteDatabase db, int arg1, int arg2) {
    Log.i("Msg", "数据库更新!");
}
}
```

数据库表的创建在 onCreate 方法中执行,该方法只有在数据库第一次创建时执行,避免了重复创建同名表的问题。使用 SQLiteDatabase 数据库时,采用如下代码:

```
SQLiteDatabase db = new MyDbHelper(this,"dbnam.db",null,1).getWritableDatabase();
```

如果仅读取数据库中的内容,可以获取只读数据库,采用如下代码:

```
SQLiteDatabase db = new MyDbHelper(this,"dbnam.db",null,1).getReadableDatabase();
```

上述方式会在手机内部存储中创建数据库,如果需要在指定位置创建数据库,可以采用下面的代码:

```
File root = Environment.getExternalStorageDirectory();
File dbf = new File(root,"bk.db");
try {
    dbf.createNewFile();
} catch (IOException e) {
    e.printStackTrace();
}
//打开数据库
SQLiteDatabase  db = new MyDbHelper(this,dbf.getAbsolutePath(),null,1).getWritableDatabase();
```

9.3 SQLite 图形化查看工具

SQLite 数据库可以在应用程序中直接创建,也可以借助相应工具软件创建,然后引入到应用程序中。查看和管理 SQLite 数据库需要借助外部工具,与 Oracle、SQL Server 等数据库一样,SQLite 数据库也支持图形化的管理软件。sqlite3.exe 位于 Android SDK 的 tools 文件夹下,是基于命令窗口的管理软件。除此之外,比较常用的管理软件还有 SQLiteSpy 和 SQLiteExpert。

SQLiteSpy 也是用于管理 SQLite 数据库的工具,只有 2MB 大小。不同于 sqlite3.exe,它是图形化操作界面,如图 9.7 所示。通过 File 菜单下的 Open Database 就可以打开

SQLite 数据库,当然也可以创建 SQLite 数据库。该款软件的使用比较简单,在此不作说明。

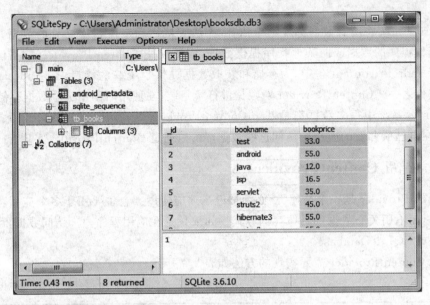

图 9.7 SQLiteSpy 软件管理界面

SQLiteExpert 管理软件比 SQLiteSpy 功能更加丰富,运行界面如图 9.8 所示,下载地址是 http://www.sqliteexpert.com/download.html。借助它可以很容易执行查看已有 SQLite 数据库、创建新数据库、新建表等操作。

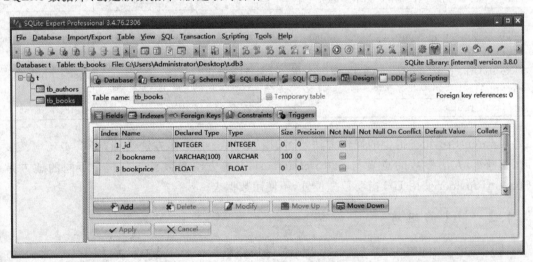

图 9.8 SQLiteExpert 软件管理界面

创建表结构时,注意 SQLite 数据库中的字段类型为 INTEGER(整形)、REAL(浮点型)、TEXT(文本)和 BLOB(二进制)类型,但也可以接受 varchar(n) 和 char(n) 类型,只不过在存储数据时会自动转换为上述类型。除了声明为 INTEGER Primary key 的主键只能为整形之外,SQLite 对类型的限制并不严格。

9.4 Content Provider

Android 系统将所有数据都规定为私有,无法直接访问应用程序之外的数据。如果需要访问其他程序的数据或向其他程序提供数据,需要使用 ContentProvider(数据供应器)。该类位于 android.content 包中,为存储和获取数据提供了统一的接口,是 Android 应用开发四大组件之一。ContentProvider 对数据进行了封装,在使用时不用关心数据存储方式。

Android 系统中很多数据都是基于 ContentProvider 方式访问,如视频、音频、图片、通讯录等,在自己所开发的应用中访问这些数据都可以通过 ContentProvider 实现。

9.4.1 使用 ContentProvider

ContentProvider 是一个抽象类,必须被继承才能使用。它主要用于多个应用程序之间访问数据,而不用关心数据的具体存储方式。如果只是在应用程序内部访问数据,可以使用文件系统或 SQLiteDatabase。

继承 ContentProvider 需要实现的方法如表 9.7 所示。

表 9.7 ContentProvider 抽象方法

方法	说明	返回值类型
onCreate()	ContentProvider 对象创建后,会自动执行该方法	abstract boolean
query(Uri uri, String[] projection, String selection, String[] selectionArgs, String sortOrder)	查询操作	abstract Cursor
insert(Uri uri, ContentValues values)	插入操作	abstract Uri
update(Uri uri, ContentValues values, String selection, String[] selectionArgs)	更新操作	abstract Uri
delete(Uri uri, String selection, String[] selectionArgs)	删除操作	abstract int
getType(Uri uri)	获取 MIME 类型	abstract String

ContentProvider 提供的操作方法与 SQLiteDatabase 操作方法类似,可以实现增、删、改、查操作,方法参数的意义与 SQLiteDatabase 中的方法参数意义一样。不同的地方是 ContentProvider 使用 Uri 指明数据位置,不使用数据表。

9.4.2 Uri 的组成

Uri 类位于 android.net 包中,代表了要操作的数据。一个 Uri 一般有 3 个组成部分,分别是访问协议、唯一性标识字符串(或访问权限字符串)、资源部分。图 9.9 是一个比较典型的 Uri,A 区域是访问协议,B 区域是唯一性标识,C 和 D 构成资源部分,其含义如下:

- A 为前缀标识,任何 Uri 都有一个固定的前缀,"content://"标明这是 ContentProvider 中使用的 Uri。
- B 为数据的权限标识,要求唯一。
- C 为路径标识,代表 Uri 所要访问的具体数据表。

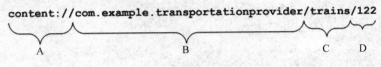

图 9.9 Uri 组成部分图例

- D 为 ID 信息，C 中数据表的某一个 ID 值。

将合法的 Uri 字符串转变为 Uri 对象，可以使用 Uri 类的静态方法 parse，如 Uri.parse("content://com.android.contacts/data")。Android 系统提供了两个用于操作 Uri 的工具类，分别为 UriMatcher 和 ContentUris，都位于 android.content 包中。

UriMatcher 类用于匹配 Uri 的检验。主要涉及方法 addURI(String authority, String path, int code) 和 match(Uri uri)。UriMatcher 检验 Uri 的代码如下：

```
//匹配检验
UriMatcher uMatcher = new UriMatcher(UriMatcher.NO_MATCH);
uMatcher.addURI("com.android.contacts", "data", 1);
int r = uMatcher.match(ContactsContract.Data.CONTENT_URI);
Log.i("Tag", "匹配结果:" + r);
```

创建 UriMatcher 对象，传入常量 NO_MATCH，当匹配不成功时会返回该常量的值。使用 addURI 方法添加匹配样式，这相当于模板，供待检验的 Uri 与此比对，可以多次调用此方法，添加多个模板。该方法有 3 个参数：第一个是 Uri 的数据权限；第二个是路径标识，可以使用通配符（*用于通配文本，#用于通配数字）；第三个参数是匹配成功时的返回值，通过与此值比对，可以判定是否匹配成功。match 方法用于匹配 Uri，若匹配成功，返回 addURI 方法中的第三个参数，若不成功，返回构造 UriMatcher 时传入的值。

ContentUris 类用于操作 Uri 路径后面的 ID 部分，它有两个比较实用的方法：

- static Uri withAppendedId(Uri contentUri, long id)，向 Uri 路径后面追加 id。
- static long parseId(Uri contentUri)，获取 Uri 中的 id 值。

9.4.3 ContentProvider 基本操作

ContentProvider 作为统一访问数据的接口，屏蔽了数据存储的差异，使得增、删、改、查操作时可以不用关注底层的具体存储方式。ContentProvider 的统一数据访问是对调用者而言的，对继承 ContentProvider，实现操作方法的类而言，数据存储方式必须被确认。

访问 ContentProvider 数据接口时，通常使用 ContentResolver 类，它位于 android.content 包中，是一个抽象类，可以通过 Context 类的 getContentResolver 方法获得实现对象。ContentResolver 类执行增、删、该、查操作的方法与 ContentProvider 类几乎一致，可以参考表 9.7。

下面通过两个应用程序项目 Proj09_6 和 Proj09_7 演示自定义 ContentProvider 的应用。在 Proj09_6 项目中创建 ContentProvider，并使用 SQLiteDatabase 实现数据的插入和查询操作。在 Proj09_7 项目中，通过 ContentResolver 访问 Proj09_6 中的数据。

1. 创建 ContentProvider，向外提供访问接口

数据存储方式可以是文件系统，也可以是数据库。在本项目中，数据存储采用

SQLiteDatabase,下面是数据库帮助类的实现代码。

📖 Proj09_6 项目 CostDbHelper.java 文件

```java
public class CostDbHelper extends SQLiteOpenHelper {
    public CostDbHelper(Context context, String name, CursorFactory factory, int version) {
        super(context, name, factory, version);
    }
    @Override
    public void onCreate(SQLiteDatabase db) {
        String sql = "create table tb_costInfo(_id integer primary key autoincrement," +
                "costPro varchar,costValue float)";
        db.execSQL(sql);
    }
    @Override
    public void onUpgrade(SQLiteDatabase db, int oldVersion, int newVersion) {
        db.execSQL("drop table if exists tb_costInfo");
    }
}
```

自定义 ContentProvider 需要继承 ContentProvider,并实现其中的抽象方法,具体代码如下。静态常量 AUTHORITY 作为 Uri 唯一性标识,与配置文件中的权限字符串一样。静态常量_ID、COSTPRO、COSTVALUE 是数据字段,它们的取值与数据库表中的字段一致。静态常量 CONTENT_URI 是访问本 ContentProvider 的 Uri。

📖 Proj09_6 项目 ContentProvider.java 文件

```java
public class CostProvider extends ContentProvider {
    //唯一性标识,作为授权标识符
    public final static String AUTHORITY = "edu.freshen.contentProvider.COSTPROVIDER";
    //数据字段
    public final static String _ID = "_id";
    public final static String COSTPRO = "costPro";
    public final static String COSTVALUE = "costValue";
    //数据 Uri
    public final static Uri CONTENT_URI = Uri.parse("content://" + AUTHORITY);
    //数据库
    SQLiteDatabase db;
    CostDbHelper dbHelper;
    //删除操作
    @Override
    public int delete(Uri uri, String selection, String[] selectionArgs) {
        return 0;
    }
    //获取类型
    @Override
    public String getType(Uri uri) {
        Log.i("Msg", "From CostProvider getType!  uri = " + uri);
        return null;
    }
```

```java
//插入操作
@Override
public Uri insert(Uri uri, ContentValues cv) {
    db = dbHelper.getWritableDatabase();
    long rowId = db.insert("tb_costInfo", null, cv);
    if(rowId > 0){
        Uri rowUri = ContentUris.withAppendedId(CONTENT_URI, rowId);
        Log.i("Msg", "CostProvider insert! rowId = " + rowId);
        return rowUri;
    }
    return null;
}
//创建数据存储
@Override
public boolean onCreate() {
    dbHelper = new CostDbHelper(this.getContext(),"costDb.db",null,1);
    return dbHelper == null?false:true;
}
//查询操作
@Override
  public Cursor query (Uri uri, String [ ] projection, String selection, String [ ] selectionArgs, String sortOrder) {
    db = dbHelper.getReadableDatabase();
    Cursor c = db.query("tb_costInfo", projection, selection, selectionArgs,
                                               null, null, sortOrder);
    return c;
}
//更新操作
@Override
public int update(Uri uri, ContentValues cv, String selection, String[] selectionArgs) {
    return 0;
}
}
```

上面的自定义 ContentProvider 仅实现了插入和查询操作，感兴趣的读者可以将更新和删除操作补充完整。

自定义的 ContentProvider 需要在配置文件中进行注册，使用标签 provider。属性 name 指明 ContentProvider 的全路径，属性 authorities 是访问标识，属性 exported 是允许外部应用访问。具体配置信息如下。

📖 Proj09_6 项目 AndroidManifest.xml 文件

```xml
< manifest xmlns:android = "http://schemas.android.com/apk/res/android"
    package = "edu.frehen.proj09_6"
    android:versionCode = "1"
    android:versionName = "1.0" >
    < uses - sdk
        android:minSdkVersion = "14"
```

```xml
            android:targetSdkVersion = "18" />
    <application
        android:allowBackup = "true"
        android:icon = "@drawable/ic_launcher"
        android:label = "@string/app_name"
        android:theme = "@style/AppTheme" >
        <activity
            android:name = "edu.frehen.proj09_6.MainActivity"
            android:label = "@string/app_name" >
            <intent-filter>
                <action android:name = "android.intent.action.MAIN" />
                <category android:name = "android.intent.category.LAUNCHER" />
            </intent-filter>
        </activity>
        <provider
            android:name = "edu.frehen.proj09_6.CostProvider"
            android:authorities = "edu.freshen.contentProvider.COSTPROVIDER"
            android:exported = "true"
            />
    </application>
</manifest>
```

运行上述项目,创建自定义 ContentProvider,等待被访问。

2. 调用项目 Proj09_6 中的 ContentProvider 对数据进行访问

项目 Proj09_7 通过 ContentResolver 访问 Proj09_6 中的 ContentProvider,布局文件是 activity_main.xml,具体代码如下,运行效果可以参考图 9.10。Spinner 控件使用 entries 属性指定下拉列表中呈现的数组,两个 Button 控件使用 onClick 属性指定监听方法。

📖 Proj09_7 项目 activity_main.xml 文件

```xml
<RelativeLayout xmlns:android = "http://schemas.android.com/apk/res/android"
    xmlns:tools = "http://schemas.android.com/tools"
    android:layout_width = "match_parent"
    android:layout_height = "match_parent"
    tools:context = ".MainActivity" >
    <Spinner
        android:id = "@+id/spinner1"
        android:layout_width = "120dp"
        android:layout_height = "wrap_content"
        android:entries = "@array/costPro"/>
    <EditText
        android:id = "@+id/et01"
        android:layout_width = "60dp"
        android:layout_height = "wrap_content"
        android:singleLine = "true"
        android:inputType = "numberDecimal"
        android:layout_toRightOf = "@id/spinner1"
        android:layout_alignBottom = "@id/spinner1" />
    <Button
```

```xml
        android:id = "@ + id/bt01"
        android:layout_width = "wrap_content"
        android:layout_height = "wrap_content"
        android:text = "增加"
        android:layout_toRightOf = "@id/et01"
        android:layout_alignBottom = "@id/et01"
        android:onClick = "saveCost"/>
    <Button
        android:id = "@ + id/bt02"
        android:layout_width = "wrap_content"
        android:layout_height = "wrap_content"
        android:text = "查询"
        android:layout_toRightOf = "@id/bt01"
        android:layout_alignBottom = "@id/et01"
        android:onClick = "queryCost" />
    <ListView
        android:id = "@ + id/listView1"
        android:layout_width = "match_parent"
        android:layout_height = "wrap_content"
        android:layout_below = "@id/spinner1" >
    </ListView>
</RelativeLayout>
```

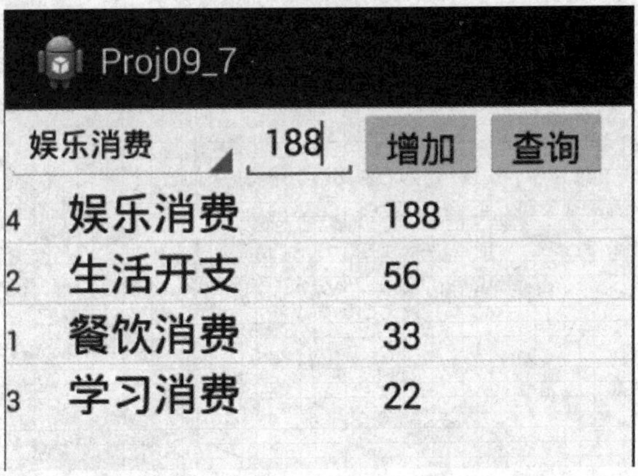

图 9.10 ContentProvider 运行界面

访问 ContentProvider 数据，Uri 使用 content://edu.freshen.contentProvider.COSTPROVIDER/，要与 Proj09_6 中对外声明的一致。具体代码如下。

📖 Proj09_7 项目 MainActivity.java 文件

```java
public class MainActivity extends Activity {
    Spinner sp;
    EditText et;
    ListView lv;
    //声明字段
    static final String _ID = "_id";
```

```java
        static final String COSTPRO = "costPro";
        static final String COSTVALUE = "costValue";
        Uri CONTENT_URI = Uri.parse("content://edu.freshen.contentProvider.COSTPROVIDER/");
        @Override
        protected void onCreate(Bundle savedInstanceState) {
            super.onCreate(savedInstanceState);
            setContentView(R.layout.activity_main);
            sp = (Spinner) findViewById(R.id.spinner1);
            et = (EditText) findViewById(R.id.et01);
            lv = (ListView) findViewById(R.id.listView1);
        }
        //插入数据
        public void saveCost(View v){
            ContentResolver cr = getContentResolver();
            ContentValues cv = new ContentValues();
            cv.put(COSTPRO, sp.getSelectedItem().toString());
            cv.put(COSTVALUE, et.getText().toString());
            Uri uri = CONTENT_URI;
            Uri rowUri = cr.insert(uri, cv);
        }
        //查询数据
        public void queryCost(View v){
            ContentResolver cr = getContentResolver();
            Uri uri = CONTENT_URI;
            //执行查找,按 COSTVALUE 字段降序排列
            Cursor cursor = cr.query(uri, new String[]{_ID,COSTPRO,COSTVALUE},
                    null,null, COSTVALUE + " DESC");
            //创建 SimpleCursorAdapter 适配器
            SimpleCursorAdapter sca = new SimpleCursorAdapter(
                    this,R.layout.list_costitem,cursor,
                    new String[]{_ID,COSTPRO,COSTVALUE},
                    new int[]{R.id.list_costitem_id,R.id.list_costitem_cost,R.id.list_costitem_value}
            );
            lv.setAdapter(sca);
        }
    }
```

运行项目 Proj09_7,执行"增加"和"查询"操作,可以对项目 Proj09_6 中的数据进行操作。最终效果如图 9.10 所示。

9.5 管理手机联系人信息

在常规 Android 应用开发中,使用 ContentProvider 的场合多是访问 Android 系统数据,如联系人信息、多媒体信息、日历信息等。在这些应用开发中,不需要自定义 ContentProvider,而是借助 ContentResolver 的 insert、delete、update 和 query 方法访问内部数据。

访问 Android 系统内部数据，需要了解这些内部 ContentProvider 的 Uri 构成，具体细节可以查看 Android API 文档，位于 android.provider 包中。本节以手机联系人信息管理为例，演示如何访问 Android 内部数据。

从 Android 2.0 SDK 开始有关联系人 ContentProvider 的类变成了 ContactsContract，虽然 android.provider.Contacts 还可以继续使用，但在 SDK 中标记为 deprecated，将其视为不推荐的方法。

ContactsContract 的子类 ContactsContract.Contacts 是一张表，代表了所有联系人的统计信息，如联系人 ID(_ID)、查询键(LOOKUP_KEY)、联系人的姓名(DISPLAY_NAME_PRIMARY)、头像的 id(PHOTO_ID)等。此外，联系人数据模型还涉及另外两个类，分别是 ContactsContact.Data 和 ContactsContact.RawContacts。前者存储通讯录中联系人的全部信息，如名字、电话、E-mail 等；后者存储联系人描述信息和唯一账号。手机联系人数据访问涉及的 Uri 如下：

ContactsContract.Contacts.CONTENT_URI
ContactsContract.RawContacts.CONTENT_URI
ContactsContract.Data.CONTENT_URI

应用程序采用的布局文件是 activity_main.xml，含有增删改查操作的 4 个按钮、相应的编辑文本框和显示联系人信息的 ListView，设计视图可以参考最终运行效果(图 9.11)。增删改查按钮分别调用 insertDate、deleteData、updateData 和 queryData 方法。

下面的代码实现查询联系人信息，将其联系人 id、姓名和第一个手机号码封装为 Map，加入集合 List < Map < String, String >> contactsData，并填充给列表控件。列表控件元素采用的布局文件是 list_contactitem.xml。

📖 Proj09_8 项目 MainActivity.java 文件

```java
//查询联系人
public void queryData(){
    Cursor contactCursor = null;
    contactsData.clear();
    //联系人 Uri
    Uri contactUri = ContactsContract.Contacts.CONTENT_URI;
    //查询指定 ID 联系人,条件查询
    if(et1.getText().toString().length()>0){
        contactCursor = getContentResolver()
                .query(contactUri,
                        new String[]{ContactsContract.Contacts._ID,
                        ContactsContract.Contacts.DISPLAY_NAME},
                        ContactsContract.Contacts._ID + " = ?",
                        new String[]{et1.getText().toString()},null);
    }else{
        //查询所有联系人
        contactCursor = getContentResolver()
                .query(contactUri,
                        new String[]{ContactsContract.Contacts._ID,
                        ContactsContract.Contacts.DISPLAY_NAME},
                        null,null,null);
    }
```

```java
            Log.i("Msg","查询联系人个数: " + contactCursor.getCount());
            //遍历查询结果
            while(contactCursor.moveToNext()){
                //封装联系人信息的集合
                Map<String,String> contactItem = new HashMap<String,String>();
                //读取联系人 id
                contactItem.put("contactId", contactCursor.getString(
                        contactCursor.getColumnIndex(ContactsContract.Contacts._ID)));
                //读取联系人姓名
                contactItem.put("contactName", contactCursor.getString(
                        contactCursor.getColumnIndex(ContactsContract.Contacts.DISPLAY_NAME)));
                //查询对应联系人的电话号码,一个联系人会有多个电话号码
                Uri phoneUri = ContactsContract.CommonDataKinds.Phone.CONTENT_URI;
                //根据联系人 id,在电话号码对应的 Uri 中条件查询电话号码
                Cursor phoneCursor = getContentResolver().query(phoneUri,
                        new String[]{ContactsContract.CommonDataKinds.Phone.NUMBER},
                        ContactsContract.CommonDataKinds.Phone.CONTACT_ID + " = ?",
                        new String[]{contactItem.get("contactId")}, null);
                //Log.i("Msg","查询联系人电话个数: " + phoneCursor.getCount());
                while(phoneCursor.moveToNext()){
                    //读取联系人电话号码,为了简化查询,此处只读取一个电话号码
                    contactItem.put("contactPhone",phoneCursor.getString(phoneCursor
                    .getColumnIndex(ContactsContract.CommonDataKinds.Phone.NUMBER)));
                    break;
                }
                phoneCursor.close();
                contactsData.add(contactItem);       //将联系人信息加入集合
            }
            contactCursor.close();
            //填充适配器
            initLv();
    }
    //填充列表控件
    private void initLv() {
        if(contactsData.size()<1)return;
        SimpleAdapter sa = new SimpleAdapter(this,
                contactsData,R.layout.list_contactitem,
                new String[]{"contactId","contactName","contactPhone"},
                new int[]{R.id.listview_item_id,R.id.listview_item_name,
                                    R.id.listview_item_phone});
        lv.setAdapter(sa);
    }
```

查询联系人的操作分两步进行,首先是查询 ContactsContract.Contacts.CONTENT_URI,获取联系人基本数据、id 和姓名,对应字段是 Contacts._ID 和 Contacts.DISPLAY_NAME。如果查询条件为输入,则查询所有联系人信息,否则查询对应联系人信息。

由于一个联系人会有多个不同类型电话号码,如移动电话、办公电话、家庭电话等,所有查询出的联系人,需要在 ContactsContract.CommonDataKinds.Phone.CONTENT_URI 查询具体电话号码数据,对应字段是 ContactsContract.CommonDataKinds.Phone.NUMBER。为了简化功能,上述代码在查询电话号码时只读取一个号码。

下面的代码实现插入联系人信息的功能。

📖 Proj09_8 项目 MainActivity.java 文件

```java
//插入联系人
public void insertData(){
    //插入联系人要分两步完成,生成联系人 id 和插入电话号码
    ContentValues cv = new ContentValues();
    Uri rawContactUri = getContentResolver().insert(
                        ContactsContract.RawContacts.CONTENT_URI, cv);
    //获取生成的联系人 id
    long rawId = ContentUris.parseId(rawContactUri);
    //步骤1: 插入姓名
    cv.put(ContactsContract.Data.RAW_CONTACT_ID, rawId);
    cv.put(ContactsContract.Data.MIMETYPE,
        ContactsContract.CommonDataKinds.StructuredName.CONTENT_ITEM_TYPE);
    cv.put(ContactsContract.CommonDataKinds.StructuredName.GIVEN_NAME,
                        et2.getText().toString());         //联系人姓名
    //执行插入
    getContentResolver().insert(ContactsContract.Data.CONTENT_URI, cv);
    //步骤2: 插入电话号码
    cv.clear();
    cv.put(ContactsContract.Data.RAW_CONTACT_ID, rawId);
    cv.put(ContactsContract.Data.MIMETYPE ,
                ContactsContract.CommonDataKinds.Phone.CONTENT_ITEM_TYPE);
    cv.put(ContactsContract.CommonDataKinds.Phone.NUMBER,
                        et3.getText().toString());         //电话号码
    cv.put(ContactsContract.CommonDataKinds.Phone.TYPE,
            ContactsContract.CommonDataKinds.Phone.TYPE_MOBILE);  //号码类型
    getContentResolver().insert(ContactsContract.Data.CONTENT_URI, cv);
    Log.i("Msg", "插入联系人号码完成");
}
```

插入联系人数据分两步,首先是在 ContactsContract.RawContacts.CONTENT_URI 插入记录,生成 RAW_CONTACT_ID,这是联系人数据的唯一标识。其次是把 RAW_CONTACT_ID、数据类型和数据封装在 ContentValues 中,插入到 ContactsContract.Data.CONTENT_URI。上面的代码实现插入联系人姓名和手机号码两组数据,需要分别实现。

下面的代码实现根据联系人 id 更新手机号码的功能。

📖 Proj09_8 项目 MainActivity.java 文件

```java
//更新联系人
public void updateData(){
    ContentValues cv = new ContentValues();
    cv.put(ContactsContract.CommonDataKinds.Phone.NUMBER,et5.getText().toString());
    int r = getContentResolver().update(ContactsContract.Data.CONTENT_URI, cv,
            ContactsContract.Data.CONTACT_ID + " = ? and "
            + ContactsContract.Data.MIMETYPE + " = ?",
            new String[]{et4.getText().toString(),
                ContactsContract.CommonDataKinds.Phone.CONTENT_ITEM_TYPE});
    Log.i("Msg", "更新联系人号码完成!r= " + r);
}
```

在 Data.CONTENT_URI 中,联系人数据的基本信息都独立记录,如姓名、联系号码、邮箱等,因此在更新数据时,条件是主键 Data.CONTACT_ID 和类型 Data.MIMETYPE。下面的代码实现根据 CONTACT_ID 删除联系人数据的功能。

📖 Proj09_8 项目 MainActivity.java 文件

```java
//删除联系人
public void delData(){
    int dr = getContentResolver().delete(ContactsContract.Data.CONTENT_URI,
            ContactsContract.Data.CONTACT_ID + " = ?",
            new String[]{et6.getText().toString()});
    int rr = getContentResolver().delete(ContactsContract.RawContacts.CONTENT_URI,
            ContactsContract.RawContacts.CONTACT_ID + " = ?",
            new String[]{et6.getText().toString()});
    Log.i("Msg","删除联系人完成! rr = " + rr + "   dr = " + dr);
}
```

删除联系人数据时,分别在 ContactsContract.Data.CONTENT_URI 和 ContactsContract.RawContacts.CONTENT_URI 进行删除操作。删除操作的条件是 CONTACT_ID。

运行该项目前,需要在 AndroidManifest.xml 文件配置如下权限:

```xml
<uses-permission android:name = "android.permission.READ_CONTACTS"/>
<uses-permission android:name = "android.permission.WRITE_CONTACTS"/>
```

项目最终运行效果如图 9.11 所示。

图 9.11 联系管理界面

9.6 习 题

1. 选择题

(1) SharedPreferences 读写的文件类型是(　　)。
   A. 数据库文件　　B. XML 文件　　C. HTML 文件　　D. 二进制文件

(2) 下列关于 Android 应用程序读写文件的说法中正确的是(　　)。
   A. Android 应用程序只能读写内部存储中的文件,无法读写外部存储文件
   B. Android 应用程序可以读写内部存储文件,只能读取外部存储文件
   C. Android 应用程序在获取相应权限后,可以读写外部存储文件
   D. Android 应用程序只能读写文本文件,无法读写二进制文件

(3) 获取 SQLiteDatabase 对象的方法不包括(　　)。
   A. 执行 Context 类中的 openOrCreateDatabase 方法
   B. 执行 SQLiteDatabase 中的 openDatabase 方法
   C. 执行 SQLiteDatabase 中的 openOrCreateDatabase 方法
   D. 继承 SQLiteDatabase 类,自定义实现方式

(4) 下列关于 SQLiteOpenHelper 说法正确的是(　　)。
   A. 它是一个抽象类,必须被继承才能使用
   B. 每次打开数据库时都会执行它的 onCreate 方法
   C. 创建它的对象时,会自动创建一个 SQLiteDatabase 对象
   D. 它只能在外部存储中创建数据库

(5) 下列关于 ContentProvider 说法有误的是(　　)。
   A. 它是一个抽象类,必须被继承才能使用
   B. 它用于向外部应用程序提供数据访问接口
   C. 它和 SQLiteDatabase 一样,可以存储数据
   D. 它使用 Uri 指明数据位置

2. 简答题

(1) 使用 SharedPreferences 实现"记住密码"功能,App 运行界面参考下图:

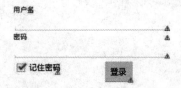

(2) 使用 SQLiteDatabase 存储数据,实现日常消费记录 App,运行界面参考下图:

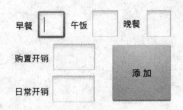

（3）使用 SQLiteDatabase 存储数据，借助 SQLiteOpenHelper 实现好友管理 App，运行界面参考下图：

第 10 章　Android 网络编程

本章学习目标
- 掌握使用 Socket 编程与 DatagramSocket 编程。
- 掌握使用 URLConnection 发起 Get 请求与 Post 请求。
- 掌握使用 HttpClient 发起 Get 请求与 Post 请求。
- 了解访问 WebService 的方法。
- 掌握解析 JSON 数据和 XML 数据的方法。

Android 网络编程涉及 Android 应用程序与 PC 或服务器通信,发送请求、获取数据的相关技术和知识。Android SDK 直接引入了 JDK 网络编程的 API,可以直接在传输层实现基于 TCP 协议的 Socket 编程和基于 UDP 协议的 DatagramSocket 编程。在应用层可以使用 URLConnection 访问网络资源,发起 HTTP 请求操作;也通过第三方 API 提供的 HttpClient 访问受保护的网络资源。在联网过程中,传输数据通常采用 JSON 或 XML 格式,因此需要对这些格式的数据进行解析。

10.1　基于传输层协议的联网

计算机网络的抽象模型 TCP/IP 协议包含了一系列构成互联网基础的网络协议。这个模型中,所有的 TCP/IP 系列网络协议都被归类到 4 个抽象的"层",由高到低分别是应用层、传输层、网络互联层(网络层)和网络接口层。每一抽象层建立在低一层提供的服务上,并且为高一层提供服务。

10.1.1　传输层协议介绍

传输层是整个 TCP/IP 协议模型的重要组成,其功能是从源主机到目的主机提供可靠的数据传输,这种数据传输与具体使用的网络或网关无关。传输层位于应用层和网络层之间,利用网络层提供的数据,向应用层提供服务。网络层解决了数据传输的点到点问题,即数据可以从一个网络节点到达另外一个网络节点。传输层解决了数据传输的端到端问题,即数据从一个网络节点上的某个端口到达另外一个网络节点的某个端口。不同设备上的应用程序借助网络实现数据互通,要经过 TCP/IP 协议模型的各层,数据流向如图 10.1 所示。

端口是传输层上很重要的一个概念,它相当于应用程序与网络互通数据的一个接口。端口都有唯一的编号,从 0 到 65525,其中前 1024 个端口号都分配给操作系统联网使用,如 HTTP 通信使用 80 端口,FTP 通信使用 21 端口。1024 之后的端口都可以分配给应用程

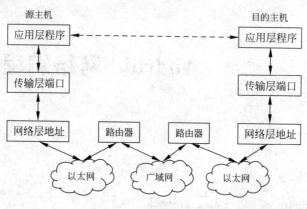

图 10.1 数据传输示意图

序使用,如 MySQL 数据库默认使用 3306 端口,SQL Server 数据库默认使用 1433 端口。一个端口只能被一个应用程序使用,否则会出现端口冲突,在选择端口时尽量避开这些常用端口。

传输层提供了两种比较基本的数据传输模式,分别使用 TCP 协议和 UPD 协议实现。TCP 实现"可靠的、面向连接"的传输机制,它提供一种可靠的字节流保证数据完整、无损且按顺序到达目的地。TCP 协议会连续不断地测试网络的负载,控制发送数据的速度,避免网络拥塞。数据会按照规定的顺序发送,这种特性导致 TCP 协议在完成实时数据传输或者遇到网络丢包率较高的情况下会成为一个缺陷。

UDP 是一种无连接的数据报协议,它实现"不可靠"的数据传输模式。所谓"不可靠"不是说 UDP 无法将数据传输到目的地,而是指 UDP 协议不会检查数据包是否已经到达目的地,并且不保证它们按顺序到达。

鉴于两种传输模式的特征,实现传输层数据联网时应根据传输要求有所选取。TCP 传输模式常用于文本类型数据传输,UDP 传输常用于实时性较高的音视频传输。

10.1.2 Socket 与 ServerSocket

使用 Java 语言实现 Socket 编程,实质上是在传输层使用 TCP 协议实现数据传输。Socket 是对 TCP 协议的一种封装,让程序员不用关注传输层对数据分组大小、分组协议的限制,使用一系列 API 函数就可以实现数据互通。

Socket 编程可以实现 C/S(Client/Server,客户/服务器)架构的数据访问模式,让众多客户端与同一个服务器中心进行通信。在软件开发时服务器端和客户端需要分别开发,采用确定的 IP 地址和端口号。

服务器端启动时,先初始化 ServerSocket,与 IP 地址和端口绑定,然后调用 accept 方法监听所绑定的端口,直到有客户端连接。当连接成功时,accept 方法返回一个 Socket 对象,实现与客户端通信连接,此时通常需要启动一个线程实现数据的读写交互。客户端发送请求数据,服务器接收请求数据并处理,然后把结果发送给客户端,客户端读取数据,最后关闭连接。编写 Socket 程序时通常需要进行的操作如下:

- 创建 Socket 和 ServerSocket。
- 打开输入输出流。

- 按照约定的协议进行读写操作。
- 关闭 Socket 连接。

ServerSocket 类常用方法如表 10.1 所示。

表 10.1 ServerSocket 常用方法

方法名及参数	说明	返回值类型
ServerSocket(int port)	构造方法，port 是绑定的端口号	
accept()	等待客户连接的阻塞方法，执行到该方法时程序将进入"阻塞"状态，不再向下执行	Socket
bind(SocketAddress localAddr)	绑定到本地地址	void
isClosed()	检测 Server 端是否关闭	boolean

Socket 类常用方法描述如表 10.2 所示。

表 10.2 Socket 常用方法

方法名及参数	说明	返回值类型
Socket(String dstName, int dstPort)	构造方法，dstName 是连接的服务器 IP 地址，dstPort 是端口号	
Socket(String dstName, int dstPort, InetAddress localAddress, int localPort)	构造方法，创建 Socket 并绑定本地地址和端口号	
close()	关闭 Socket，输入输出流都会被关闭	void
getInputStream()	获取输入流	InputStream
getOutputStream()	获取输出流	OutputStream
getRemoteSocketAddress()	返回 Socket 连接的 IP 地址，如果未连接则返回 null	SocketAddress

Android 应用程序之间或 Android 应用程序与服务器之间都可以实现 Socket 通信，下面通过一个实例演示 Android 应用与服务器端通信。

1. 编写服务器端程序

在本实例中服务器启动后，监听 8090 端口，有客户端连接时，开启子线程处理与客户端的通信。读取客户端发送的数据，追加服务器时间后返回给客户端。服务端应用程序可以是一个 Java Project 项目，直接由 main 方法启动即可。

ServerSocket 作为响应客户端请求的对象，应该可以为众多的客户端服务，其具体通信任务应在子线程中实现。当 accept 方法接收到客户端连接后，获取 Socket 对象，交给子线程 SocketThread。详细代码如下。

📖 Proj10_2 项目 ServerClient.java 文件

```
public class ServerClient {
    ServerSocket ss;              //服务器端 ServerSocket
    public ServerClient(){
        try {
            System.out.println("服务器端开启,等待连接……");
            ss = new ServerSocket(8090);
```

```
                    while(true){
                        Socket s = ss.accept();
                        System.out.println("有客户端已经连接!" + s.getRemoteSocketAddress());
                        //启动子线程
                        new Thread(new SocketThread(s)).start();
                    }
                } catch (IOException e) {
                    e.printStackTrace();
                }
            }
    ...
```

子线程处理通信的逻辑比较简单,当连接没有关闭时,开启读写流,读取客户端发送来的信息。如果不是 Quit 信息,则在原信息之后追加服务器时间,再发送给客户端;否则退出信息处理,线程结束。详细代码如下。

📖 Proj10_2 项目 ServerClient.java 文件

```
...
//处理通信的线程类
class SocketThread implements Runnable{
    Socket socket;
    public SocketThread(Socket socket) {
        super();
        this.socket = socket;
    }
    @Override
    public void run() {
        BufferedReader br = null;          //输入流
        PrintWriter pw = null;              //输出流
        //格式化时间
        SimpleDateFormat sdf = new SimpleDateFormat("yyyy-MM-dd HH:mm:ss");
        try {
            while(!socket.isClosed()){
                //获取输入流
                br = new BufferedReader(new InputStreamReader(socket.getInputStream()));
                //获取输出流
                pw = new PrintWriter(socket.getOutputStream());
                String txt;
                //读取客户端信息,并返回
                while( (txt = br.readLine())!= null){
                    if(txt.trim().equals("Quit")){
                        socket.close();
                        break;
                    }
                    pw.println(txt + "[服务器时间: " + sdf.format(new Date()) + "]");
                    pw.flush();
                }
            }
```

```
        } catch (IOException e) {
            e.printStackTrace();
        }
        System.out.println("客户端连接关闭!");
    }
}
...
```

多次客户端连接、通信、退出后,服务器端程序在控制台输出信息如图 10.2 所示。两次连接的客户端程序与服务器端程序位于同一台电话,因此 IP 地址相同。

```
服务器端开启,等待连接……
有客户端已经连接! /192.168.99.115:64253
[客户端:]hello server
[客户端:]show time
[客户端:]Quit
客户端连接关闭!
有客户端已经连接! /192.168.99.115:64332
[客户端:]client 2
[客户端:]Quit
客户端连接关闭!
```

图 10.2 服务器端输出信息

2. 编写客户端程序

客户端程序操作界面比较简单,通过按钮的单击将输入框(EditText)中的内容发送给服务器,读取服务器的返回数据,显示在文本视图(TextView)。由于 Android 系统要求所有涉及网络连接的代码必须写在子线程,因此客户端启动后,需要开启子线程,创建 Socket 连接。在本实例中,Activity 执行 onResume 方法时启动子线程,连接服务器,读取服务器发送回来的数据。代码如下。

📖 Proj10_1 项目 MainActivity.java 文件

```
...
public class MainActivity extends Activity {
    //声明控件
    EditText et;
    TextView tv;
    //声明 Socket
    Socket socket;
    //输入输出流
    PrintWriter pw;
    BufferedReader br;
    //handler 更新界面
    Handler handler = new Handler(){
        public void handleMessage(android.os.Message msg) {
            tv.append(msg.obj + "\n");
        }
    };
```

```java
@Override
protected void onCreate(Bundle savedInstanceState) {
    super.onCreate(savedInstanceState);
    setContentView(R.layout.activity_main);
    et = (EditText) findViewById(R.id.editText1);
    tv = (TextView) findViewById(R.id.content);
}
@Override
protected void onResume() {
    super.onResume();
    //开启线程
    new Thread(new MsgThread()).start();
}
...
```

联网子线程由内部类 MsgThread 实现，主要完成创建 Socket 连接，获取输入输出流的功能。Socket 连接服务器端程序所运行的主机 IP 地址 192.168.99.115（本实例运行时的主机 IP 地址），并指定服务器监听的端口 8090。要查看主机 IP 地址，可以在开始菜单的命令行中输入 cmd，打开命令行窗口，然后输入 ipconfig 命令。请勿使用 127.0.0.1，手机模拟器会将其视为自身 IP 地址，无法连接服务器程序。

如果 Socket 连接创建成功，获取输入输出流，以备通过该链接发送数据和接收数据。BufferedReader 类 readLine 方法是阻塞方法，它将"挂起"后续程序，一直等到服务器返回数据。若读取到服务器数据，则将数据发送给 Handler 对象，通知它更新界面，把数据追加到 TextView 中。详细代码如下。

📖 Proj10_1 项目 MainActivity.java 文件

```java
...
//子线程,读取服务器端返回的数据
class MsgThread implements Runnable {
    @Override
    public void run() {
        try {
            //ip 指定为服务器端的 IP
            socket = new Socket("192.168.99.115",8090);
            if(socket!= null){
                pw = new PrintWriter(socket.getOutputStream());
                br = new BufferedReader(new InputStreamReader
                                        (socket.getInputStream()));
            }
            Log.i("Msg","连接建立");
            while(socket.isConnected()){
                //读取服务器数据
                String txt = br.readLine();
                Message msg = new Message();
                msg.obj = txt;
```

```
                    handler.sendMessage(msg);
                }
            } catch (IOException e) {
                e.printStackTrace();
            }
        }
    }
...
```

对布局文件中的按钮,添加 android:onClick="sendMsg"属性,设定监听方法。通过输出流向服务器发送数据。代码如下。

📖 Proj10_1 项目 MainActivity.java 文件

```
public void sendMsg(View v){
    if(socket == null){
        Toast.makeText(this, "服务器连接失败.", Toast.LENGTH_LONG).show();
        return ;
    }
    //发送信息
    pw.println(et.getText().toString());
    pw.flush();
    et.setText("");
}
```

当客户端退出时,向服务器发送最后一次数据,约定为 Quit 字符串,标识客户端将退出,随后关闭 Socket 连接。本实例将连接的创建和关闭分别放在 onResume 方法和 onPause 方法中,该功能也可以分别放在 onCreate 和 onStop 方法中。

📖 Proj10_1 项目 MainActivity.java 文件

```
@Override
protected void onPause() {
    super.onPause();
    if(socket!= null)
        try {
            //通知服务器,客户端退出,Quit 为约定退出的标识
            pw.println("Quit");
            pw.flush();
            socket.close();
            Log.i("Msg", "连接关闭!");
        } catch (IOException e) {
            e.printStackTrace();
        }
}
```

应用程序连接网络,需要声明联网权限,在 Manifest.xml 文件中添加如下代码:

```
<uses-permission android:name="android.permission.INTERNET"/>
```

在服务器端程序启动之后,运行客户端程序,向服务器发送信息,会收到服务器返回带有时间的信息,如图10.3所示。

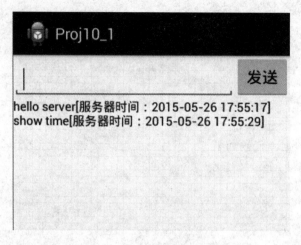

图10.3 客户端程序运行界面

上述实例主要演示Socket通信的基本过程,在此基础上可以很容易实现客户端的实时信息互通。服务器端维持所有在线客户端的链接,根据客户端选择广播信息或是单播信息,将数据转发给对应的其他链接,实现类似于微信的在线文本聊天功能。

10.1.3 DatagramSocket 与 DatagramPacket

与采用TCP协议的Socket编程不同,采用UDP协议的DatagramSocket,没有客户端与服务器端之分。数据的发送端和接收端都采用DatagramSocket,需要传输的数据被封装在DatagramPacket中。

使用UDP协议的通信过程一般可以总结为以下两个步骤:

(1) 发送端把需要发送的数据封装成一个DatagramPacket对象,标明发送目的地IP地址和端口号,使用DatagramSocket把数据包发送出去。

(2) 接收端使用DatagramSocket接收某端口数据包,并将数据包对应成一个DatagramPacket对象。

DatagramPacket是UDP协议中数据包的抽象表示,它的常用方法如表10.3所示。

表10.3 DatagramPacket常用方法

方法	说明	返回值类型
DatagramPacket(byte[] data, int length)	构造方法,创建接收数据的对象。data是保存读取到的数据的缓存,length是数据缓存的大小	
DatagramPacket(byte[] data, int length, InetAddress host, int port)	构造方法,创建发送数据的对象,data是需要发送的数据缓存,length是缓存大小,host是目的地址,port是目的端口号	
getAddress()	得到数据包的目的地址或源地址	InetAddress
getData()	得到数据包中的数据缓存	byte[]
getLength()	得到缓存数据的有效长度	int

续表

方法	说 明	返回值类型
getPort()	得到数据包的目的地址或源地址端口号	int
setAddress(InetAddress addr)	设置数据包的目的地址	void
etData(byte[] buf)	设置数据包的数据缓存区域	void
setPort(int aPort)	设置数据包的目的端口号	void

DatagramSocket 在 UDP 协议通信中起到发送数据包、接收数据包的作用,它的常用方法如表 10.4 所示。

表 10.4 DatagramSocket 常用方法

方法	说 明	返回值类型
DatagramSocket()	构造方法,创建 UDP 数据包接收对象	
DatagramSocket(int aPort)	构造方法,创建接收对象,指定接收数据的端口号	
close()	关闭接收	void
receive(DatagramPacket pack)	接收数据包,并将其存入 pack 中	void
send(DatagramPacket pack)	发送数据包,pack 是等待发送的数据包	void

UDP 通信常用于对实时性要求较高,可以接受较小数据传输错误的场合。下面介绍一个实例,演示同一个网络中两部手机借助 UDP 协议实时通信。文本的实时通信并不是 UDP 所擅长的,即使较小的丢包率,也会导致传递文字信息错误,本实例实现文本的通信只是为了简化功能代码。

应用程序布局文件比较简单,包括两个 EditText 分别用于设定目标 IP 地址和输入聊天内容,一个 Button 用于发送数据,两个 TextView 分别用于提示输入和呈现聊天内容,具体可以参考运行效果。初始化控件,添加按钮监听器,启动发送数据和接收数据的代码如下。

📖 Proj10_3 项目 MainActivity.java 文件

```java
...
public class MainActivity extends Activity {
    private EditText et1,et2;              //确定 IP 地址和发送文本的输入框
    private Button sendBt;                 //发送按钮
    private TextView tv;                   //显示内容的文本
    private boolean isQuit;                //标识程序是否退出
    private Handler handler = new Handler(){
        public void handleMessage(android.os.Message msg) {
            tv.append("From:" + msg.obj + " [" + sdf.format(new Date()) + "]\n");
        };
    };
    //格式化时间
    SimpleDateFormat sdf = new SimpleDateFormat("yyyy-MM-dd HH:mm:ss");
    @Override
    protected void onCreate(Bundle savedInstanceState) {
```

```
            super.onCreate(savedInstanceState);
            setContentView(R.layout.activity_main);
            //初始化控件
            et1 = (EditText) findViewById(R.id.editText1);
            et2 = (EditText) findViewById(R.id.editText2);
            sendBt = (Button) findViewById(R.id.button1);
            tv = (TextView) findViewById(R.id.textView2);
            //按钮监听器
            sendBt.setOnClickListener(new OnClickListener(){
                @Override
                public void onClick(View arg0) {
                    //启动发送数据线程
                    new Thread(new SendMsgThread()).start();
                    tv.append("To:" + et2.getText().toString() + " [" + sdf.format(new Date()) + "]\n");
                }}
            );
        }
        @Override
        protected void onResume() {
            super.onResume();
            isQuit = false;
            //启动接收数据线程
            new Thread(new ReceiveMsgThread()).start();
            Log.i("Msg","启动线程");
        }
        @Override
        protected void onStop() {
            super.onStop();
            isQuit = true;
        }
    ...
```

数据的发送由线程 SendMsgThread 类实现,发送数据包的目的地 IP 地址由输入框确定,端口号统一设定为 8001。待发送的字符串需要转型为字节数组,然后封装到 DatagramPacket 中。数据发送完成,线程即结束。在封装待传数据大小时,一般应控制小于 8192B(8KB),否则多余的字节将被丢弃,这是由于 UDP 协议的数据包大小有限制。若数据比较多,可以拆分成几组,依次发送。

📖 Proj10_3 项目 MainActivity.java 文件

```
    ...
        //发送数据的线程
        class SendMsgThread implements Runnable{
            @Override
            public void run() {
                String txt = et2.getText().toString();         //需要发送的信息
                try {
                    //目标地址
```

```
            InetAddress ipAddr = InetAddress.getByName(et1.getText().toString());
            //待发送的数据包,标明目标地址和目标端口
            DatagramPacket dp = new DatagramPacket(txt.getBytes(),
                                    txt.getBytes().length, ipAddr, 8001);
            //准备发送数据包
            DatagramSocket ds = new DatagramSocket();
            ds.send(dp);              //发送
            Log.i("Msg", "数据包已经发出!" + txt.getBytes().length);
            ds.close();
        } catch (UnknownHostException e) {
            e.printStackTrace();
        } catch (SocketException e) {
            e.printStackTrace();
        } catch (IOException e) {
            e.printStackTrace();
        }
    }
}
...
```

线程 ReceiveMsgThread 实现接收数据的功能,缓存字节数组为 1024B,这是目前收发端约定的大小,即发送端不会发送超过 1024B 的数据。实际项目中,这种逻辑并不合理,应实现更具通用性的数据传输。接收端通过 receive 方法接收数据,该方法是阻塞方法,会"挂起"后续代码,直至接收到数据。

📖 Proj10_3 项目 MainActivity.java 文件

```
...
    //接收数据的线程
    class ReceiveMsgThread implements Runnable{
        @Override
        public void run() {
            byte buffer [] = new byte[1024];         //准备接收数据的缓存
            //接收数据包
            DatagramPacket dp = new DatagramPacket(buffer, buffer.length);
            try {
                //准备接收
                DatagramSocket ds = new DatagramSocket(8001);
                while(!isQuit){
                    //接收数据
                    ds.receive(dp);
                    String msg = new String(dp.getData(),0,dp.getLength());
                    //向 handler 发送消息,通知更新界面
                    Message message = new Message();
                    message.obj = msg;
                    handler.sendMessage(message);
                    Log.i("Msg","收到信息: " + msg + " " + dp.getLength());
                }
```

```
                ds.close();
            } catch (SocketException e) {
                e.printStackTrace();
            } catch (IOException e) {
                e.printStackTrace();
            }
        }
    }
...
```

应用程序连接网络,需要声明联网权限,在 Manifest.xml 文件中添加如下代码:

```
< uses - permission android:name = "android.permission.INTERNET"/>
```

如果借助一台设备测试该项目,可以在"目标主机 IP 地址"栏输入 127.0.0.1,此时应用程序会把数据包发送给本机,接收线程将收到数据。本教程中的实例是通过两台真机测试运行,"目标主机 IP 地址"分别设定为对方的 IP 地址,运行效果如图 10.4 所示。手机 IP 地址可以在"设置"→"网络"→WiFi 栏中查找,不同机型,查看路径可能有所区别。两台真机 IP 地址应属于同一个网段。

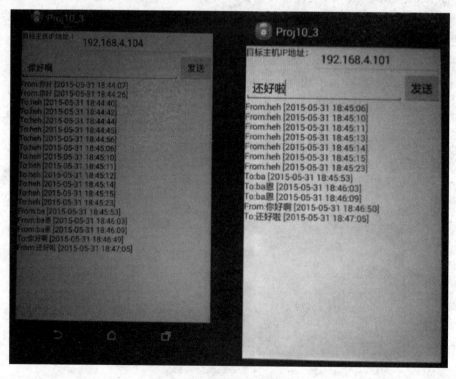

图 10.4 真机模拟通信界面

10.1.4 Android 对联网代码的限制

在 Android 2.3.X 版本及之前版本,联网代码可以写在主 UI 线程中,该版本之后各版本都不允许直接在 UI 线程中建立网络连接,访问网络资源,这些操作需要写在子线程中。这种硬性规定在很大程度上提高了用户使用的体验。

由于网络连接需要很长的时间,一般 2~5s,甚至更长的时间才能返请求的数据。如果连接动作直接写在主线程,也就是 UI 线程中,整个程序会处于等待状态,无法响应用户的其他操作,界面似乎是"卡死"了。联网代码放置到单独线程中运行,可以避免阻塞 UI 线程,不会对主线程有任何影响。

在新版本的 SDK 中,如果主线程中出现与联网相关的代码,如创建 Socket、执行 URLConnection 连接等,则会抛出异常信息 NetworkOnMainThreadException,它属于运行时异常。

无论采用 WiFi 联网还是 GPS 联网,都需要申请联网权限,该权限配置在 Manifest.xml 文件中:< uses-permission android:name="android.permission.INTERNET"/>。与网络相关的权限还有 android.permission.ACCESS_NETWORK_STATE、android.permission.ACCESS_WIFI_STATE 和 android.permission.CHANGE_WIFI_STATE。

10.2 基于应用层协议的联网

应用层是 TCP/IP 模型的最上一层,包含所有的高层协议,如虚拟终端协议(TELNET)、文件传输协议(FTP)、电子邮件传输协议(SMTP)、域名服务(DNS)、超文本传送协议(HTTP)等。其中 HTTP 协议主要用于从 WWW 服务器传输超文本到本地浏览器,也可以用于其他因特网/内联网应用系统之间的通信,从而实现各类应用资源超媒体访问的集成。通过 HTTP 或者 HTTPS(增加安全通道的 HTTP 协议)请求的资源由统一资源定位符(Uniform Resource Locator,URL)来标识。本节主要介绍在 Android 平台下如何通过 HTTP 协议访问 Web 服务器,向服务器提交请求数据,获取服务器返回的数据。

10.2.1 URL 介绍

统一资源定位符(URL)相当于一个文件名在网络范围的扩展,是与网络相联的机器上的任何可访问对象的一个指针。通常,URL 的一般形式是:< URL 的访问协议>://<主机>:<端口>/<路径>[? 请求字符串]。常用的访问协议有 file(读取文件)、ftp(文件传输协议)、http(超文本传输协议)、https(安全 http)、jar(读取 jar 文件)。

JDK 提供了 URL 类,位于 java.net 包中,Android 平台可以直接使用。URL 类提供了较多的构造方法,可以根据不同的参数创建 URL 对象。URL 类的常用方法如表 10.5 所示,其中 openConnection 或 openStream 都可以获得输入流,进而以"流"的方式访问资源。

URL 类构造方法如下:
- URL(String spec),根据 String 表示形式创建 URL 对象。
- URL(String protocol, String host, int port, String file),根据指定 protocol、host、port 和 file 创建 URL 对象。
- URL(String protocol, String host, int port, String file, URLStreamHandler handler),根据指定的 protocol、host、port、file 和 handler 创建 URL 对象。
- URL(String protocol, String host, String file),根据指定的 protocol、host 和 file

创建 URL。
- URL(URL context, String spec), 通过在指定的上下文中对给定的 spec 进行解析来创建 URL。
- URL(URL context, String spec, URLStreamHandler handler), 通过在指定的上下文中用指定的处理程序对给定的 spec 进行解析来创建 URL。

表 10.5　URL 常用方法

方　法	说　　明	返回值类型
getFile()	返回 URL 中的资源名称	String
getHost()	返回 URL 中的主机名或 IP 地址	String
getPath()	返回 URL 中的路径部分字符串	String
getPort()	返回 URL 中的端口号	int
getProtocol()	返回 URL 所采用的访问协议	String
getQuery()	返回 URL 的请求字符串,即"?"后面的字符串	String
openConnection()	返回到 URL 指定资源的连接对象,类似于数据库连接的 Connection	URLConnection
openStream()	返回到 URL 指定资源的输入流,此输入流也可以通过 URLConnection 的 getInputStream 方法获取	InputStream

新建项目 Proj10_4 使用 URL 获取输入流,并读取资源。布局文件运行效果参考图 10.5。EditText 用于输入图片 URL(网址),Button 用于触发下载事件,ImageView 用于显示下载的图片。

当按钮单击时,获取输入框中的图片网址字符串,创建图片下载线程对象 LoadPicThread。在下载线程中创建 URL 对象,调用 openStream 方法,获取字节输入流,再使用 BitmapFactory.decodeStream 方法,把输入流转为位图 Bitmap,随后调用 Handler,更新 UI。

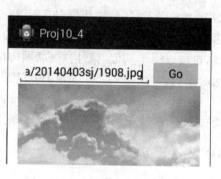

图 10.5　显示联网图片

📖 Proj10_4 项目 MainActivity.java 文件

```
...
public class MainActivity extends Activity {
    private EditText et;
    private ImageView iv;
    private Bitmap pic;                         //下载到的图片
    //如果下载完成,更新界面
    Handler handler = new Handler(){
        public void handleMessage(android.os.Message msg) {
            if(pic!= null)
                iv.setImageBitmap(pic);
        };
    };
```

```java
@Override
protected void onCreate(Bundle savedInstanceState) {
    super.onCreate(savedInstanceState);
    setContentView(R.layout.activity_main);
    et = (EditText) findViewById(R.id.editText1);
    iv = (ImageView) findViewById(R.id.imageView1);
}
//通过指定的 URL 获取图片
public void loadImage(View v){
    String urlStr = et.getText().toString();
    if(urlStr == null || urlStr.trim().length()<1)return;
    //开启下载图片的线程
    new Thread(new LoadPicThread(urlStr)).start();
}
/**
 * 下载图片的线程
 * */
class LoadPicThread implements Runnable{
    String urlStr;
    //构造方法,接收 URL 字符串
    public LoadPicThread(String urlStr) {
        this.urlStr = urlStr;
    }
    @Override
    public void run() {
        try {
            URL url = new URL(urlStr); //创建 URL 对象
            //获取输入流,并解析成位图
            pic = BitmapFactory.decodeStream(url.openStream());
            //通知界面更新
            handler.sendEmptyMessage(0x100);
        } catch (MalformedURLException e) {
            e.printStackTrace();
        } catch (IOException e) {
            e.printStackTrace();
        }
    }
}
```

10.2.2 GET 请求和 POST 请求

直接使用 URL 获取联网资源,实现代码比较简单,但存在很大的局限性。例如,在访问资源时,必须先向服务器提交请求信息,或者需要保持登录状态才能访问资源。对于这样的资源请求,需要了解客户端如何与服务器交互,使用稍微复杂一些的实现方案。HTTP 协议定义了客户端与服务器交互的不同方法,最基本的方法有 4 种,分别是 GET、POST、

PUT、DELETE。这 4 种操作相当于对 URL 资源的查、改、增、删 4 个操作。GET 一般用于获取/查询资源信息,POST 一般用于更新资源信息,PUT 和 DELETE 的操作都可以借助 POST 实现。但是在实际开发中,很少按照 HTTP 协议的严格规定实现。有时为了方便,更新资源也使用 GET,因为 POST 必须用到 FORM(表单),这样会麻烦一点。GET 和 POST 二者都可以完成数据传递,但它们具有很大的区别。

首先,GET 将表单中数据的按照 variable=value 的形式添加到 action 所指向的 URL 后面,并且两者使用"?"连接,而各个变量之间使用"&"连接;POST 是将数据放在 FORM 表单中,按照变量和值相对应的方式传递到 action 所指向 URL。

其次,GET 是不安全的,因为在传输过程中,数据被放在请求的 URL 中,是可见的(当然也可以对其加密),而现有的很多服务器、代理服务器或者用户代理都会将请求 URL 记录到日志文件中,然后放在某个地方,这样就可能会有一些隐私的信息被第三方看到。另外,用户也可以在浏览器上直接看到提交的数据,一些系统内部消息将会一同显示在用户面前。POST 的所有操作对用户来说都是不可见的。

第三,GET 传输的数据量小,这主要是因为受 URL 长度的限制,不同浏览器大小限制有所不同,一般不超过 1KB 都不会出问题;而 POST 可以传输大量的数据,所以在上传文件只能使用 POST。

第四,GET 和 POST 的编码格式不同。Tomcat 服务器在处理 POST 请求时会使用 request.setCharacterEncoding 方法所设置的编码来处理,如果未设置,则使用默认的 iso-8859-1 编码。处理 GET 请求则不同,Tomcat 对于 GET 请求并不会考虑使用 request.setCharacterEncoding 方法设置的编码,而会永远使用 iso-8859-1 编码。在向服务器提交数据时,如果请求串中含有中文,应该注意编码格式的统一和转换。

10.2.3 使用 HttpURLConnection 联网

使用 URL 可以获取相应的网络资源,但无法获取更多的控制信息,如果需要配置网络连接,需要借助 URLConnection,该类位于 java.net 包中,它相当于 URL 资源与应用程序之间的连接桥梁,作用与 JDBC 中的 Connection 较为类似。借助它可以向 URL 发送请求数据,配置连接信息(如超时时间等),读取返回的资源。HttpURLConnection 是 URLConnection 的子类,主要是用于建立 HTTP 协议上的连接,访问网页资源,可以直接使用 HttpURLConnection 类。

使用 URLConnection 或 HttpURLConnection 建立网络资源连接时,可以参考以下步骤:

(1)调用 URL 类中的 OpenConnection 方法,获取 URLConnection 对象,如果 URL 创建时采用 HTTP 协议,则可以直接强制转换为 HttpURLConnection 对象。

(2)配置连接属性。

(3)调用 connect 方法,正式建立连接。

(4)建立连接后,使用 getHeaderFields 和 getHeaderFieldKey(int posn)方法可以查询头信息。

(5)获取输入流,读取返回资源,该输入流与在 URL 中获取的输入流一致。

HttpURLConnection 类的常用方法如表 10.6 所示。

表 10.6 HttpURLConnection 常用方法

方　法	说　明	返回值类型
getResponseCode()	返回服务器的响应代码	int
getResponseMessage()	返回服务器的响应信息	String
getRequestMethod()	返回请求方法	String
setRequestMethod(String method)	设置请求方法	void
setDoInput(boolean newValue)	设置连接的 doInput 请求头字段	void
setDoOutput(boolean newValue)	设置连接的 doOutput 请求头字段	void
setRequestProperty(String field, String newValue)	设置连接的请求字段值,如 setRequestProperty("accept","*/*")	void
getContent()	获取连接的内容	Object
getHeaderField(String header)	获取指定响应头字段	String
getInputStream()	获取连接的输入流	InputStream
getOutputStream()	获取连接的输出流,用于发送请求参数	OutputStream
getContentEncoding()	获取响应头 content-encoding 对应的值	String
getContentLength()	获取响应头 content-length 对应的值	String

下面新建 Proj10_5 项目,演示如何向 Web 站点发送 GET 请求和 POST 请求,并获取站点的返回信息。该项目的布局文件比较简单,提供 URL 输入框、GET 请求按钮、POST 请求按钮和显示返回信息的 TextView。

服务器端程序项目是 Ser10_5,需要部署到服务器才可以访问,如 Tomcat。关于 Java Web 项目的开发请参考其他资料。服务器端提供 LoginServlet 类,实现了 doGet 和 doPost 方法,分别响应 GET 请求和 POST 请求,返回对应的字符串。

📖 Proj10_5 项目 MainActivity.java 文件核心代码

```java
...
//Get 按钮单击事件
public void getRequest(View v){
    final String urlStr = et.getText().toString();
    new Thread(new Runnable(){
        @Override
        public void run() {
            //创建 GET 请求,并获得返回数据
            String resultStr = new RequestHelper().sendGet(urlStr,
                        "loginName = admin&loginPwd = admin");
            Message msg = new Message();
            msg.obj = resultStr;
            handler.sendMessage(msg);
        }}).start();
}
//Post 按钮单击事件
public void postRequest(View v){
    final String urlStr = et.getText().toString();
    new Thread(new Runnable(){
        @Override
```

```java
        public void run() {
            //创建POST请求,并获得返回数据
            String resultStr = new RequestHelper().sendPost(urlStr,
                            "loginName = admin&loginPwd = admin");
            Message msg = new Message();
            msg.obj = resultStr;
            handler.sendMessage(msg);
        }}).start();
    }
...
```

上面的程序中,getRequest 和 postRequest 方法分别对应两个按钮的单击事件,执行创建线程,请求资源,并调用 Handler 更新 UI 的功能。联网代码由工具类 RequestHelper 实现,sendGet 方法和 sendPost 方法的写法基本类似,但请求参数的处理不同,GET 请求需要把请求参数以"?"拼接在请求地址之后;POST 请求需要借助输出流。

📖 Proj10_5 项目 RequestHelper.java 文件

```java
...
    public String sendGet(String urlStr,String params){
        BufferedReader br = null;
        StringBuffer sbuff = new StringBuffer();
        try {
            //建立URL对象,参数以"?"连接在地址后面
            URL url = new URL(urlStr + "?" + params);
            //获取连接
            HttpURLConnection conn = (HttpURLConnection) url.openConnection();
            //设置连接
            conn.setRequestProperty("accept", "*/*");
            conn.setRequestProperty("Charset", "UTF-8");
            conn.setConnectTimeout(5000);
            conn.setRequestMethod("GET");
            conn.connect();
            //获取输入流
            br = new BufferedReader(new InputStreamReader(url.openStream()));
            Log.i("Msg", "Get over");
            String line;
            //读取返回信息
            while( (line = br.readLine())!= null ){
                sbuff.append(line);
            }
            br.close();
            //断开连接
            conn.disconnect();
        } catch (MalformedURLException e) {
            e.printStackTrace();
        } catch (IOException e) {
            e.printStackTrace();
        }
```

①

```
        return sbuff.toString();
    }
...
```

POST 请求时需要配置连接属性,请求方法必须指定为 POST,否则默认为 GET,同时需要设置 setDoOutput。请求参数串借助输出流提交到服务器端,如下面程序 2 处所示,这不同于 GET 方式代码中的①。

📖 Proj10_5 项目 RequestHelper.java 文件

```
public String sendPost(String urlStr,String params){
    StringBuffer sbuff = new StringBuffer();
    URL url;
    HttpURLConnection conn = null;
    try {
        //建立 URL 对象,不需要添加请求参数
        url = new URL(urlStr);
        //获取 HTTP 连接
        conn = (HttpURLConnection) url.openConnection();
        //指定使用 POST 请求方式
        conn.setRequestMethod("POST");
        //向连接中写入数据,读取数据
        conn.setDoInput(true);
        conn.setDoOutput(true);
        //禁止缓存
        conn.setUseCaches(false);
        //自动执行 HTTP 重定向
        conn.setInstanceFollowRedirects(true);
        //设置内容类型
        conn.setRequestProperty("Content-Type","application/x-www-form-urlencoded");
        //获取输出流
        DataOutputStream out = new DataOutputStream(conn.getOutputStream());
        out.writeBytes(params);                                                     ②
        out.flush();
        out.close();
        // 判断服务器是否响应成功
        if (conn.getResponseCode() == HttpURLConnection.HTTP_OK) {
            //获取输入流对象
            BufferedReader buffer = new BufferedReader(
                        new InputStreamReader(conn.getInputStream()));
            String inputLine = null;
            while ((inputLine = buffer.readLine()) != null) {
                sbuff.append(inputLine);
            }
            buffer.close();                   //关闭字符输入流
        }
        //断开连接
        conn.disconnect();
    } catch (MalformedURLException e) {
```

```
            e.printStackTrace();
        } catch (IOException e) {
            e.printStackTrace();
        }
        return sbuff.toString();
    }
```

程序连接网络需要赋予联网权限，在 AndroidManifest.xml 文件中配置如下授权代码：

```
<uses-permission android:name="android.permission.INTERNET"/>
```

所访问的服务器端地址 http://192.168.4.103:8080/Proj10_6/servlet/LoginServlet 需要根据具体 Web 配置信息修改，一般 IP 地址更改为服务器主机地址即可。请勿使用 localhost，否则虚拟设备会将此视为自身，无法访问站点服务器。程序运行效果如图 10.6 所示。

图 10.6　GET 和 POST 请求

10.2.4　使用 HttpClient 联网

使用 URLConnection 或 HttpURLConnection 可以实现访问网络资源和提交请求数据的功能，但如果需要访问一些受保护的资源时（登录后才能访问，需要处理 Session、Cookie 等），处理起来比较麻烦。HttpClient 位于 org.apache.http.client 包中，是 Apache 开源组织提供的项目，用于处理客户端与服务器的连接，核心任务是发送 HTTP 请求，接收 HTTP 响应，管理连接，维持连接状态，但是不会对接收的内容进行解析。HttpClient 可以比较方便地实现前面所说的很多功能，它可以被视为 HttpURLConnection 的增强版本，除了实现 HttpURLConnection 的功能，还提供对连接状态的管理。

Android 操作系统完全集成了 HttpClient，可以直接使用。使用 HttpClient 发起 GET 请求和 POST 的步骤基本一致，只是提交参数的方式不同。使用 HttpClient 访问网络资源可以按照以下步骤。相对于 HttpURLConnection，HttpClient 联网更加严格和规范，熟练掌握操作步骤之后，建议读者将其封装成一个工具类。在 Android5.0(API23)中 Google 已经移除了 Apache HttpClient 相关的类，如果继续使用需要下 libs 添加 org.apache.http.legacy.jar。也可以寻求第三方框架，如 Volley、OKHttp、xUtils3 等，实现联网功能。

(1) 创建 HttpClient 对象，HttpClient client = new DefaultHttpClient()。由于 HttpClient 是一个接口，所以在创建时，只能选择它的实现类 DefaultHttpClient 或 AndroidHttpClient。

(2) 创建 HttpGet 对象执行 GET 请求或者创建 HttpPost 对象执行 POST 请求，二者都是接口 HttpUriRequest 的实现类。

- HttpGet get = new HttpGet(String url)，发起 GET 请求的对象，Get 请求的参数

可以直接连接在后面。
- HttpPost post=new HttpPost(String url),创建 HttpPost,此处 url 不包括请求串信息。给 HttpPost 设置参数可以借助 BasicNameValuePair 类,该类与 Map 类类似,以键-值对形式存放请求参数:BasicNameValuePair pair = new BasicNameValuePair(String name,String value)。由于一次请求的参数对有多个,需要把这些对象添加到一个集合中,通常是 List 集合,再将集合创建为 HttpEntity 对象:UrlEncodedFormEntity entity = new UrlEncodedFormEntity(List < NameValuePair > list, String encoding),创建 HttpEntity 实体,UrlEncodedFormEntity 实现了 HttpEntity 接口,在创建时可以指定编码格式。最后一步把 HttpEntity 对象设置给 HttpPost 请求:post.setEntity(entity)。

(3) 执行请求,获得服务器响应对象:HttpResponse response = client.execute(HttpUriRequest)。请求后的响应对象可以获取服务器响应状态和信息。例如获取服务器响应码:int code = response.getStatusLine().getStatusCode。实际上 getStatusLine 方法会返回 StatusLine 对象,通过该对象的方法可以查看协议版本、响应代码等,常用的就是查看响应代码。HttpResponse 对象还提供了 getAllHeaders 和 getHeaders(String name)这样的方法用于查询头部信息。

(4) 获取服务器返回数据:InputStream in = response.getEntity().getContent(),获取返回的内容。getEntity 方法会返回 HttpEntity 对象,该对象封装了服务器的响应内容。通过 HttpEntity 类的 getContent 方法可以获取 InputStream,读取返回的内容。

项目 Proj10_6 演示了如何使用 HttpClient 请求服务器访问受保护的资源。服务器端项目是 Ser10_6,需要部署到 Tomcat。服务器端提供了两个 Servlet:LoginServlet 和 QueryServlet,分别实现登录和查询功能。执行 QueryServlet 时,会检查 Session 中是否有登录。两个 Servlet 的访问地址如下(IP 需要修改为 Tomcat 服务器所在地址):

http://192.168.4.100:8080/Ser10_6/servlet/LoginServlet

http://192.168.4.100:8080/Ser10_6/servlet/QueryServlet

客户端布局比较简单,提供地址输入、按钮和文本视图。EditText 默认是上面的网址,按钮分别对应登录事件和查询事件。登录时,HttpClient 采用 POST 请求方式;查询时,HttpClient 采用 GET 方式。具体实现代码如下。

📖 Proj10_6 项目 MainActivity.java 核心代码

```
...
    protected void onCreate(Bundle savedInstanceState) {
        super.onCreate(savedInstanceState);
        setContentView(R.layout.activity_main);
        et1 = (EditText) findViewById(R.id.editText1);
        et2 = (EditText) findViewById(R.id.editText2);
        tv = (TextView) findViewById(R.id.textView1);
        //第 1 步,创建访问对象                                        ①
        httpClient = new DefaultHttpClient();
    }
    //登录事件
    public void login(View v){
```

```java
            final String urlStr = et1.getText().toString();
            new Thread(new Runnable(){
                @Override
                public void run() {
                    try {
                        //第2步,创建访问方式并设置请求参数
                        HttpPost post = new HttpPost(urlStr);
                        NameValuePair p1 = new BasicNameValuePair("loginName","admin");
                        NameValuePair p2 = new BasicNameValuePair("loginPwd","admin");
                        List<NameValuePair> nps = new ArrayList<NameValuePair>();
                        nps.add(p1);
                        nps.add(p2);
                        UrlEncodedFormEntity enEntity = new UrlEncodedFormEntity(nps);
                        post.setEntity(enEntity);
                        //第3步,执行请求,获取响应对象
                        HttpResponse resp = httpClient.execute(post);
                        Message msg = new Message();
                        if(resp.getStatusLine().getStatusCode() == 200){
                            //第4步,获取服务器返回的数据
                            HttpEntity entity = resp.getEntity();
                            //获得服务器返回的字符串,获取借助输入流读取
                            String resStr = EntityUtils.toString(entity, "utf-8");
                            Log.i("Msg", "登录,服务器返回 + " + resStr);
                            msg.obj = resStr;
                        }else{
                            msg.obj = "网络访问失败!";
                        }
                        handler.sendMessage(msg);
                    } catch (ClientProtocolException e) {
                        e.printStackTrace();
                    } catch (IOException e) {
                        e.printStackTrace();
                    }
                }
            }).start();
        }
        //查询事件
        public void query(View v){
            final String urlStr = et2.getText().toString();
            new Thread(new Runnable(){
                @Override
                public void run() {
                    //第2步,创建访问方式,并设置请求参数
                    HttpGet get = new HttpGet(urlStr + "?account=65001201");
                    try {
                        //第3步,执行请求,获取响应对象
                        HttpResponse resp = httpClient.execute(get);
                        Message msg = new Message();
                        if(resp.getStatusLine().getStatusCode() == 200){
```

```
                //第4步,获取服务器返回的数据
                HttpEntity entity = resp.getEntity();
                //获得服务器返回的字符串,获取借助输入流读取
                String resStr = EntityUtils.toString(entity, "utf-8");
                Log.i("Msg", "查询,服务器返回" + " " + resStr);
                msg.obj = resStr;
            }else{
                msg.obj = "网络访问失败!";
            }
            handler.sendMessage(msg);
        } catch (ClientProtocolException e) {
            e.printStackTrace();
        } catch (IOException e) {
            e.printStackTrace();
        }
    }
}).start();
...
```

POST 请求和 GET 请求只是在处理请求参数时不同,其他步骤一样。两次请求使用的 HttpClient 是同一个,即在上面的代码中①处创建,请勿使用不同的 HttpClient 对象。注意,程序运行时需要添加网络访问权限。上述项目的运行效果如图 10.7 所示。

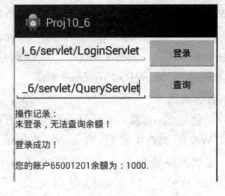

图 10.7 访问受保护的资源

先单击"查询"按钮,访问 QueryServlet,服务器检查 Session,没有登录信息,提示登录。登录操作,访问 LoginServlet 后,将登录信息 loginName=admin,loginPwd=admin,发送给服务器端,Session 记录登录信息,再次查询时便可以通过。此处的账号信息是固定的,有兴趣的读者可以修改项目,需要用户输入账号信息。

10.3 访问 Web Service

Web Service 是一种远程调用标准,通过它可以将不同操作系统、不同语言、不同技术整合到一起;是一种基于 Web 的服务,也是一个应用程序,它向外界暴露出一个能够通过 Web 进行调用的 API,通过编程的方法就可以调用这个应用程序。一般把能调用这个 Web Service 的应用程序叫做客户端程序。本节主要讨论如何在 Android 系统中调用 Web Service 服务,实现一些比较实用的功能,如天气预报功能、票务查询功能等。如果对发布 Web Service 感兴趣可以查阅其他资料。

10.3.1 WSDL 和 SOAP

WSDL(Web Service Description Language, Web Service 描述语言)使用 XML 方式描述 Web Service。它提供 Web Service 的位置信息、能完成的功能以及通信方式，使用它就可以掌握访问 Web Service 的方法。

通常，WSDL 需要描述 3 个方面的信息。使用 < type />、< message />、< portType /> 标签定义 Web Service 能完成的方法、数据格式详情和访问 Web Service 需要使用的协议，使用 < service /> 标签定义 Web Service 网络地址。

在 Android 平台调用 Web Service 可以使用第三方类库 KSOAP2，它是一个 SOAP Web Service 客户端开发包，主要用于资源受限制的 Java 环境，如 Applets 或 J2ME 应用程序(CLDC/ CDC/MIDP)。通常情况下，Android 开发中都是使用 KSOAP2 Android。它是 Android 平台上一个高效的、轻量级的 SOAP 开发包，等同于 Android 平台上的 KSOAP2 的移植版本。

KSOAP2 当前版本是 3.0.0(ksoap2-android-assembly-3.0.0-jar-with-dependencies.jar)，无法下载的读者可以使用本节项目中的 jar 文件。

使用 KSOAP2 Android 调用 Web Service 的步骤通常可以概括为以下几步：

(1) 创建 HttpTransportSE 对象，通过 HttpTransportSE 类的构造方法可以指定 WebService 的 WSDL 文档的 URL，该 URL 由 WebService 提供。

(2) 创建 SoapSerializationEnvelope 对象，通过构造方法设置 SOAP 协议的版本号(一般为 SoapEnvelope.VER11)。该版本号需要根据服务端 Web Service 的版本号设置。在创建 SoapSerializationEnvelope 对象后，还需要设置 SoapSerializationEnvelope 类的 bodyOut 属性。

创建 SoapObject 对象，该类构造方法的第 1 个参数表示 Web Service 的命名空间，可以从 WSDL 文档中找到，第 2 个参数表示要调用的 Web Service 方法名。

(3) 传送参数(一般情况都需要)，使用 SoapObject 类的 addProperty(String key, Object value)方法，其中 key 是 Web Service 要求的参数，不能随意写。调用 SoapSerializationEnvelope 类的 setOutputSoapObject 方法或者直接使用 bodyOut 属性，将 SoapObject 对象附加给 SoapSerializationEnvelope，以明确请求参数。

(4) 调用 HttpTransportSE 的 call 方法，执行 Web Service 访问。call 方法有两个参数，第一个是 Web Service 的 SOAPAction(这可以从 Web Service 提供的文档找到)；第二个是 SoapSerializationEnvelope 对象。

(5) 处理返回的数据，通过 SoapSerializationEnvelope 对象的 bodyIn 属性可以获取 SoapObject 对象，该对象是 Web Service 返回的信息，解析该对象就能得到服务器的返回值。

10.3.2 调用 Web Service

项目 Proj10_7 演示通过执行 Web Service 访问实现手机号码归属地的查询功能。手机号码的归属地查询需要借助第三方平台，本项目中访问的第三方平台是 webxml 网站，注册后就可以开通测试。webxml 平台对手机号码归属地查询的 SOAP 描述信息如下：

```
POST /WebServices/MobileCodeWS.asmx HTTP/1.1
Host: webservice.webxml.com.cn
Content-Type: text/xml; charset = utf-8
Content-Length: length
SOAPAction: "http://WebXml.com.cn/getMobileCodeInfo"
<?xml version = "1.0" encoding = "utf-8"?>
<soap:Envelope xmlns:xsi = http://www.w3.org/2001/XMLSchema-instance
                xmlns:xsd = "http://www.w3.org/2001/XMLSchema"
                xmlns:soap = "http://schemas.xmlsoap.org/soap/envelope/">
  <soap:Body>
    <getMobileCodeInfo xmlns = "http://WebXml.com.cn/">
      <mobileCode>string</mobileCode>
      <userID>string</userID>
    </getMobileCodeInfo>
  </soap:Body>
</soap:Envelope>
```

通过分析上述描述信息,可以获取使用 SOAP 的全部信息。如 Web Service 的 URL 地址(Pos 连接 Host, http://webservice.webxml.com.cn/WebServices/MobileCodeWS.asmx)、编码格式、命名空间(http://WebXml.com.cn/)、访问方法(getMobileCodeInfo)、需要提交的参数(mobileCode 和 userID,测试时 userID 可以不写)等。

📖 Proj10_7 项目 MainActivity.java

```java
public void query(View v){
    final String phoneNum = et.getText().toString();
    new Thread(new Runnable(){
        @Override
        public void run() {
            String url = "http://webservice.webxml.com.cn/WebServices/MobileCodeWS.asmx";
            //第1步,创建 HttpTransportSE 对象
            HttpTransportSE httpSE = new HttpTransportSE(url);
            //第2步,创建 SoapSerializationEnvelope 对象
            SoapSerializationEnvelope soapEn = new
                                SoapSerializationEnvelope(SoapEnvelope.VER11);
            soapEn.dotNet = true;   // 是否调用 DotNet 开发的 Web Service
            //第3步,准备传送参数
            SoapObject soapObj = new
                    SoapObject("http://WebXml.com.cn/","getMobileCodeInfo");
            //参数名要与 Web Service 的描述一致
            soapObj.addProperty("mobileCode",phoneNum);
            soapObj.addProperty("userID", "");
            soapEn.bodyOut = soapObj;
            try {
                //第4步,调用 HttpTransportSE 的 call 方法,执行 Web Service 访问
                httpSE.call("http://WebXml.com.cn/getMobileCodeInfo", soapEn);
                //第5步,处理返回的数据
                if(soapEn.getResponse()!= null){
                    SoapObject soapOut = (SoapObject) soapEn.bodyIn;
```

```
                    int n = soapOut.getPropertyCount();
                    Message msg = new Message();
                    msg.obj = soapOut.getProperty(0).toString();
                    handler.sendMessage(msg);
                }else{
                    Log.i("Msg","服务器无返回信息");
                }
            } catch (IOException e) {
                e.printStackTrace();
            } catch (XmlPullParserException e) {
                e.printStackTrace();
            }
        }
    }).start();
}
```

程序需要联网,在 AndroidManifest.xml 文件中赋予相应权限。程序运行效果如图 10.8 所示。如果未获得第三方平台的访问权限或该平台不再提供服务,将无法得到结果。读者也可以尝试使用其他的第三方平台,如"聚合数据"等。

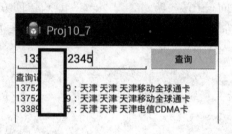

图 10.8 手机归属地查询

10.4 解析网络传输中的数据

手机终端无论以何种方式联网,访问服务器通常都会得到相应的返回数据。在移动应用开发中,这些返回数据多是平台无关性的(为了兼容 Android 和 iOS)。前几节的项目中,客户端获取的服务器数据都是字符串形式,这不适用于处理比较复杂的数据请求,如航班信息、天气信息等。对于这些有一定格式的数据,需要按照一定规则进行封装,JSON 和 XML 是目前应用比较广泛的、平台无关的、封装传输数据的格式。本节介绍在 Android 平台上如何解析收到的 JSON 数据和 XML 数据。

10.4.1 解析 JSON 格式数据

JSON(JavaScript Object Notation)是一种轻量级的数据交换格式,常用于 JavaScript 语言的数据交换,现在已经逐渐成为一种语言无关、平台无关的数据交换格式,这一点类似于 XML,但它要比 XML 更加轻量级。JSON 常应用于表示对象和数组,二者有不同的表示方法。

JSON 字符串在表示对象数据时采用下面的规则：
- 一个对象以"{"（左括号）开始，"}"（右括号）结束。
- 每个"名称"后跟一个":"（冒号）。
- "名称/值对"之间使用","（逗号）分隔。

例如 JSON 串{"firstName":"Freshen","lastName":"Green","age":20}表示一个对象拥有 3 个属性，即 fristName、lastName 和 age，其值分别是 Freshen、Green 和 20。

JSON 字符串在表示数组时采用下面的规则：
- 数组是值（value）的有序集合。
- 一个数组以"["（左中括号）开始，"]"（右中括号）结束。
- 值之间使用","（逗号）分隔。

例如 JSON 串

```
{"people":[
{"firstName":"Freshen","lastName":"Arthur","age":20},
{"firstName":"Ann","lastName":"Carl","age":18},
{"firstName":"Bob","lastName":"Steven","age":33}
]}
```

表示一个集合 people，有 3 个对象类型的元素。

Android 操作系统集成了 JSON 的 Jar 包，与 JSON 解析相关的类位于 org.json 包中，主要有 JSONArray（数组形式，数组元素可以是对象）、JSONObject（对象形式）、JSONString 等。通过这些类就可以很轻松地实现 JSON 字符串与 JSONObject 和 JSONArray 直接的相互转换。

项目 Proj10_8 演示如何解析 JSON 数据，访问"聚合数据"平台（需要注册，并申请 API 接口，获取对应的 appKey，网址是 http://www.juhe.cn/），调用"航线查询"功能，获取航班基础信息的 JSON 数据。在 API 帮助中可以查找到 URL 地址、需要提交的请求参数、请求方式（当下航线查询仅支持 GET）。以 GET 方式访问航班信息的核心代码如下。

📖 Proj10_8 项目 MainActivity.java 文件

```
...
//查询事件
    public void queryLines(View v){
        new Thread(new Runnable(){
            @Override
            public void run() {
                String urlStr = "http://apis.juhe.cn/plan/bc";
                HttpURLConnection httpConn = null;
                try {
                    String scity = et1.getText().toString();
                    String dcity = et2.getText().toString();
                    URL url = new URL(urlStr + "?start = " + scity + "&end = " + dcity +
                            "&key = 替换为申请的 Key"
                    );
                    httpConn = (HttpURLConnection) url.openConnection();
```

①

```java
                    httpConn.connect();
                    Log.i("Msg", "服务器返回码: " + httpConn.getResponseCode());
                    if(httpConn.getResponseCode() == 200){
                        StringBuffer sbuf = new StringBuffer();
                        String line;
                        BufferedReader br = new BufferedReader(new
                                InputStreamReader(httpConn.getInputStream()));
                        while( (line = br.readLine())!= null){
                            sbuf.append(line);
                        }
                        //解析 JSON
                        JSONObject  jsonObj = new JSONObject(sbuf.toString());
                        String rs = jsonObj.getString("result");
                        parseAirLine(rs);                        //解析航班
                        handler.sendEmptyMessage(0x100);         //更新 UI
                    }
                } catch (MalformedURLException e) {
                    e.printStackTrace();
                } catch (IOException e) {
                    e.printStackTrace();
                } catch (JSONException e) {
                    e.printStackTrace();
                }finally{
                    httpConn.disconnect();
                }
            }
        }).start();
    }
...
```

注意代码中的①处需要替换为自己申请的 app key,否则服务器不予响应。以北京(代码 PEK)和上海(代码 PVG,各城市代码可以调用"聚合数据"的"城市列表"功能获取)为出发地和目的地,成功访问数据之后,会返回如下 JSON 字符串:

```
{"resultcode":"200",
"reason":"查询成功",
"result":[
{"FlightNum":"HU7609",  "AirlineCode":"HU","Airline":"海南航空","DepCity":"北京首都",
"ArrCity":"上海浦东","DepCode":"PEK","ArrCode":"PVG","OnTimeRate":"97.5%", "
DepTerminal":"T1","ArrTerminal":"T2","FlightDate":"2015 - 08 - 12","PEKDate":"2015 - 08 -
12",  "DepTime":"06:35","ArrTime":"08:45",  "Dexpected":"07:18","Aexpected":"09:
00"},
{"FlightNum":"CA155","AirlineCode":"CA","Airline":"中国航空","DepCity":"北京首都","
ArrCity":"上海浦东","DepCode":"PEK","ArrCode":"PVG","OnTimeRate":"","DepTerminal":"
T3","ArrTerminal":"T2","FlightDate":"2015 - 08 - 12","PEKDate":"2015 - 08 - 12","DepTime":"
07:20","ArrTime":"09:35","Dexpected":"07:33","Aexpected":"09:17"},
...
```

result 对应的是航班信息的结果集,根据项目要求,将其解析为对象集合或键值对集合。本项目中将航班信息解析为 Map,对应的方法是 parseAirLine,解析完成后,通过 Handler 更新 UI,将数据填充到 ListView 列表。

📖 Proj10_8 项目 MainActivity.java

```java
…
//解析JSON串,转换为航线键值对集合
public void parseAirLine(String jsonStr){
    try {
        JSONArray  jsonArr = new JSONArray(jsonStr);
        airLines.clear();
        for(int i = 0 ;i< jsonArr.length();i++){
            JSONObject jobj = jsonArr.getJSONObject(i);
            Map< String,String > item = new HashMap< String,String >();
            item.put("FlightNum", jobj.getString("FlightNum"));        //航班号
            item.put("Airline", jobj.getString("Airline"));            //航空公司
            item.put("DepCity", jobj.getString("DepCity"));            //出发城市
            item.put("ArrCity", jobj.getString("ArrCity"));            //到达城市
            item.put("DepTerminal", jobj.getString("DepTerminal"));    //出发航站楼
            item.put("ArrTerminal", jobj.getString("ArrTerminal"));    //到达航站楼
            item.put("FlightDate", jobj.getString("FlightDate"));      //日期
            item.put("DepTime", jobj.getString("DepTime"));            //出发时间
            item.put("ArrTime", jobj.getString("ArrTime"));            //到达时间
            airLines.add(item);                                        //填充到集合中
        }
        Log.i("Msg", "共解析航班数目 " + airLines.size());
    } catch (JSONException e) {
        e.printStackTrace();
    }
}
…
```

由于结果集 JSON 串是一个数组(以[ ]为标志),在解析时要创建 JSONArray,数组中的元素会自动转换为 JSONObject 对象,再一一提取需要的属性即可。ListView 元素的布局文件可以查看 listview_airline_item.xml 文件。该项目有联网代码,需要赋予联网的权限,最终运行效果如图 10.9 所示。

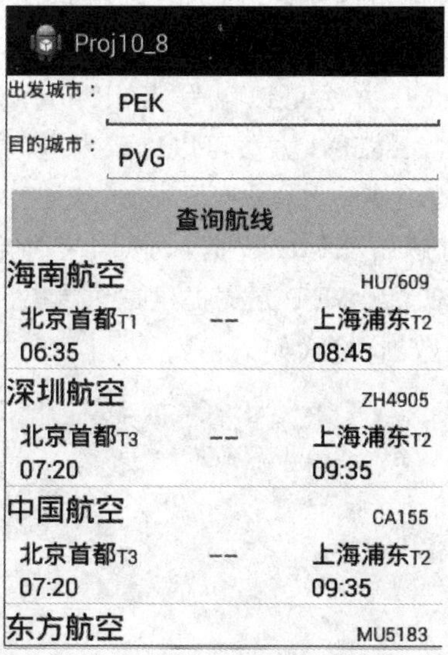

图 10.9 解析 JSON 数据

10.4.2 解析 XML 格式数据

XML 是可扩展标记语言(eXtensible Markup Language),是一种标记语言。XML 用来传送及携带数据信息,不用来表现或展示数据,这不同于 HTML 语言,HTML 语言重在表现数据。XML 用途的焦点是它要说明的数据是什么以及携带数据信息的格式。由于 XML 数据以纯文本格式进行存储,因此它与 JSON 一样,也提供了独立于软件和硬件的数据存储方法,让不同应用程序共享数据变得更加容易。

在 Android 系统中,常见的 XML 解析器分别是 DOM 解析器、SAX 解析器和 Pull 解析器。其中,Pull 解析器是 Android 附带的解析器,其工作方式类似于 SAX,是基于事件的模式。Pull 解析器小巧轻便,解析速度快,简单易用,非常适合在移动设备中使用。Android 系统内部在解析各种 XML 时也使用 Pull 解析器,这也是官方推荐的解析技术。Pull 解析技术属于第三方开源技术,它同样可以应用于 JavaSE 开发。

Android 系统中和 Pull 方式相关的包是 org.xmlpull.v1,在这个包中提供了 Pull 解析器的工厂类 XmlPullParserFactory 和 Pull 解析器 XmlPullParser。XmlPullParserFactory 对象通过调用 newPullParser 方法创建 XmlPullParser 解析器对象。

XmlPullParser 对象有两种创建方法。

(1) 使用工厂类创建:

```
XmlPullParserFactory pullFactory = XmlPullParserFactory.newInstance();
XmlPullParser xmlPullParser = pullFactory.newPullParser();
```

(2) 使用 Android 提供的工具类 android.util.Xml 创建

```
XmlPullParser xmlPullParser = Xml.newPullParser();
```

创建 XmlPullParser 对象之后,可以调用 setInput(InputStream inputStream, String inputEncoding)方法,设置需要解析的文件流,然后调用 getEventType 方法获取元素,通过判断文档事件是否为 START_DOCUMENT、END_DOCUMENT、START_TAG、END_TAG、TEXT 进行相应解析,next 方法可以触发 XmlPullParser 向下一个文档事件移动,getName 方法可以获取标签名,getText 方法可以读取内容区域。Pull 方式比较简单,而且可以根据判断停止解析。

项目 Proj10_9 演示如何解析 XML 数据,访问"聚合数据"航空城市列表,网址是 http://apis.juhe.cn/plan/city。以 GET 方式提供请求参数 key 和 dtype,当 dtype=xml,则服务器会以 XML 格式返回数据,格式如下:

```
<?xml version = "1.0" encoding = "utf-8" ?>
<root>
    <resultcode>
        200
    </resultcode>
    <reason>
        SUCCESSED!
```

```
        </reason>
        <result>
            <item>
                <city>
                    R
                </city>
                <spell>
                </spell>
            </item>
            <item>
                <city>
                    亚特兰大
                </city>
                <spell>
                    ATL
                </spell>
            </item>
...
```

在上面的 XML 数据中，<item>标签包含一个城市信息（城市名和拼写），<city>标签包含城市名，<spell>标签包含城市的拼写。XmlPullParser 解析的原理，是按照从上到下的顺序（不可逆）依次扫描各标签和内容区域，这类似与数据库中游标读取数据。遇到标签开始、标签结束、内容区域，都对应响应的事件，根据不同事件，执行响应的逻辑就可以将其中的数据解析出来。

📖 Proj10_9 项目 MainActivity.java 访问城市列表的代码

```java
...
    //按钮事件
    public void queryCitys(View v){
        new Thread(new Runnable(){
            @Override
            public void run() {
                HttpURLConnection httpConn = null;
                String urlStr = "http://apis.juhe.cn/plan/city";
                String params = "key=更换为申请的key&dtype=xml";     ①
                try {
                    URL url = new URL(urlStr + "?" + params);
                    httpConn = (HttpURLConnection) url.openConnection();
                    httpConn.connect();
                    Log.i("Msg", "服务器返回码 " + httpConn.getResponseCode());
                    if(httpConn.getResponseCode() == 200){
                        parseXml(httpConn.getInputStream());     //解析数据
                        handler.sendEmptyMessage(0x100);
                    }
                } catch (MalformedURLException e) {
                    e.printStackTrace();
                } catch (IOException e) {
                    e.printStackTrace();
```

```
            }
        }
    }).start();
}
...
```

注意代码中的①处提交的请求参数,dtype 如果不提供,服务器默认返回 JSON 格式的数据。获取服务返回输入流后,由方法 parseXml 解析。解析的过程也就是判定 XML 文档元素事件和标签内容的过程。

📖 Proj10_9 项目 MainActivity.java 解析 XML 数据的代码

```
...
//解析 XML
public void parseXml(InputStream is){
    //解析器
    XmlPullParser parser = Xml.newPullParser();
    try {
        parser.setInput(is, "UTF-8");
        //XML 文档事件
        int type = parser.getEventType();
        //存放一个城市信息
        Map<String,String> item = null;
        //只要没到文件结尾,就一直遍历文件
        while( (type = parser.next())!= XmlPullParser.END_DOCUMENT ){
            //读取标签名
            String tag = parser.getName();
            //遇到 item 标签开始,则创建一个城市信息
            if(type == XmlPullParser.START_TAG && tag.equalsIgnoreCase("item")){
                item = new HashMap<String,String>();
            }
            //如果遇到 city 标签开始
            if(type == XmlPullParser.START_TAG && tag.equalsIgnoreCase("city")){
                //从标签移动到内容区域,处理内容区域
                type = parser.next();
                String city = parser.getText();
                if(city!= null&&city.length()>0)
                    item.put("city", city);
            }else if(type == XmlPullParser.START_TAG && tag.equalsIgnoreCase("spell")){
                //处理内容区域
                type = parser.next();
                String spell = parser.getText();
                if(spell!= null&&spell.length()>0)
                    item.put("spell", spell);
            }
            //遇到 item 标签结束,将城市信息添加到集合中
            if(type == XmlPullParser.END_TAG && tag.equalsIgnoreCase("item")){
                citys.add(item);
                Log.i("Msg", "城市信息: " + item.get("city") + ":" + item.get("spell"));
            }
        }
        Log.i("Msg", "共解析城市信息个数: " + citys.size());
```

```
        } catch (XmlPullParserException e) {
            e.printStackTrace();
        } catch (IOException e) {
            e.printStackTrace();
        }
    }
...
```

XML 文档的解析过程要针对具体的 XML 文件，掌握 XML 文件标签的构成和元素的含义，才能正确解析出需要的数据。上述解析过程比较简单，整个文档中的主要标签是<item>，它对应一个完整的城市信息，它的子标签<city>和<spell>是具体的城市信息项。解析过程可以总结为 3 步：首先，如果扫描到<item>标签开始，则新建 Map 对象，准备存放一个城市信息；其次，扫描到子标签<city>和<spell>，读取其文本内容；最后，扫描到<item>标签结束，将一个城市信息添加到集合中，继续下一个城市信息的扫描。程序运行时需要赋予联网权限，运行效果如图 10.10 所示。

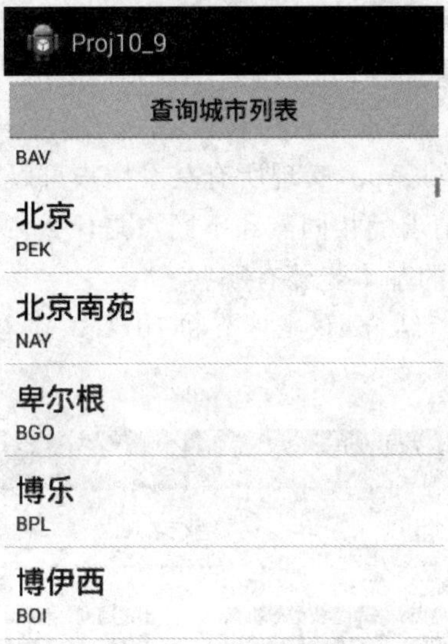

图 10.10 解析 XML 数据

10.5 习   题

1. 选择题

（1）以下关于 Socket 编程的说法有误的是（　　）。
    A. Socket 是客户端类，ServerSocket 是服务器端类
    B. 服务器端应先启动，等待客户端的连接
    C. accept 方法是非阻塞方法，不会影响 ServerSocket 后续代码的执行
    D. Socket 编程是基于 TCP 层的通信技术

（2）一个完成 URL 的组成部分可以不包括（　　）。

A. 访问协议　　　B. 端口号　　　C. 主机地址　　　D. 资料路径

(3) 以下关于 GET 请求的说法正确的是(　　)。
   A. GET 请求提交请求数据时，以"?"连接在 URL 之后
   B. GET 请求默认采用 UTF-8 编码格式
   C. GET 请求数据大小没有限制
   D. GET 请求不能替换成 POST 请求

(4) 以下关于 HttpURLConnection 执行 POST 请求的说法正确的是(　　)。
   A. HttpURLConnection 默认采用 POST 方式连接网络
   B. HttpURLConnection 执行 POST 访问，无须设定 doInput
   C. HttpURLConnection 可以访问受保护的资源
   D. HttpURLConnection 必须先用输出流，再用输入流

(5) 下列关于 HttpClient 执行 POST 请求的描述说法有误的是(　　)。
   A. 每次访问都应创建一个新的 HttpClient 对象
   B. 提交请求参数时使用 NameValuePair 封装
   C. 执行请求的方法参数是 HttpUriRequest 的实现类
   D. HttpClient 执行请求后，可以得到 HttpResponse 对象

2. 简答题

(1) 使用 Socket 通信，编写 App，实现所有登录用户可以相互聊天的功能。

(2) GET 请求和 POST 请求的相同点和不同点是什么？

(3) HttpClient 连接网络的基本步骤有哪些？

(4) 使用 HttpClient 编程，结合 10.4.1 节和 10.4.2 节，实现查询指定城市之间航线信息的功能，如下图所示。

```
查询到有57班次飞机

深圳                      北京

厦门航空    航班号：MF1059
出发地点：深圳宝安国际机场    出发时间：10:30
到达地点：北京首都国际机场    到达时间：13:35
每周班次：123456日           机型：    333

深圳航空    航班号：ZH9959
出发地点：深圳宝安国际机场    出发时间：07:30
到达地点：北京首都国际机场    到达时间：10:45
每周班次：123456日           机型：    739

中国国航    航班号：CA3310
出发地点：深圳宝安国际机场    出发时间：11:55
到达地点：北京首都国际机场    到达时间：14:55
每周班次：135日              机型：    737

中国联航    航班号：KN5852
出发地点：深圳宝安国际机场    出发时间：11:55
到达地点：北京南苑机场        到达时间：14:55
每周班次：135日              机型：    737

南方航空    航班号：CZ3193
出发地点：深圳宝安国际机场    出发时间：12:15
到达地点：北京首都国际机场    到达时间：15:30
每周班次：123456日           机型：    333
```

第 11 章　传感器应用与蓝牙通信

本章学习目标

- 掌握使用 SensorManager 获取传感器对象的方法。
- 掌握使用 Sensor 获取传感器数据的方法。
- 了解 Android 系统中常用的传感器。
- 了解 Android 系统中蓝牙通信的基本使用方法。

传感器(sensor)是一种器件或装置,可以按照一定的规律测量物理信号,并将测量结果转换为电信号输出。传感器通常由敏感元件和转换元件组成。中国物联网校企联盟认为,传感器的存在和发展让物体有了"视觉""听觉"和"嗅觉"等感官,让物体慢慢变得"活了起来"。

Android 系统中内置了多个传感器,主要包括加速度传感器(accelerometer)、陀螺仪(gyroscope)、环境光照传感器(light)、磁力传感器(magnetic field)、方向传感器(orientation)、压力传感器(pressure)、距离传感器(proximity)和温度传感器(temperature)。本章主要介绍 Android 系统中的内置传感器,讲解调用传感器的基础知识。

11.1　Android 中的传感器

传感器在智能手机中的使用越来越普及,功能越来越丰富,借助传感器可以开发很多新奇的应用程序。但并不是所有型号的手机中都内置了传感器,因此传感器应用程序不具有通用性。在开发传感器应用程序之前,需要先检测手机中是否支持相应的传感器。

11.1.1　传感器概述

在 Android 系统中使用传感器主要与 SensorManager 类和 Sensor 类相关,二者都位于 android.hardware 包。SensorManager 是各种传感器的管理类,通过它可以获得具体的传感器对象。要获取 SensorManager 对象,可以使用 Context.getSystemService 方法传入参数 SENSOR_SERVICE。

SensorManager 中的常用方法如表 11.1 所示。

表 11.1 SensorManager 常用方法

方法	说明	返回值类型
getDefaultSensor(int type)	获取指定类型的默认传感器	Sensor
getDefaultSensor(int type, boolean wakeUp)	获取指定类型和启动状态的传感器	Sensor
getSensorList(int type)	获取指定类型的传感器列表	List < Sensor >
registerListener（SensorEventListener listener, Sensor sensor, int samplingPeriodUs)	注册传感器监听器，建议在 Activity.onResume 方法中执行	boolean
unregisterListener(SensorEventListener listener)	解除监听器，建议在 Activity.onPause 方法中执行	void

在获取传感器时需要指定类型，在 Android 系统中常用的传感器类型如下：
- Sensor.TYPE_ACCELEROMETER，表示加速度传感器。
- Sensor.TYPE_MAGNETIC_FIELD，表示磁力传感器。
- Sensor.TYPE_ORIENTATION，表示方向传感器，该类型已经过时，建议使用 SensorManager 类中的方法 static float[] getOrientation（float[] R，float[] values)获取方向数据。
- Sensor.TYPE_GYROSCOPE，表示陀螺仪传感器。
- Sensor.TYPE_LIGHT，表示环境光照传感器。
- Sensor.TYPE_PRESSURE，表示压力传感器。
- Sensor.TYPE_AMBIENT_TEMPERATURE，表示温度传感器。
- Sensor.TYPE_PROXIMITY，表示距离传感器。
- Sensor.TYPE_ALL，表示获取全部传感器，可以在 getSensorList(int type)中使用。

通过传感器可以实时获取大量原始数据，这将是一个非常耗电的功能，合理控制传感器的启动、停止以及获取数据的速率尤为重要。传感器启动之后，不会因为应用程序退出或者屏幕关闭就停止工作，因此不要忘记显式调用 unregisterListener 方法解除注册监听程序。

注册传感器监听器时，可以设定获取传感器数据的速率 samplingPeriodUs。该设置只是对传感器系统的一个建议，不保证传感器一定会按照这个采样率工作。常用的采样率参数如下：
- SensorManager.SENSOR_DELAY_FASTEST，最快级别采样率，最低延迟。一般不是特别敏感的应用不推荐使用，这种模式会造成手机电力大量消耗。
- SensorManager.SENSOR_DELAY_GAME，游戏级别的采样率。通常实时性较高的游戏使用该级别。
- SensorManager.SENSOR_DELAY_NORMAL，普通级别采样率，标准延迟。对于一般的应用程序或益智类游戏可以使用，过低的采样率可能对一些实时性较高的程序（如赛车类游戏）造成跳帧现象。
- SensorManager.SENSOR_DELAY_UI，界面级别采样率。一般对于屏幕方向自动旋转使用，相对节省电能和逻辑处理。

获取传感器对象后，可以调用 Sensor 类中的方法获取传感器基本信息。Sensor 中的常用方法如表 11.2 所示。

表 11.2　Sensor 常用方法

方　　法	说　　　　明	返回值类型
getName()	获取传感器设备名称	String
getType()	获取传感器类型	int
getStringType()	获取传感器类型的字符串，API level 20 可用	String
getVendor()	获取传感器提供商	String
getVersion()	获取传感器当前版本	int

11.1.2　测试传感器应用程序

测试手机上的传感器需要借助真机或者 SensorSimulator 软件。前者可以比较真实地反映传感器的工作情况，但受限于机型，无法将所有传感器都测试一遍；后者是一个开源免费的测试工具，借助它可以在模拟器中调试传感器应用程序。

项目 Proj11_1 演示了如何查看手机上所支持的传感器，将传感器信息显示在 TextView 中。代码如下。

📖 Proj11_1 项目 MainActivity.java 文件

```java
public class MainActivity extends Activity {
    TextView tx1;
    SensorManager sm;              // 传感器管理对象
    @Override
    protected void onCreate(Bundle savedInstanceState) {
        super.onCreate(savedInstanceState);
        setContentView(R.layout.activity_main);
        tx1 = (TextView) findViewById(R.id.tv);
        sm = (SensorManager) getSystemService(Context.SENSOR_SERVICE);
        List<Sensor> sensors = sm.getSensorList(Sensor.TYPE_ALL);
        for (Sensor s : sensors) {
            String ts = "设备名称:" + s.getName() + "\n" + "供应商:"
                + s.getVendor() + "\n" + "设备版本:" + s.getVersion() +
                "\n---------------------\n";
            switch (s.getType()) {
                case Sensor.TYPE_ACCELEROMETER:
                    tx1.append("加速度传感器 accelerometer" + ts);
                    break;
                case Sensor.TYPE_GYROSCOPE:
                    tx1.append("陀螺仪传感器 gyroscope" + ts);
                    break;
                case Sensor.TYPE_LIGHT:
                    tx1.append("环境光线传感器 light" + ts);
                    break;
                case Sensor.TYPE_MAGNETIC_FIELD:
                    tx1.append("电磁场传感器 magnetic field" + ts);
                    break;
                case Sensor.TYPE_ORIENTATION:
                    tx1.append("方向传感器 orientation" + ts);
```

```
                    break;
                case Sensor.TYPE_PRESSURE:
                    tx1.append("压力传感器 pressure" + ts);
                    break;
                case Sensor.TYPE_PROXIMITY:
                    tx1.append("距离传感器 proximity" + ts);
                    break;
                case Sensor.TYPE_AMBIENT_TEMPERATURE:
                    tx1.append("温度传感器 temperature" + ts);
                    break;
                default:
                    tx1.append("未知传感器"  + ts);
                    break;
            }
        }
        tx1.append("\n传感器数目: " + sensors.size());
    }
}
```

在真机上运行项目,输出结果如图 11.1 所示。在不同品牌、型号的手机上运行上述项目,输出结果不同。

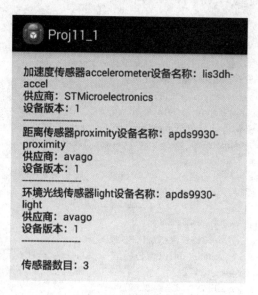

图 11.1　传感器列表

11.2　加速度传感器

加速度传感器是比较常见的传感器,几乎在每种型号的手机中都可以使用。这里的加速度特指重力加速度,当手机静止时,加速度传感器返回的值与重力加速度的值接近,约为 $g=9.8\text{m/s}^2$。加速度传感器有 3 个方向的值,分别表示三维空间中 X、Y 和 Z 轴的加速度。

在传感器中使用的坐标系统与 View 视图中的坐标系统不同。将手机竖直向上放置,如图 11.2 所示。平行于屏幕,水平方向从左到右为 X 轴正方向;平行于屏幕,竖直方向从下到上为 Y 轴正方向;垂直屏幕,从内到外为 Z 轴正方向。

参考上述坐标系统,传感器返回数值与手机状态的关系如下:

- 当设备竖直朝上放置时,Y 轴的返回值接近 g 值。
- 当设备竖直朝下放置时,Y 轴的返回值接近负 g 值。
- 当设备左边朝下放置时,X 轴的返回值接近 g 值。
- 当设备右边朝下放置时,X 轴的返回值接近负 g 值。
- 当设备的屏幕朝上平放在桌面上时,Z 轴的返回值接近 g 值。
- 当设备的屏幕朝下平放在桌面上时,Z 轴的返回值接近负 g 值。

图 11.2 传感器中坐标系统

当设备以图 11.2 的姿态竖直向上以 $5m/s^2$ 加速度向上加速移动时,Y 轴的返回值是 $g+5$。当设备以图 11.2 的姿态自由落体时,Y 轴的返回值约为 0。

监听传感器数值的变化需要借助监听器接口 SensorEventListener,它有两个方法,分别处理传感器精度变化和传感器数值变化。方法的声明如下:

- abstract void onAccuracyChanged(Sensor sensor, int accuracy),当注册的监听器精度发生变化时调用。
- abstract void onSensorChanged(SensorEvent event),当传感器数值发生变化时调用,参数是 SensorEvent。

SensorEvent 是封装了传感器数值(float[] values 属性)、时间戳(long timestamp 属性)、精度(int accuracy 属性)和传感器(Sensor sensor)信息的对象。其中 values 属性中存放的就是传感器返回的数据。不同的类型的传感器,返回的数据条数和意义不同。对于加速度传感器,values 数组中会有 3 个值,分别表示 X、Y、Z 轴的加速度。对于光线传感器,values 数组中只有一个值,表示光照强度。values 数组中的数据含义,可以查阅 SensorEvent 类的 API 文档。

项目 Proj11_2 演示了加速度传感器的使用。布局文件中有 3 个 TextView 控件,实时显示 X、Y、Z 轴的加速度值。当某个方向的数值大于 12 时,启用设备的振动功能。代码如下。

📖 Proj11_2 项目 MainActivity.java 文件

```java
public class MainActivity extends Activity implements SensorEventListener {
    SensorManager sm;              //传感器管理
    Sensor s;                      //传感器
    TextView tv1,tv2,tv3;          //显示控件
    Vibrator vibrator;             //振动器
    @Override
    protected void onCreate(Bundle savedInstanceState) {
        super.onCreate(savedInstanceState);
        setContentView(R.layout.activity_main);
```

```java
            tv1 = (TextView) findViewById(R.id.tv1);
            tv2 = (TextView) findViewById(R.id.tv2);
            tv3 = (TextView) findViewById(R.id.tv3);
            //初始化振动器对象
            vibrator = (Vibrator) getSystemService(Context.VIBRATOR_SERVICE);
            //获取传感器管理对象
            sm = (SensorManager) getSystemService(Context.SENSOR_SERVICE);
            //获取加速度传感器
            s = sm.getDefaultSensor(Sensor.TYPE_ACCELEROMETER);
        }
        @Override
        protected void onResume() {
            super.onResume();
            if(s!= null)                    //注册传感器
                sm.registerListener(this, s, SensorManager.SENSOR_DELAY_NORMAL);
        }
        @Override
        protected void onPause() {
            super.onPause();
            if(s!= null)                    //解除注册关系
                sm.unregisterListener(this);
        }
        //实现传感器监听接口 SensorEventListener 中的方法
        @Override
        public void onAccuracyChanged(Sensor arg0, int arg1) {
        }
        @Override
        public void onSensorChanged(SensorEvent arg0) {
            //获取3个方向的传感器数值
            float xr = arg0.values[0];
            float yr = arg0.values[1];
            float zr = arg0.values[2];
            tv1.setText("X 轴加速度：" + xr);
            tv2.setText("Y 轴加速度：" + yr);
            tv3.setText("Z 轴加速度：" + zr);
            //启用振动的逻辑判定
            if(xr > 12||yr > 12||zr > 12)
                vibrator.vibrate(200);
        }
}
```

在真机上运行上述项目，只要设备中有加速度传感器，有振动马达，就可以通过"摇一摇"实现振动效果。有兴趣的读者可以尝试将"摇一摇"效果融合到应用程序中，实现"摇一摇"发送信息，"摇一摇"切换播放音乐文件等。

要使用振动效果，需要开启如下权限：

```xml
< uses - permission android:name = "android.permission.VIBRATE"/>
```

11.3 光线传感器

光线传感器一般位于手持设备屏幕上方,它能获取手持设备目前所处的光线亮度,通常用于自动调节手持设备屏幕亮度,给使用者带来最佳的视觉效果。例如,手持设备屏幕背光灯在弱光下会自动变暗,在强光下则自动变亮,否则无法看清屏幕中的内容。

光线传感器中 SensorEvent 对象的属性 values 数组只有第一个元素(values[0])有意义,它表示光线的强度,最大的值为 120000.0f。Android SDK 将光线强度分为不同的等级,每一个等级的最大值由一个常量表示,都定义在 SensorManager 类中。这些常量的信息如下:

public static final float LIGHT_SUNLIGHT_MAX =120000.0f;
public static final float LIGHT_SUNLIGHT=110000.0f;
public static final float LIGHT_SHADE=20000.0f;
public static final float LIGHT_OVERCAST= 10000.0f;
public static final float LIGHT_SUNRISE= 400.0f;
public static final float LIGHT_CLOUDY= 100.0f;
public static final float LIGHT_FULLMOON= 0.25f;
public static final float LIGHT_NO_MOON= 0.001f;

上述常量只是临界值,不具备代表意义。在使用光线传感器时要根据实际情况确定一个范围。例如,当太阳逐渐升起时,values[0]的值很可能会超过 LIGHT_SUNRISE,当 values[0]的值逐渐增大时,就会逐渐越过 LIGHT_OVERCAST 而达到 LIGHT_SHADE,当然,如果天特别好,也可能会达到 LIGHT_SUNLIGHT 甚至更高。

项目 Proj11_3 演示了光线传感器的使用,获取传感器数据,修改屏幕亮度。布局文件比较简单,在 ScrollView 中嵌套 TextView 以方便滚动查看光照数据。

📖 Proj11_3 项目 MainActivity.java 文件

```java
public class MainActivity extends Activity implements SensorEventListener {
    //声明所需组件
    TextView tv;
    SensorManager sm;
    Sensor s;
    @Override
    protected void onCreate(Bundle savedInstanceState) {
        super.onCreate(savedInstanceState);
        setContentView(R.layout.activity_main);
        tv = (TextView) findViewById(R.id.tv);
        sm = (SensorManager) getSystemService(Context.SENSOR_SERVICE);
        //获取光照传感器
        s = sm.getDefaultSensor(Sensor.TYPE_LIGHT);
    }
    @Override
    protected void onResume() {
```

```
        super.onResume();
        if(s!= null)                    //注册
            sm.registerListener(this, s, SensorManager.SENSOR_DELAY_UI);
    }
    @Override
    protected void onPause() {
        super.onPause();
        if(s!= null)                    //解除注册
            sm.unregisterListener(this);
    }
    @Override
    public void onAccuracyChanged(Sensor arg0, int arg1) {
    }
    @Override
    public void onSensorChanged(SensorEvent arg0) {
        float lig = arg0.values[0];     //获取光照数据
        setScreenBrightness(lig);       //修改屏幕亮度
        tv.append("光照强度: " + lig + " (lux)\n");
    }
    //设置屏幕亮度
    public void setScreenBrightness(float v) {
        //获取当前窗口属性
        WindowManager.LayoutParams params = getWindow().getAttributes();
        params.screenBrightness = v / 255f;
        getWindow().setAttributes(params);
    }
}
```

运行上述项目,通过改变光线强弱来测试光线传感器的数据采集,如图 11.3 所示。根据光照强度修改屏幕亮度时,应该划分范围,或者采用 SensorManager 中的光线强度等级常量,当光照强度发生较大变化时再调整屏幕亮度,没有必要像代码中一样实时调节屏幕亮度。注意,如果屏幕亮度调节不起作用,可以尝试关闭系统设置中的"自动调整屏幕亮度"选项。

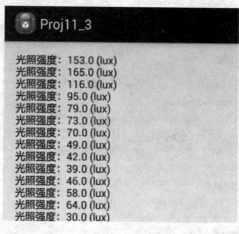

图 11.3 光线传感器采集数据

11.4 距离传感器

距离传感器通过发射特别短的光脉冲来测量此光脉冲从发射到被物体反射回来的时间，通过时间长短来计算设备与物体之间的距离。移动设备中的距离传感器一般都位于设备顶端，便于检测是否靠近脸部，是否正在打电话。

距离传感器在手机中的应用表现在：打电话时脸部靠近屏幕，屏幕灯会熄灭，并自动锁屏，防止误操作；当打完电话，移开手机后，屏幕灯会自动开启，并且自动解锁。

距离传感器中 SensorEvent 对象的属性 values 数组，只有第一个元素(values[0])有意义。当靠近手机时(只有向传感器位置靠近时才会起作用)，属性值为 0，离开时属性值为其他(具体数据与手机型号有关)。

项目 Proj11_4 演示了距离传感器的应用，获取传感器数据，控制屏幕电源的启动和关闭。布局文件比较简单，在 ScrollView 中嵌套 TextView 以方便滚动查看距离数据。

📖 Proj11_4 项目 MainActivity.java 文件

```java
public class MainActivity extends Activity implements SensorEventListener {
    TextView tv;
    SensorManager sm;
    Sensor s;
    //屏幕开关
    private PowerManager pm = null;                    //电源管理对象
    private PowerManager.WakeLock pmLock = null;       //电源锁
    @Override
    protected void onCreate(Bundle savedInstanceState) {
        super.onCreate(savedInstanceState);
        setContentView(R.layout.activity_main);
        tv = (TextView) findViewById(R.id.tv);
        sm = (SensorManager) getSystemService(Context.SENSOR_SERVICE);
        s = sm.getDefaultSensor(Sensor.TYPE_PROXIMITY);
        //获取电源管理器
        pm = (PowerManager) getSystemService(Context.POWER_SERVICE);
        pmLock = pm.newWakeLock(32, "Msg");//第一个参数为电源锁级别    }
    @Override
    protected void onResume() {
        super.onResume();
        sm.registerListener(this, s, SensorManager.SENSOR_DELAY_UI);
    }
    @Override
    protected void onDestroy() {
        super.onDestroy();
        if(sm != null){
            pmLock.release();                           //释放电源锁
            //如果不释放就结束这个 Acitivity,仍然会有自动锁屏的效果,不信可以试一试
            sm.unregisterListener(this);                //注销传感器监听
        }
    }
```

```
    @Override
    public void onAccuracyChanged(Sensor arg0, int arg1) {
    }
    @Override
    public void onSensorChanged(SensorEvent arg0) {
        float v = arg0.values[0];
        tv.append("距离传感器：" + v + "\n");
        // 贴近手机
        if (v == 0.0) {
            System.out.println("准备锁屏……");
            if (pmLock.isHeld()) return;
            else pmLock.acquire();            // 申请设备电源锁
        } else {                              // 远离手机
            System.out.println("解除锁屏……");
            if (pmLock.isHeld()) return;
            else{
                pmLock.setReferenceCounted(false);
                pmLock.release();             // 释放设备电源锁
            }
        }
        tv.append("距离传感器：" + v + "\n");
    }
}
```

运行上述项目前,需要在 AndroidManifest.xml 文件中配置如下权限:

```
<uses-permission android:name = "android.permission.DEVICE_POWER"/>
<uses-permission android:name = "android.permission.WAKE_LOCK"/>
```

项目运行效果如图 11.4 所示。当靠近距离传感器时,数值为 0;当离开距离传感器时,数值为 9。需要注意的是,并非所有设备都按照上述数据输出。

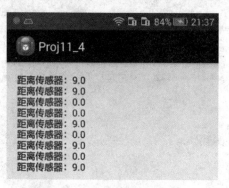

图 11.4　距离传感器采集数据

11.5　蓝牙通信技术应用

蓝牙是一种无线近距离通信技术,可以应用在智能手机、平板电脑、笔记本电脑、音响、耳机等众多设备中。Android 系统包含了对蓝牙通信协议的支持,并对其进行了封装,使用相应的 API 就可以完成蓝牙设备的发现、连接和通信。本节主要介绍如何在 Android 系统中使用蓝牙通信完成基本的数据交换功能,对蓝牙协议栈的知识不作介绍,要了解这部分内容的读者可以参考其他资料。

11.5.1 近距离通信技术介绍

随着物联网技术的发展与成熟,短距离无线通信应用步伐不断加快,逐渐走向成熟。所谓短距离通信,一般距离限制在较短范围,从几厘米到几十米。在短距离通信中,根据对传输速度、距离和耗电量等指标的要求不同,出现了很多各具特点的技术。目前,国内应用较为普遍的短距离通信技术有 RFID、WiFi、ZigBee、蓝牙和 NFC。

RFID(Radio Frequency Identification,无线射频识别)主要是通过无线电信号识别特定目标并读写相关数据。这种技术广泛应用在门禁系统、食品安全溯源、仓储货物管理等场合。

WiFi 是一种允许将设备连接到一个无线局域网(WLAN)的技术,是目前使用最广的一种无线网络传输技术,几乎所有智能手机、平板电脑和笔记本电脑都支持 WiFi 连接。使用这种技术,需要借助无线路由器把有线信号转换成 WiFi 信号。

ZigBee 是一种短距离、低功耗的无线通信技术。在低耗电待机模式下,2 节 5 号干电池可支持 1 个节点工作 6~24 个月甚至更长,这是 ZigBee 的突出优势。相比而言,蓝牙节点可工作数周,WiFi 节点仅能工作数小时。ZigBee 技术在智能家居、智能楼宇、智能小区建设中逐渐发挥作用。

蓝牙技术发展到现在已经历很多年,先后出现了很多通信标准。蓝牙最初由电信巨头爱立信公司于 1994 年创制,当时是作为 RS232 数据线的替代方案。蓝牙 4.0 是 2012 年推出的蓝牙版本,是 3.0 的升级版本,与 3.0 版本相比,具有更省电、成本低、3ms 低延迟、超长有效连接距离、AES-128 加密等新特点。

NFC 技术由 RFID 演变而来,是飞利浦半导体(现恩智浦半导体公司)、诺基亚和索尼共同研制开发的,其基础是 RFID 及互连技术。它是一种短距高频的无线通信技术,通信距离在 10cm 以内。可以通过带有 NFC 功能的手机实现支付、签到、刷公交卡或门票,或者和别人交换名片、传输文件等功能。

11.5.2 Android 系统中的蓝牙组件

Android 系统中与蓝牙通信相关的类位于 android.bluetooth 包中,提供管理蓝牙基本功能的各种类,实现如扫描蓝牙设备、与其他设备连接、完成设备之间的数据传输等基础功能。Android 系统中的蓝牙模块同时支持传统蓝牙通信和低功耗蓝牙通信,即蓝牙 4.0 标准。

使用 Android 系统提供的蓝牙 API 实现蓝牙设备之间的通信,主要包括 4 个基本步骤:设置蓝牙设备并寻找匹配的设备,请求连接匹配,完成设备间配对,使用 I/O 在设备之间传输数据,如图 11.5 所示。在 Android 系统中实现蓝牙通信的编程模式,可以参考 TCP 协议中的 Socket 通信模式,二者的思路基本一致。

使用蓝牙通信时常用的类有 BluetoothAdapter、BluetoothDevice、BluboothServerSocket 和 BluetoothSocket。

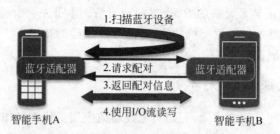

图 11.5 蓝牙通信过程

BluetoothAdapter 类代表一个本地的蓝牙适配器,它是所有蓝牙交互的入口点。使用 BluetoothAdapter 可以实现发现其他蓝牙设备、查询已经配对的设备功能,使用已知 MAC 地址实例化一个 BluetoothDevice 对象,建立 BluetoothServerSocket(作为服务器端,相当于 Socket 编程中的 ServerSocket)来监听来自其他设备的连接。

获取 BluetoothAdapter 对象的方式是调用该类的静态方法 getDefaultAdapter,这也是蓝牙通信编程中的第一步。该类的常用方法如表 11.3 所示。

表 11.3 BluetoothAdapter 常用方法

方 法	说 明	返回值类型
cancelDiscovery()	取消当前设备扫描附近蓝牙设备的进程	boolean
checkBluetoothAddress(String address)	验证蓝牙地址的有效性,MAC 地址中字母需要大写	static boolean
disable()	关闭蓝牙适配器	boolean
enable()	打开蓝牙适配器	boolean
getAddress()	获取本地蓝牙适配器地址,如 00:43:A8:23:10:F0	String
getName()	获取本地蓝牙设备的名称	String
getBondedDevices()	返回已经配对的蓝牙设备集合	Set<BluetoothDevice>
getRemoteDevice(String address)	获取指定地址的蓝牙设备对象,地址必须合法	BluetoothDevice
getState()	获取本地蓝牙适配器的状态,如 STATE_OFF、STATE_TURNING_ON、STATE_ON、STATE_TURNING_OFF	int
listenUsingInsecureRfcommWithServiceRecord(String name, UUID uuid)	以非安全的方式创建服务器端蓝牙套接对象,uuid 是通用唯一识别码	BluetoothServerSocket
listenUsingRfcommWithServiceRecord(String name, UUID uuid)	以安全的方式创建服务器端蓝牙套接对象	BluetoothServerSocket
setName(String name)	设置本地蓝牙适配器名称	boolean
startDiscovery()	开启蓝牙设备扫描进程	boolean

需要注意的是蓝牙通信中使用的 MAC 地址,与 WLAN 中的 MAC 地址是不同的,二

者没有必然联系。如无特殊说明,本节中 MAC 地址所指的是蓝牙 MAC 地址。

BluetoothDevice 类代表蓝牙设备,使用它可以连接远端蓝牙设备,获取远端蓝牙设备的名称、地址、种类和绑定状态。BluetoothDevice 类的常用方法如表 11.4 所示。

表 11.4 BluetoothDevice 常用方法

方法	说明	返回值类型
createBond()	开始匹配进程	boolean
createInsecureRfcommSocketToServiceRecord (UUID uuid)	以非安全的方式创建蓝牙套件对象	BluetoothSocket
createRfcommSocketToServiceRecord(UUID uuid)	创建 RFCOMM 蓝牙套件,并以安全方式连接	BluetoothSocket
getAddress()	获取地址信息	String
getName()	获取名称	String
getBondState()	获取绑定(配对)状态,如 BOND_NONE、BOND_BONDING、BOND_BONDED	int
setPin(byte[] pin)	设置 pin 码	boolean

BluboothServerSocket 类是蓝牙连接的服务端监听对象,等待可能到来的连接请求。要实现连接两个蓝牙设备互联,则必须有一个设备作为服务器,打开一个服务套接字,即 BluboothServerSocket。当远端设备发起连接请求并且已经连接成功时,该对象将会返回一个 BluetoothSocket 对象。这类似于 TCP 通信中 ServerSocket 使用阻塞方法 accept 等待连接,如果有连接到来,则返回 Socket 对象。BluboothServerSocket 常用方法如表 11.5 所示。

表 11.5 BluboothServerSocket 常用方法

方法	说明	返回值类型
accept(int timeout)	阻塞方法,在指定时间内等待连接	BluetoothSocket
accept()	阻塞方法,等待连接建立	BluetoothSocket
close()	关闭断开,并释放相关资源	void

BluetoothSocket 类是一个蓝牙连接的套接字对象,类似于 TCP 中的 Socket 套接字。通过 BluetoothSocket 对象,可以开启 I/O 流,实现与其他蓝牙设备的通信。BluetoothSocket 的常用方法如表 11.6 所示。

表 11.6 BluetoothSocket 常用方法

方法	说明	返回值类型
close()	关闭连接,释放资源	void
connect()	阻塞方法,开始连接其他设备,直到连接建立或失败	void
getInputStream()	获取输入流	InputStream
getOutputStream()	获取输出流	OutputStream
getRemoteDevice()	获取该端口正在连接或已经连接的设备	BluetoothDevice
isConnected()	检测是否连接	boolean

项目Proj11_5演示了如何在智能手机上开启蓝牙,扫描蓝牙设备。布局文件是activity_main.xml,使用两个Button控件和一个TextView控件,布局比较简单,可以参考项目的运行效果图。

当单击"设置蓝牙可见"按钮时,执行如下代码。

📖 Proj11_5项目MainActivity.java文件的"设置蓝牙可见"部分代码

```java
...
//开启蓝牙可见,允许被其他蓝牙设备发现
private void setVisibale(){
    tv1.append("设置蓝牙设备显示!\n");
    //创建一个Intent对象,设置蓝牙设备为可见状态
    Intent intent = new Intent(BluetoothAdapter.ACTION_REQUEST_DISCOVERABLE);
    //指定可见状态的持续时间,若大于300s,就认为是300s
    intent.putExtra(BluetoothAdapter.EXTRA_DISCOVERABLE_DURATION, 300);
    startActivity(intent);
}
...
```

当单击"扫描连接的蓝牙设备"按钮时,执行如下代码:

📖 Proj11_5项目MainActivity.java文件的"扫描连接的蓝牙设备"部分代码

```java
...
//扫描已经连接的蓝牙设备
    private void scannBT(){
        tv1.append("开始扫描蓝牙设备……\n");
        //得到BluetoothAdapter对象
        BluetoothAdapter adapter = BluetoothAdapter.getDefaultAdapter();
        //判断BluetoothAdapter对象是否为空,如果为空,则表明本机没有蓝牙设备
        if(adapter != null){
            tv1.append("本机拥有蓝牙设备\n");
            //调用isEnabled方法判断当前蓝牙设备是否可用
            if(!adapter.isEnabled()){
                //如果蓝牙设备不可用,提示用户启动蓝牙适配器
                Intent intent = new Intent(BluetoothAdapter.ACTION_REQUEST_ENABLE);
                startActivity(intent);
            }
            //得到所有已经配对的蓝牙适配器对象
            Set<BluetoothDevice> devices = adapter.getBondedDevices();
            if(devices.size()>0) {
                //迭代已配对的设备
                tv1.append("已经连接蓝牙设备信息如下:\n");
                for(Iterator iterator = devices.iterator();iterator.hasNext();){
                    //得到BluetoothDevice对象,也就是得到配对的蓝牙适配器
                    BluetoothDevice device = (BluetoothDevice)iterator.next();
                    //得到远程蓝牙设备的地址
                    tv1.append("设备名:" + device.getName() + "地址:" + device.getAddress() + "\n");
                }
```

```
            }
        }else{
            tv1.append("本机没有蓝牙设备\n");
        }
    }
...
```

运行该项目之前,需要在配置文件中声明如下权限:

```
<uses-permission android:name="android.permission.BLUETOOTH" />
<uses-permission android:name="android.permission.BLUETOOTH_ADMIN" />
```

为测试该项目的运行效果,现采用两部手机进行演示(分别是华为和 HTC,都具有蓝牙通信功能)。在华为手机上运行上述项目,单击"设置蓝牙可见"按钮,时间为 300s,效果如图 11.6 所示。

使用 HTC 手机连接华为手机,根据提示设置配对信息,完成配对过程,如图 11.7 所示。已经配对成功的设备会存储在手机中,以便后续使用。配对成功是连接的前提,但配对成功与蓝牙设备建立通信连接是两个不同的过程。前者相当于认证过程,后者才是真正的数据交换。

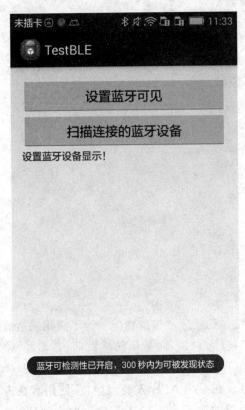

图 11.6　设置蓝牙可见

图 11.7　手机配对请求

因为任何无线通信技术都存在被监听和破解的可能，所以为了保证蓝牙通信的安全性，需要先采用认证的方式确认连接，再进行数据交互。同时为了保证使用的方便性，以配对的形式完成两个蓝牙设备之间的首次通信认证，经配对之后，随后的通信连接就不必每次都要连接确认了。所以不进行配对，两个设备之间便无法建立认证关系，无法进行连接及其之后的操作，配对在一定程度上保证了蓝牙通信的安全。

当蓝牙配对成功完成后，单击"扫描连接的蓝牙设备"按钮，会显示相应的连接信息，如图 11.8 所示，将已配对的 HTC 手机和 MAC 地址显示在 TextView 控件中。

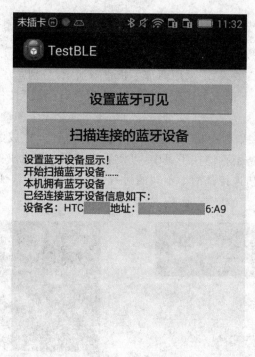

图 11.8　已配对蓝牙设备信息

11.5.3　蓝牙设备间的通信

在 11.5.2 节的基础上，本节实现两部智能手机借助蓝牙通信的功能。为了简化通信流程，项目 Proj11_6 只实现了单向通信，即由一部手机（客户端）向另一部手机（服务器端）发送数据，后者将接收到的数据显示在文本控件中。

在该项目中，将客户端功能和服务器端功能融合在一个 APK 安装包中，由使用者在进行蓝牙组网时确定哪个手机作为服务器端。感兴趣的读者也可以将其分为两个 APK 来开发。

该项目的功能相对简单，主要由 MainActivity.java、TalkActivity.java 和两个布局文件

activity_main.xml、activity_talk.xml组成。MainActivity作为蓝牙通信的主界面,实现启动蓝牙设备、发现其他设备、启动服务器、接收并显示客户端发送的信息的功能。TalkActivity作为客户端发送信息的界面,实现连接服务器、发送数据的功能。

activity_main.xml文件的代码如下,ListView控件用于呈现已匹配的蓝牙设备,3个Button控件依次实现设置蓝牙可见、扫描匹配设备、启动服务器功能,TextView控件显示连接信息和服务器收到的客户端数据。布局效果可以参考图11.10。

📖 Proj11_6项目中activity_main.xml文件的主要控件

```xml
…
<ListView
    android:id = "@ + id/listView1"
    android:layout_width = "match_parent"
    android:layout_height = "wrap_content" >
</ListView>
<ScrollView
    android:id = "@ + id/scrollView1"
    android:layout_width = "wrap_content"
    android:layout_height = "wrap_content"
    android:layout_below = "@id/listView1" >
    <LinearLayout
        android:layout_width = "match_parent"
        android:layout_height = "match_parent"
        android:orientation = "vertical" >
        <Button
            android:id = "@ + id/button1"
            android:layout_width = "match_parent"
            android:layout_height = "wrap_content"
            android:text = "设置蓝牙可见" />
        <Button
            android:id = "@ + id/button2"
            android:layout_width = "match_parent"
            android:layout_height = "wrap_content"
            android:text = "扫描连接的蓝牙设备" />
        <Button
            android:id = "@ + id/button3"
            android:layout_width = "match_parent"
            android:layout_height = "wrap_content"
            android:text = "启动服务器" />
        <TextView
            android:id = "@ + id/textview1"
            android:layout_width = "match_parent"
            android:layout_height = "wrap_content" />
    </LinearLayout>
</ScrollView>
…
```

3个按钮控件的监听器方法如下,让MainActivity实现OnClickListener接口。在监听器方法中执行不同的功能逻辑。

📖 Proj11_6 项目中 MainActivity.java 文件的按钮单击监听方法

```java
@Override
public void onClick(View arg0) {
    switch(arg0.getId()){
    case R.id.button1:setVisibale();break;
    case R.id.button2:scannBT();break;
    case R.id.button3:startBTSer();break;
    }
}
```

(1) 设置蓝牙可见, 在 11.5.2 节中已经介绍过;
(2) 发现匹配设备与 11.5.2 节中介绍的基本一致, 只是在迭代设备时将设备信息呈现在 ListView 中。

📖 Proj11_6 项目中 MainActivity.java 文件的扫描匹配设备方法

```java
private void scannBT(){
    adapter = BluetoothAdapter.getDefaultAdapter();
    if(adapter != null){
        if(!adapter.isEnabled()){
            Intent intent = new Intent(BluetoothAdapter.ACTION_REQUEST_ENABLE);
            startActivity(intent);
        }
        Set<BluetoothDevice> devices = adapter.getBondedDevices();
        data = new ArrayList<String>();
        if(devices.size()>0) {
            for(Iterator iterator = devices.iterator();iterator.hasNext();){
                BluetoothDevice device = (BluetoothDevice)iterator.next();
                tv1.append("设备名: " + device.getName() + "地址: "
                    + device.getAddress() + "\n");
                //List<String> data 属性, 存放匹配设备的 MAC 地址
                data.add(device.getAddress());
            }
            ArrayAdapter aa = new ArrayAdapter(MainActivity.this,
                        android.R.layout.simple_list_item_1,data);
            lv.setAdapter(aa);
            lv.setOnItemClickListener(this);       //ListView 监听器
        }
    }else{
        tv1.append("本机没有蓝牙设备\n");
    }
}
```

ListView 控件的监听器通过 MainActivity 实现 OnItemClickListener 接口完成。当单击某个 MAC 地址时, 即把它当成服务器地址, 跳转到 TalkActivity, 建立蓝牙通信。

📖 Proj11_6 项目中 MainActivity.java 文件的 ListView 监听器方法

```java
@Override
public void onItemClick(AdapterView<?> arg0, View arg1, int arg2, long arg3) {
    // String serMac 属性,表示服务器地址
    serMac = data.get(arg2);
    tv1.append("当前点击: " + serMac);
    Intent intent = new Intent(MainActivity.this,TalkActivity.class);        //跳转
    //服务器地址传入 TalkActivity
    intent.putExtra("mac", serMac);
    startActivity(intent);
}
```

(3) 启动服务器的功能代码如下。在哪个手机点击该功能按钮,就会使它成为服务器端。服务器启动之后,需要一直监听客户端的链接,因此需要启动一个子线程,在子线程中使用 BluetoothServerSocket 的 accept 方法监听客户端连接。

📖 Proj11_6 项目中 MainActivity.java 文件的启动服务器方法

```java
private void startBTSer() {
    serFlag = true;
    new Thread(new BTServer()).start();          //启动子线程
}
private class BTServer implements Runnable{
    private BluetoothServerSocket bss;           //服务端连接对象
    public BTServer(){
        try {                                    //创建非安全模式的服务器连接
            bss = adapter.listenUsingInsecureRfcommWithServiceRecord("BTSer", uuid);
        } catch (IOException e) {
            e.printStackTrace();
        }
    }
    @Override
    public void run() {
        BluetoothSocket bs = null;
        Log.i("Msg", "服务器等待……");
        while(serFlag){
            try {
                bs = bss.accept();               //等待连接
                if(bs!= null){
                    Log.i("Msg","有客户端连接!" + bs.getRemoteDevice().getName() + "\n");
                    //将连接对象交给其他子线程,让其处理具体通信事务
                    new Thread(new TalkingThread(bs)).start();
                }else{
                    Log.i("Msg","无有效连接\n");
                }
            } catch (IOException e) {
                e.printStackTrace();
            }
        }
    }
}
```

服务器启动时使用的 UUID 对象是指定的蓝牙串口服务字符串：

```
private static final UUID uuid = UUID.fromString ( " 00001101 - 0000 - 1000 - 8000 -
    00805F9B34FB")
```

在客户端连接服务器端时也要采用该 UUID。

为了可以控制子线程的生命周期，要借助属性变量 serFlag，当 Activity 执行了 onStop 方法时，该属性值为 false。

BluetoothServerSocket 类的 accept 方法是阻塞方法，即程序执行到该方法时将"停住"等待连接。当客户端连接发起后，该方法会返回 BluetoothSocket 对象，这是双方设备互相通信的连接对象。为了方便处理两个设备相互通信，交互数据，此处又开启了 TalkingThread 线程，负责读写数据。TalkingThread 线程代码如下。

📖 Proj11_6 项目中 MainActivity.java 文件的读取数据的线程类

```java
private class TalkingThread implements Runnable{
    BluetoothSocket bs;
    public TalkingThread(BluetoothSocket bs) {
        super();
        this.bs = bs;
    }
    @Override
    public void run() {
        try {
            while(serFlag){
                Log.i("Msg", "socket state " + bs.isConnected());
                //获取输入流
                InputStream is = bs.getInputStream();
                byte buf [] = new byte[1024];
                int n = is.read(buf);
                String str = new String(buf, 0,  n);    //创建字符串对象
                Message msg = new Message();
                msg.obj = bs.getRemoteDevice().getName() + ":" + str;
                handler.sendMessage(msg);
            }
        } catch (IOException e) {
            e.printStackTrace();
        }
    }
}
```

TalkingThread 线程读取数据的方法处理得比较简单，此处不适合处理复杂数据通信或文件传输。有需要的读者可以参考 I/O 流读写的知识修改该部分代码。由于子线程无法修改 UI，因此需要借助 Handler 对象将读取到的数据显示在 TextView 控件中。Handler 对象的代码如下。

📖 Proj11_6 项目中 MainActivity.java 文件的 Handler 类

```java
Handler handler = new Handler(){
```

```java
    public void handleMessage(android.os.Message msg) {
        tv1.append(msg.obj + "\n");
    };
};
```

以上是 MainActivity 界面需要实现的功能,下面介绍 TalkActivity 实现的功能和代码。当在 MainActivity 中单击 ListView 列表中的服务器地址时,会切换到 TalkActivity 界面。TalkActivity 实现与服务器的连接,并向服务器发送聊天数据。布局文件 activity_talk.xml 比较简单,可以参考图 11.9。

初始化连接的方法如下,serMac 是在 Activity 跳转时由 MainActivity 添加的,表示需要连接的蓝牙服务器地址。

📖 Proj11_6 项目中 TalkActivity.java 文件的初始化连接方法

```java
private void initConn(){
    // BluetoothAdapter adapter;属性
    adapter = BluetoothAdapter.getDefaultAdapter();
    dname = adapter.getName();
    Log.i("Msg", "连接地址" + serMac);
    if(serMac == null)return;
    // BluetoothDevice bd;属性,表示蓝牙设备
    bd = adapter.getRemoteDevice(serMac);
    try {
        // BluetoothSocket bs;属性,表示连接对象
        bs = bd.createInsecureRfcommSocketToServiceRecord(uuid);
        bs.connect();
        if(bs.isConnected()){
            tv.append("BluetoothSocket 连接 OK\n");
        }else{
            tv.append("BluetoothSocket 连接失败\n");
        }
    } catch (IOException e) {
        e.printStackTrace();
    }
}
```

发送按钮的监听器方法如下,当单击该按钮时,将 EditText 中的内容发送给服务器端。发送数据需要获取 BluetoothSocket 类的输出流,然后将内容直接输出即可。

📖 Proj11_6 项目中 TalkActivity.java 文件的初始化连接方法

```java
public void onClick(View arg0) {
    if(!bs.isConnected()){
        tv.append("BluetoothSocket 连接未建立\n");
        return;
    }
    String msg = et.getText().toString();
    tv.append(dname + ":" + msg + "\n");
```

```
if(msg.length()<1)return;
try {
    if(!bs.isConnected())bs.connect();
    OutputStream os = bs.getOutputStream();
    os.write(msg.getBytes());
} catch (IOException e) {
    e.printStackTrace();
}
}
```

在运行该项目时，需要申请对应的蓝牙权限，具体可以参考 11.5.2 节的内容。为说明蓝牙通信的过程，现使用两部手机（HTC 和华为）演示该项目的运行效果。在两部手机上都运行该项目，分别执行"设置蓝牙可见"和"扫描连接的蓝牙设备"，此时 MainActivity 界面的 ListView 控制将显示它们已经配对的蓝牙设备。在 HTC 手机上看到是的华为手机的 MAC 地址，在华为手机上看到的是 HTC 手机的 MAC 地址。此时设置华为手机为服务器端，因此它需要执行"启动服务器"功能。当服务器成功启动后，在 HTC 手机上单击华为手机的 MAC 地址列表，会切换到 TalkActivity 界面，显示"BluetoothSocket 连接 OK"，否则需要返回 MainActivity 重新连接（读者也可以修改初始化连接的方式）。

在 TalkActivity 显示连接成功后，就可以向服务器端发送信息了，同时，发送的信息也会呈现在上方的聊天区域内，如图 11.9 所示。服务器端接收到连接后，启动 TalkingThread 线程读取客户端数据，并显示在服务器端的聊天区域内，如图 11.10 所示。到此为止，使用蓝牙通信模拟简单聊天的程序就完成了，有兴趣的读者可以在此基础上实现双向通信、"群聊"或"私聊"等功能。

图 11.9　HTC 手机客户端界面

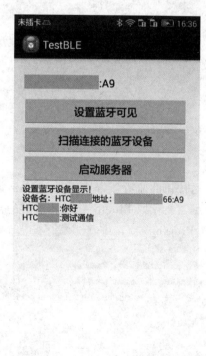

图 11.10　华为手机服务器端界面

11.6 习 题

1. 选择题

(1) 以下表示磁力传感器的是(   )。

    A. Sensor.TYPE_ACCELEROMETER

    B. Sensor.TYPE_MAGNETIC_FIELD

    C. Sensor.TYPE_ORIENTATION

    D. Sensor.TYPE_GYROSCOPE

(2) 以下关于加速度传感器的坐标系统的说法中有误的是(   )。

    A. 加速度传感器所采用的坐标系统与 Canvas 采用的坐标系统一致

    B. 手机竖直向上放置时,平行于屏幕,水平方向从左到右为 X 轴正方向

    C. 手机竖直向上放置时,平行于屏幕,竖直方向从下到上为 Y 轴正方向

    D. 手机竖直向上放置时,垂直屏幕,从内到外为 Z 轴正方向

(3) 以下关于光线传感器的说法中正确的是(   )。

    A. 获取光线传感器需要借助常量 Sensor.TYPE_AMBIENT_TEMPERATURE

    B. 获取光线传感器需要借助常量 Sensor.TYPE_PROXIMITY

    C. SensorEvent 对象的属性 values 数组只有第一个元素(values[0])有意义

    D. SensorEvent 对象的属性 values 数组有且只有一个元素

(4) 以下关于 Android 系统蓝牙通信的说法中有误的是(   )。

    A. 蓝牙通信之前需要先进行认证

    B. 蓝牙设备允许被其他设备发现的最长时间是 120s

    C. 蓝牙通信中需要借助蓝牙模块的 MAC 地址建立连接

    D. 蓝牙技术被广泛应用于近距离通信

2. 简答题

(1) 如何检验 Android 设备是否支持某种传感器设备?

(2) 简要说明加速度传感器中的坐标系统。

(3) 请列举 5 种使用传感器提升用户体验的应用场景。

第 12 章　校园 App 项目案例

本章学习目标
- 了解校园 App 项目设计过程。
- 了解客户端与服务器端通信技术。
- 了解 App 用户导航控制的基本知识。

设计和开发 Android App 产品,不但需要扎实的编程基础,熟练掌握开发知识,还需要从产品的功能需要、用户定位、系统架构、功能设计、UI 与交互设计、数据库设计等其他方面了解相关知识和技术。本章介绍校园 App 项目的设计与开发过程,旨在让初学者了解项目开发过程中涉及的相关技术。

12.1　校园 App 项目介绍

校园 App 的使用者主要是在校学生,用于阅读校园新闻、管理学习课程、查询作业与考试等功能。该项目由移动客户端和服务器端两部分组成,前者又分为 Android 客户端实现和 iOS 客户端实现;后者使用 JSP+Struts2+Hibernate4 技术实现。对该服务器端开发技术不了解的读者可以先学习 Java Web 开发,了解 JSP 技术和 SSH 技术。

本章仅介绍校园 App Android 客户端和服务器端的部分功能,主要包括客户端欢迎界面,客户端侧滑屏导航和底部标签导航,客户端新闻列表实时获取,客户端以 HTML5 网页查看新闻,客户端本地模拟聊天气泡,服务器分页查询新闻列表,服务器以 HTML5 标准呈现新闻内容页面。

1. 客户端欢迎界面

该界面是 Android 客户端运行后的第一个界面,实现过程比较简单,运行效果如图 12.1 所示。

2. 客户端主界面

主界面底部是 3 个导航标签,可以切换"校园新闻""应用中心"和"班级聊天"3 个子界面,如图 12.2 所示。

3. 侧滑栏界面

在客户端主界面向右侧滑屏,可以打开左侧导航界面,如图 12.3 所示。

4. 新闻内容界面

在主界面的"校园新闻"子界面下,单击任意新闻标题,会打开新闻内容界面,如图 12.4 所示。

图 12.1 欢迎界面　　　　　　图 12.2　App 主界面

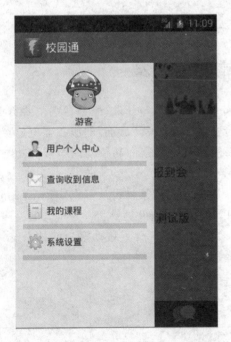

图 12.3　侧滑栏导航　　　　　　图 12.4　新闻内容界面

5. 应用中心界面

通过主界面底部导航标签可以打开应用中心子界面,如图 12.5 所示。该界面列出了校园 App 的各独立功能模块。

6. 聊天界面

通过主界面的右侧标签可以打开聊天界面,实现班级内聊天。

图 12.5　应用中心界面

图 12.6　聊天界面

12.2　服务器端功能开发

完整的移动端 App 设计与开发都要包括服务器端的设计与开发,服务器端主要负责业务逻辑的实现和数据的持久化处理,相对于"前端"的 App 来说,"后端"的服务器更显重要。在该项目中,服务端的功能并不复杂,只实现了新闻列表分页查询和新闻内容查询功能,所用到的组件有 JSP、Action 和 Hibernate。

与传统的 Web 项目相比,带有移动客户端的 Web 项目的架构设计基本一致,遵循分层设计思想。原有的系统架构可以保持不变,也可以增加一些组件提供 JSON 格式的返回,这种特点可以让以前的 Web 项目很容易扩展出移动端应用。整合了移动客户端的系统架构如图 12.7 所示。

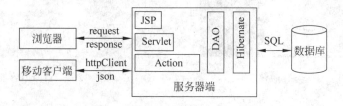

图 12.7　移动客户端与 J2EE 架构

该项目服务器端使用 Myeclipse 2015 开发,测试服务器是 Tomcat 8.0。

12.2.1 数据库表

该项目中使用 MySQL 数据库,编码格式为 UTF-8。与校园新闻功能模块相关的表有两个,分别是 tb_artcletype 和 tb_articleinfo,创建表的 SQL 语句如下。

📖 CampusSer 项目,创建表的 SQL 语句

```sql
-- Table structure for `tb_articletype`
CREATE TABLE `tb_articletype` (
  `tid` int(11) NOT NULL AUTO_INCREMENT,
  `typeName` varchar(255) DEFAULT NULL,
  `typeState` int(11) DEFAULT NULL,
  PRIMARY KEY (`tid`)
)
-- Table structure for `tb_articleinfo`
CREATE TABLE `tb_articleinfo` (
  `aid` int(11) NOT NULL AUTO_INCREMENT,
  `articleName` varchar(255) DEFAULT NULL,
  `articleAuthor` varchar(100) DEFAULT NULL,
  `articleContent` text,
  `articleType` int(11) DEFAULT NULL,
  `articleImagePath` varchar(255) DEFAULT NULL,
  `articleDate` date DEFAULT NULL,
  `readNum` int(11) DEFAULT NULL,
  `state` int(11) DEFAULT NULL,
  PRIMARY KEY (`aid`),
  KEY `articleFK` (`articleType`),
  CONSTRAINT `articleFK` FOREIGN KEY (`articleType`) REFERENCES `tb_articletype` (`tid`)
)
```

tb_articletype 存放校园新闻类型,如"教务新闻""学工新闻"等,字段类型含义如表 12.1 所示。

表 12.1 tb_articletype 表字段说明

字 段 名	类 型	说 明
tid	int(11)	新闻类型编号,主键,自增
typeName	varchar(255)	类型名称
typeState	int(11)	状态

tb_articleinfo 表存放校园新闻详细信息,字段含义如表 12.2 所示。

表 12.2 tb_articleinfo 表字段说明

字 段 名	类 型	说 明
aid	int(11)	新闻编号,主键,自增
articleName	varchar(255)	新闻类型名称
articleAuthor	varchar(100)	新闻作者
articleContent	text	新闻内容

续表

字 段 名	类 型	说 明
articleType	int(11)	新闻所属类型,外键,关联 tb_articletype 表
articleImagePath	varchar(255)	新闻图片存放路径
articleDate	date	新闻发布日期
readNum	int(11)	阅读次数
state	int(11)	新闻状态,0 表示"软删除",1 表示正常

12.2.2 实体类

本项目中的实体类通过 Hibernate"逆向工程"实现,并通过"注解"完成映射关系,也可以使用 mapping 文件实现映射关系。

因主键都采用"自增"模式,在配置主键映射时,需要指明主键的生成策略。如果在 ArticleType 实体中生成了"一对多"的映射,可以直接删除,保留如下代码即可。

📖 CampusSer 项目,edu/freshen/entity/ArticleType.java 实体类

```java
@Entity
@Table(name = "tb_articletype" , catalog = "campusdb")
public class ArticleType implements java.io.Serializable {
    private Integer tid;
    private String typeName;
    private Integer typeState;
    //映射主键,自增模式
    @GenericGenerator(name = "generator", strategy = "increment")
    @Id @GeneratedValue(generator = "generator")
    @Column(name = "tid", unique = true, nullable = false)
    public Integer getTid() {
        return this.tid;
    }
    public void setTid(Integer tid) {
        this.tid = tid;
    }
    @Column(name = "typeName")
    public String getTypeName() {
        return this.typeName;
    }
    public void setTypeName(String typeName) {
        this.typeName = typeName;
    }
    @Column(name = "typeState")
    public Integer getTypeState() {
        return this.typeState;
    }
    public void setTypeState(Integer typeState) {
        this.typeState = typeState;
    }
}
```

ArticleInfo 实体与 ArticleType 实体有多对一的关系，在指定加载模式时使用 FetchType.EAGER，这是一种积极的加载模式，即在访问 ArticleInfo 时，立即查询与之关联的 ArticleType。否则无法将与 ArticleInfo 关联的 ArticleType 提取到 JSON 串中。

📖 CampusSer 项目，edu/freshen/entity/ArticleInfo.java 实体类

```java
@Entity
@Table(name = "tb_articleinfo", catalog = "campusdb")
public class ArticleInfo implements java.io.Serializable {
    private Integer aid;
    private ArticleType articleType;
    private String articleName;
    private String articleAuthor;
    private String articleContent;
    private String articleImagePath;
    private Date articleDate;
    private Integer readNum;
    private Integer state;
    //配置主键映射
    @GenericGenerator(name = "generator", strategy = "increment")
    @Id @GeneratedValue(generator = "generator")
    @Column(name = "aid", unique = true, nullable = false)
    public Integer getAid() {
        return this.aid;
    }
    public void setAid(Integer aid) {
        this.aid = aid;
    }
    //配置多对一映射,使用积极加载模式,方便生成 JSON 串
    @ManyToOne(fetch = FetchType.EAGER)
    @JoinColumn(name = "articleType")
    public ArticleType getArticleType() {
        return this.articleType;
    }
    public void setArticleType(ArticleType articleType) {
        this.articleType = articleType;
    }
    @Column(name = "articleName")
    public String getArticleName() {
        return this.articleName;
    }
    public void setArticleName(String articleName) {
        this.articleName = articleName;
    }
    @Column(name = "articleAuthor", length = 100)
    public String getArticleAuthor() {
        return this.articleAuthor;
    }
    public void setArticleAuthor(String articleAuthor) {
        this.articleAuthor = articleAuthor;
```

```
        }
        @Column(name = "articleContent", length = 65535)
        public String getArticleContent() {
            return this.articleContent;
        }
        public void setArticleContent(String articleContent) {
            this.articleContent = articleContent;
        }
        @Column(name = "articleImagePath")
        public String getArticleImagePath() {
            return this.articleImagePath;
        }
        public void setArticleImagePath(String articleImagePath) {
            this.articleImagePath = articleImagePath;
        }
        @Temporal(TemporalType.DATE)
        @Column(name = "articleDate", length = 10)
        public Date getArticleDate() {
            return this.articleDate;
        }
        public void setArticleDate(Date articleDate) {
            this.articleDate = articleDate;
        }
        @Column(name = "readNum")
        public Integer getReadNum() {
            return this.readNum;
        }
        public void setReadNum(Integer readNum) {
            this.readNum = readNum;
        }
        @Column(name = "state")
        public Integer getState() {
            return this.state;
        }
        public void setState(Integer state) {
            this.state = state;
        }
}
```

12.2.3 DAO 层

Hibernate 配置文件描述数据库连接信息，必须具备的配置信息有：connection.url，表示数据库服务器位置信息；connection.username，表示数据库服务器的登录名；connection.password，表示数据库服务器端登录密码；connection.driver_class，表示数据库服务器的驱动类。通过注解实现的映射类需要在该配置文件中引入，否则无法找到实体类。

  📖 CampusSer 项目，hibernate.cfg.xml 配置文件

```
<?xml version = '1.0' encoding = 'UTF - 8'?>
```

```xml
<!DOCTYPE hibernate-configuration PUBLIC
        "-//Hibernate/Hibernate Configuration DTD 3.0//EN"
        "http://www.hibernate.org/dtd/hibernate-configuration-3.0.dtd">
<hibernate-configuration>
<session-factory>
    <property name="dialect">org.hibernate.dialect.MySQLDialect</property>
    <property name="connection.url">jdbc:mysql://localhost:3306/campusDB</property>
    <property name="connection.username">root</property>
    <property name="connection.password">123456</property>
    <property name="connection.driver_class">com.mysql.jdbc.Driver</property>
    <property name="myeclipse.connection.profile">mysqlConn</property>
    <property name="show_sql">true</property>
    <mapping class="edu.freshen.entity.ArticleInfo" />
    <mapping class="edu.freshen.entity.ArticleType" />
</session-factory>
</hibernate-configuration>
```

DAO层的实现相对灵活，既可以借助接口，也可以直接借助继承类。在该项目中采用后者。CommDao类是所有DAO类的父类，通过"泛型"实现共有的"增删改查"通用方法。此处仅实现了通过主键查询实体的方法。

📖 CampusSer项目，edu/freshen/dao/CommDao.java

```java
public class CommDao<T> {
    protected Session session = HibernateSessionFactory.getSession();
    public T findEntityById(Class<T> entityClass, int id){
        return (T) session.get(entityClass, id);
    }
}
```

ArticleInfoDao具体实现校园新闻功能的持久化处理，应包括该功能的所有读写操作。此处仅实现分页查询ArticleInfo实体的功能。

📖 CampusSer项目，edu/freshen/dao/ArticleInfoDao.java

```java
public class ArticleInfoDao extends CommDao<ArticleInfo> {
    public List<ArticleInfo> findArticleByPage(int start, int size){
        String hql = "from ArticleInfo a where a.state=1 order by a.articleDate Desc";
        Query q = session.createQuery(hql);
        q.setFirstResult(start);
        q.setMaxResults(size);
        return q.list();
    }
}
```

12.2.4 Action层

Action层负责客户端的请求控制和资源跳转，在该项目中仅包含一个Action，用于实现与新闻内容相关的查询业务。

📖 CampusSer 项目，edu/freshen/action/ArticleAction.java

```java
public class ArticleAction extends ActionSupport {
    private int aid;                              //新闻编号
    private ArticleInfo articleInfo;              //新闻实体
    private List<ArticleInfo> articleInfos;       //新闻实体集合
    private int currentPage;                      //当前页
    private int pageSize = 10;                    //每页包含记录条数
    //通过新闻编号查询新闻实体
    public String findArticleById(){
        articleInfo = new ArticleInfoDao().findEntityById(ArticleInfo.class, aid);
        return "findArticleById";
    }
    //分页查询新闻实体
    public String findArticleByPage(){
        articleInfos = new ArticleInfoDao().findArticleByPage(currentPage * pageSize, pageSize);
        return "findArticleByPage";
    }
    …//省略 get 和 set 方法
}
```

ArticleAction 的配置信息如下：struts.i18n.encoding 用于设置编码格式，struts.serve.static.browserCache 用于设置浏览器缓存模式，struts.devMode 用于设置开发模式。Action 配置采用通配符 *，返回结果 findArticleById 时前往 detail.jsp 页面，返回结果 findArticleByPage 时作为 JSON 串返回客户端。

📖 CampusSer 项目，struts.xml 配置文件

```xml
<?xml version="1.0" encoding="UTF-8" ?>
<!DOCTYPE struts PUBLIC "-//Apache Software Foundation//DTD Struts Configuration 2.1//EN" "http://struts.apache.org/dtds/struts-2.1.dtd">
<struts>
    <constant name="struts.i18n.encoding" value="UTF-8"></constant>
    <constant name="struts.serve.static.browserCache" value="false"></constant>
    <constant name="struts.devMode" value="true"></constant>
    <package name="mypg" extends="json-default">
        <!-- 新闻文章 action -->
        <action name="articleAct_*" class="edu.freshen.action.ArticleAction" method="{1}">
            <result name="findArticleById">/detail.jsp</result>
            <result name="findArticleByPage" type="json">
                <param name="root">articleInfos</param>
            </result>
        </action>
    </package>
</struts>
```

detail.jsp 页面是呈现新闻内容的页面，采用 HTML5 标准，可以在不同尺寸屏幕下自

适应,做到"响应式布局"。该页面引用了相关的js文件和css文件,此处不再介绍。

  📖 CampusSer 项目,detail.jsp 页面代码

```jsp
<%@ page language="java" import="java.util.*" pageEncoding="UTF-8"%>
<%@taglib prefix="c" uri="http://java.sun.com/jsp/jstl/core" %>
<%
String path = request.getContextPath();
String basePath = request.getScheme() + "://"
                + request.getServerName() + ":" + request.getServerPort() + path + "/";
%>
<!DOCTYPE html>
<head>
    <meta charset="utf-8" />
    <meta name="viewport" content="width=device-width, initial-scale=1, maximum-scale=1" />
    <title>校园通</title>
    <meta name="description" content="" />
    <meta name="author" content="" />
    <link rel="stylesheet" href="css/style.css" media="screen" />
    <link rel="stylesheet" href="css/skeleton.css" media="screen" />
    <link rel="stylesheet" href="js/jquery.fancybox.css" media="screen" />
    <script type="text/javascript" src="js/modernizr.custom.js"></script>
    <meta http-equiv="Content-Type" content="text/html; charset=utf-8" />
</head>
<body class="menu-1 h-style-1 text-1">
<div class="wrap">
    <div class="main">
    <!---------------- Container ------------------->
    <section class="container sbl clearfix">
    <!---------------- Content ------------------->
    <section id="content" class="twelve columns">
    <article class="entry clearfix single">
    <h2 class="title">${articleInfo.articleName}</h2>
    <ul class="entry-meta">
        <li><b>Date:</b> <a href="#">${articleInfo.articleDate}</a></li>
        <li><b>Author:</b> <a href="#">${articleInfo.articleAuthor}</a></li>
        <li class="tags"><b>Tags:</b>  ${articleInfo.articleType.typeName}</li>
    </ul>
    <div class="entry-body">
        <img class="entry-image" alt="" src="${articleInfo.articleImagePath}" />
        <p>
        <c:out value="${articleInfo.articleContent}" escapeXml="true"/>
        </p>
    </div>
    </article>
    </section>
    <!-- / #content -->
    </section>
    </div>
```

```
        </div>
    </body>
</html>
```

将 Web 项目部署到服务器,并在数据库中录入相应的测试数据,在浏览器地址栏输入 http://localhost:8080/CampusSer/articleAct_findArticleById?aid=1,可以直接查看新闻页面效果,如图 12.8 所示。对比前面的移动端运行界面,该页面的表现效果不同。

图 12.8　浏览器访问校园新闻效果

12.3　Android 客户端开发

校园 App 的客户端首先要合理组织预设功能,方便用户查找需要的功能,这被称作"用户导航"设计;其次是增强用户交互性,提升用户使用体验;最后是采用合理的系统组件或自定义组件将内容呈现出来。前两者主要涉及 UI 与 UE 的相关知识,在本章不做深入介绍。该项目中所采用的背景图片、图标可以在项目代码中找到。

12.3.1　欢迎界面与标题栏样式

FlashActivity 是校园 App 的欢迎界面,使用 Handler 延迟 2s 跳转到主界面。Activity 切换时使用系统动画效果,欢迎界面完成后直接执行 finish 销毁。

📖 Proj12 项目,edu/freshen/proj12/FlashActivity.java

```java
public class FlashActivity extends Activity {
    @Override
    protected void onCreate(Bundle savedInstanceState) {
        super.onCreate(savedInstanceState);
        //requestWindowFeature(Window.FEATURE_NO_TITLE);
        setContentView(R.layout.activity_flash);
        new Handler().postDelayed(new Runnable(){
            @Override
            public void run() {
                Intent intent = new Intent(FlashActivity.this,MainActivity.class);
                startActivity(intent);
                overridePendingTransition(android.R.anim.fade_in, android.R.anim.fade_out);
```

```
                FlashActivity.this.finish();    //本Activity结束
        }},2000);
    }
}
```

FlashActivity 采用的布局文件是 activity_flash.xml，主要显示应用程序 Logo 和版本信息，布局较为简单，可以参考图 12.1 所示的运行效果。

修改应用程序的样式文件，重新定义 android：actionBarStyle 的背景颜色和字体样式，代码如下。在配置文件中设定 android：theme＝"@style/AppTheme"，就可以让所有 Activity 具备相同的样式。

  📖 Proj12 项目，styles.xml 样式文件

```xml
<resources>
    <style name="AppBaseTheme" parent="android:Theme.Light">
    </style>
    <!-- Application theme. -->
    <style name="AppTheme" parent="AppBaseTheme">
        <item name="android:actionBarStyle">@style/myActionBarStyle</item>
    </style>
    <style name="myActionBarStyle" parent="android:Widget.ActionBar">
        <item name="android:background">#D01839</item>
        <item name="android:titleTextStyle">@style/AcBar_titleStyle</item>
    </style>
    <style name="AcBar_titleStyle">
        <item name="android:textSize">18sp</item>
        <item name="android:textColor">#FFFFFF</item>
    </style>
</resources>
```

12.3.2　主界面 Activity

MainActivity 是客户端主界面，集成了客户端的所有功能模块，其中与用户和系统相关的功能模块处于侧滑栏中；新闻列表、应用中心和聊天记录位于默认显示的内容区域。该项目中，侧滑栏功能采用了 Android Support v4（让低版本系统具备新的控件特性）中的 DrawerLayout 布局。

DrawerLayout 布局管理器分为侧边菜单和主内容区两部分，侧边菜单可以根据手势展开与隐藏，这是 DrawerLayout 自身的特性，无须编写代码；主内容区的内容可以随着菜单的点击而变化，这需要使用者自己实现。

MainActivity 采用的布局文件代码如下，保持两个 FrameLayout 的 id 不变，内容区域引入布局文件 activity_main_content.xml，侧滑栏区域引入布局文件 activity_main_leftmenu.xml。

  📖 Proj12 项目，activity_main.xml 布局文件

```xml
<android.support.v4.widget.DrawerLayout
```

```xml
xmlns:android = "http://schemas.android.com/apk/res/android"
android:id = "@ + id/drawer_layout"
android:layout_width = "match_parent"
android:layout_height = "match_parent">
<FrameLayout
    android:id = "@ + id/content_frame"
    android:layout_width = "match_parent"
    android:layout_height = "match_parent">
    <include
        android:visibility = "visible"
        android:layout_width = "fill_parent"
        android:layout_height = "fill_parent"
        layout = "@layout/activity_main_content"/>
</FrameLayout>
<!-- 左侧边栏 -->
<FrameLayout
    android:id = "@ + id/left_drawer"
    android:layout_width = "220dp"
    android:layout_height = "match_parent"
    android:layout_gravity = "start"
    android:background = "#EEE">
    <include
        android:visibility = "visible"
        android:layout_width = "fill_parent"
        android:layout_height = "fill_parent"
        layout = "@layout/activity_main_leftmenu"/>
</FrameLayout>
</android.support.v4.widget.DrawerLayout>
```

内容区域的布局代码如下，主体区域是 FrameLayout(id 是 fragmentContent)包含的自定义 Fragment 组件；底部是 LinearLayout 包含的 3 个 ImageButton 控件，用于切换主体区域 fragmentContent 中的 Fragment 组件。3 个自定义的 Fragment 分别是 LeftFragment、MiddleFragment 和 RightFragment，将在后面给出详细代码。内容区域的布局效果可以参考图 12.2。

📖 Proj12 项目，activity_main_content.xml 布局文件

```xml
<?xml version = "1.0" encoding = "utf-8"?>
<RelativeLayout xmlns:android = "http://schemas.android.com/apk/res/android"
    android:layout_width = "match_parent"
    android:layout_height = "match_parent" >
    <FrameLayout
        android:layout_width = "match_parent"
        android:layout_height = "match_parent"
        android:id = "@ + id/fragmentContent"
        android:paddingBottom = "40dp" >
        <fragment class = "edu.freshen.proj12.LeftFragment"
            android:layout_width = "match_parent"
```

```xml
            android:layout_height = "match_parent"/>
    </FrameLayout>
    <LinearLayout
        android:layout_width = "match_parent"
        android:layout_height = "40dp"
        android:background = "#966"
        android:layout_alignParentBottom = "true"  >
        <ImageButton
            android:id = "@ + id/imageButton1"
            android:layout_width = "40dp"
            android:layout_height = "40dp"
            android:padding = "2dp"
            android:scaleType = "fitCenter"
            android:layout_weight = "1"
            android:background = "@drawable/selector_btbar"
            android:src = "@drawable/bt_l" />
        <ImageButton
            android:id = "@ + id/imageButton2"
            android:layout_width = "40dp"
            android:layout_height = "40dp"
            android:padding = "2dp"
            android:scaleType = "fitCenter"
            android:layout_weight = "1"
            android:background = "@drawable/selector_btbar"
            android:src = "@drawable/bt_m" />
        <ImageButton
            android:id = "@ + id/imageButton3"
            android:layout_width = "40dp"
            android:layout_height = "40dp"
            android:padding = "2dp"
            android:scaleType = "fitCenter"
            android:layout_weight = "1"
            android:background = "@drawable/selector_btbar"
            android:src = "@drawable/bt_r" />
    </LinearLayout>
</RelativeLayout>
```

侧滑栏的具体布局代码如下，运行效果可以参考图12.3。侧边栏的底部是ListView控件，id是leftmenu_list，用于设置侧边栏中的列表菜单项，菜单列表会在代码中被初始化。

📖 Proj12项目，activity_main_leftmenu.xml布局文件

```xml
<?xml version = "1.0" encoding = "utf - 8"?>
<RelativeLayout xmlns:android = "http://schemas.android.com/apk/res/android"
    android:layout_width = "match_parent"
    android:layout_height = "match_parent"
    android:paddingTop = "12dp">
    <ImageView
        android:layout_width = "60dp"
        android:layout_height = "60dp"
```

```
            android:id = "@ + id/leftmenu_face"
            android:layout_centerHorizontal = "true"
            android:src = "@drawable/f6_02" />
        <TextView
            android:id = "@ + id/leftmenu_name"
            android:layout_width = "wrap_content"
            android:layout_height = "wrap_content"
            android:layout_centerHorizontal = "true"
            android:layout_below = "@id/leftmenu_face"
            android:layout_marginTop = "8dp"
            android:text = "游客"
            android:textAppearance = "?android:attr/textAppearanceSmall" />
        <View
            android:layout_width = "match_parent"
            android:layout_height = "1px"
            android:layout_below = "@id/leftmenu_name"
            android:layout_marginLeft = "6dp"
            android:layout_marginRight = "6dp"
            android:background = "#669966" />
        <ListView
            android:id = "@ + id/leftmenu_list"
            android:layout_width = "match_parent"
            android:layout_height = "match_parent"
            android:layout_below = "@id/leftmenu_name"
            android:layout_marginTop = "12dp"
            android:choiceMode = "singleChoice"
            android:layout_marginLeft = "12dp"
            android:layout_marginRight = "12dp"
            android:dividerHeight = "8dp" />
</RelativeLayout>
```

MainActivity 主要完成侧滑栏的菜单项初始化功能和 Fragment 区域的切换功能,详细代码如下。创建侧滑栏列表使用 ListView 项布局文件 list_item_leftmenu,该布局文件比较简单,使用 ImageView 显示菜单图标,TextView 显示菜单标题,布局效果可以参考图 12.3,布局代码此处不再给出。

📖 Proj12 项目,edu/freshen/proj12/MainActivity.java

```java
public class MainActivity extends Activity implements OnClickListener {
    private Fragment lf, mf, rf;                    //左中右 Fragment
    private ImageButton lb, mb, rb;                 //左中右按钮
    private FrameLayout fragmentContent;            //Fragment 内容显示区域
    @Override
    protected void onCreate(Bundle savedInstanceState) {
        super.onCreate(savedInstanceState);
        setContentView(R.layout.activity_main);
        //初始化左侧边栏
        initLeftMenu();
        //初始化底部按钮
```

```java
        initBottomBt();
        fragmentContent = (FrameLayout) findViewById(R.id.fragmentContent);
        //创建自定义Fragment
        lf = new LeftFragment();
        mf = new MiddleFragment();
        rf = new RightFragment();
    }
    private void initBottomBt() {
        lb = (ImageButton) findViewById(R.id.imageButton1);
        mb = (ImageButton) findViewById(R.id.imageButton2);
        rb = (ImageButton) findViewById(R.id.imageButton3);
        lb.setOnClickListener(this);
        mb.setOnClickListener(this);
        rb.setOnClickListener(this);
    }
    //初始化侧边栏中的导航菜单项
    private void initLeftMenu() {
        ListView lv = (ListView) findViewById(R.id.leftmenu_list);
        List<Map<String,Object>> data = new ArrayList();
        Map<String,Object> item1 = new HashMap();
        item1.put("img", R.drawable.m_user);
        item1.put("name", "用户个人中心");
        data.add(item1);
        Map<String,Object> item2 = new HashMap();
        item2.put("img", R.drawable.m_mail);
        item2.put("name", "查询收到信息");
        data.add(item2);
        Map<String,Object> item3 = new HashMap();
        item3.put("img", R.drawable.m_course);
        item3.put("name", "我的课程");
        data.add(item3);
        Map<String,Object> item4 = new HashMap();
        item4.put("img", R.drawable.m_set);
        item4.put("name", "系统设置");
        data.add(item4);
        SimpleAdapter sa = new SimpleAdapter(MainActivity.this, data,
                R.layout.list_item_leftmenu, new String[]{"img","name"},
                new int[]{R.id.list_item_leftmenu_icon, R.id.list_item_leftmenu_name});
        lv.setAdapter(sa);
    }
    //切换Fragment
    private void changeFragment(int vid, Fragment frag){
        FragmentTransaction ft = getFragmentManager().beginTransaction();
        ft.replace(vid, frag);
        ft.commit();
    }
    @Override
    public void onClick(View arg0) {
        switch(arg0.getId()){
        case R.id.imageButton1:changeFragment(R.id.fragmentContent,lf);break;
```

```
            case R.id.imageButton2:changeFragment(R.id.fragmentContent,mf);break;
            case R.id.imageButton3:changeFragment(R.id.fragmentContent,rf);break;
        }
    }
}
```

12.3.3 自定义 Fragment

MainActivity 界面的内容区域是 3 个 Fragment 切换实现的,每个 Fragment 都对应各自的布局文件,LeftFragment 实现"校园新闻",MiddleFragment 实现"应用中心",RightFragment 实现"聊天记录"。Fragment 的切换原理如图 12.9 所示。

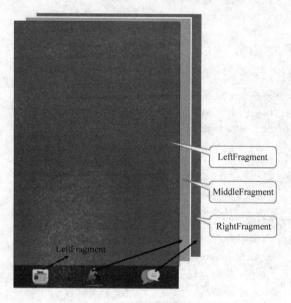

图 12.9 Fragment 切换示意图

1. LeftFragment 代码与布局文件

LeftFragment 主要完成显示新闻列表的功能,它的布局文件比较简单,上面是 ImageView 控件,显示广告图片;下面是 ListView 控件,显示具体的新闻列表信息。布局文件代码如下。

📖 Proj12 项目,fragment_left.xml

```
<?xml version = "1.0" encoding = "utf - 8"?>
<RelativeLayout xmlns:android = "http://schemas.android.com/apk/res/android"
    android:layout_width = "match_parent"
    android:layout_height = "match_parent"
    android:background = "#DDD">
    <ImageView
        android:id = "@ + id/lf_banner"
        android:layout_width = "match_parent"
        android:layout_height = "80dp"
```

```xml
            android:scaleType = "centerCrop" />
    <ListView
            android:id = "@ + id/lf_lv"
            android:layout_width = "match_parent"
            android:layout_height = "wrap_content"
            android:layout_below = "@ + id/lf_banner"
            android:layout_marginTop = "4dp" >
    </ListView>
</RelativeLayout>
```

在初始化广告图片和新闻列表时，都需要从服务器端获取，因此将服务器端地址信息定义在 UrlUtil 类中，方便在代码中使用。注意，服务器端地址不能使用 localhost，应明确写出 IP 地址或域名地址。

📖 Proj12 项目，edu/freshen/util/ UrlUtil.java

```java
public class UrlUtil {
    public static final String webRoot = "http://192.168.4.100:8080/CampusSer/";
}
```

为简化网络访问和图片加载，该项目中使用 Xutil 第三方框架，主要包括 DbUtils 模块、ViewUtils 模块、HttpUtils 模块和 BitmapUtils 模块。关于该框架的 API 可以参考 http://xutilsapi.oschina.mopaas.com。

服务器返回的数据列表是 JSON 格式，借助 Gson 类，可以很容易实现 JSON 字符串到应用类的转换。Gson 是 Google 提供的用来在 Java 对象和 JSON 数据之间进行映射的 Java 类库，对应资源可以从 https://github.com/google/gson 下载。

📖 Proj12 项目，edu/freshen/proj12/ LeftFragment.java

```java
public class LeftFragment extends Fragment {
    private ImageView banner;              //首页新闻图片或广告图片
    private ListView lv;                   //新闻列表
    List < ArticleInfo > data = null;      //新闻列表数据
    @Override
    public View onCreateView(LayoutInflater inflater, ViewGroup container,
            Bundle savedInstanceState) {
        View v = inflater.inflate(R.layout.fragment_left, null);
        return v;
    }
    @Override
    public void onActivityCreated(Bundle savedInstanceState) {
        super.onActivityCreated(savedInstanceState);
        banner = (ImageView) getView().findViewById(R.id.lf_banner);
        lv = (ListView) getView().findViewById(R.id.lf_lv);
        initBanner();                      //初始化广告图标
        initLv();                          //初始化文章列表
    }
    //初始化新闻列表
```

```java
        private void initLv() {
            if(data!= null && data.size()>0){
                setAdapter(data);return;
            }
            //使用Xutils访问服务器,获取JSON数据
            HttpUtils http = new HttpUtils();
            http.send(HttpMethod.GET,
                UrlUtil.webRoot + "articleAct_findArticleByPage?token = " + System.currentTimeMillis(),
                new RequestCallBack<String>(){
                    @Override
                    public void onFailure(HttpException arg0, String arg1) {
                        Toast.makeText(getActivity(), "未获取数据", Toast.LENGTH_SHORT).show();
                    }
                    @Override
                    public void onSuccess(ResponseInfo<String> arg0) {
                        //创建Gson对象
                        Gson gson = new Gson();
                        Log.i("Msg", "校园通服务器,新闻列表返回码: " + arg0.statusCode);
                        try {
                            if(arg0.statusCode == 200){
                                //将JSON数据直接解析,映射为List<ArticleInfo>集合
                                data = gson.fromJson(arg0.result,
                                    new TypeToken<List<ArticleInfo>>(){}.getType());
                                Log.i("Msg", "解析出文章数量: " + data.size());
                                setAdapter(data);      //设置适配器
                            }else{
                                Toast.makeText(getActivity(), "校园通知加载中...",
                                    Toast.LENGTH_SHORT).show();
                            }
                        } catch (Exception e) {
                            Toast.makeText(getActivity(), "获取服务数据异常!",
                                Toast.LENGTH_SHORT).show();
                        }
                    }
                });
        }
        //采用自定义适配器初始化列表数据
        private void setAdapter(final List<ArticleInfo> data){
            ArticleAdapter aa = new ArticleAdapter(getActivity(), R.layout.list_item_news, data);
            lv.setAdapter(aa);
        }
        //初始化广告图片
        private void initBanner() {
            //Xutils框架的图片加载模块
            BitmapUtils bu = new BitmapUtils(this.getActivity());
            bu.display(banner, UrlUtil.webRoot + "upload/201610/02.jpg");
        }
    }
```

新闻列表填充数据时使用自定义适配器 ArticleAdapter，代码如下。列表项布局文件 list_item_news 可以参考图 12.2。完成列表项数据初始化之后，需要对列表项添加监听器，实现新闻内容跳转功能，并将新闻编号传递到新闻内容 Activity。显示新闻内容的 Activity 是 ArticleActivity，在下面有详细说明。

📖 Proj12 项目，edu/freshen/util/ ArticleAdapter.java

```java
public class ArticleAdapter extends BaseAdapter {
    private Context context;
    private int layout;                         //列表项布局
    private List<ArticleInfo> data;             //待填充数据
    private BitmapUtils bu;                     //Xutil 的图片加载模块
    public ArticleAdapter(Context context, int layout, List<ArticleInfo> data) {
        super();
        this.context = context;
        this.layout = layout;
        this.data = data;
        bu = new BitmapUtils(context);
    }
    @Override
    public int getCount() {   return data.size();   }
    @Override
    public Object getItem(int arg0) {   return data.get(arg0);   }
    @Override
    public long getItemId(int arg0) {   return arg0;   }
    @Override
    public View getView(int index, View arg1, ViewGroup arg2) {
        final ArticleInfo ai = data.get(index);
        LayoutInflater lf = LayoutInflater.from(context);
        View v = lf.inflate(layout, null);
        ImageView iv = (ImageView) v.findViewById(R.id.list_item_news_icon);
        //加载图片
        bu.display(iv, UrlUtil.webRoot + ai.getArticleImagePath());
        iv.setFocusable(false);
        TextView name = (TextView) v.findViewById(R.id.list_item_news_name);
        name.setText(ai.getArticleName());
        TextView author = (TextView) v.findViewById(R.id.list_item_news_author);
        author.setText(ai.getArticleAuthor());
        TextView date = (TextView) v.findViewById(R.id.list_item_news_date);
        date.setText(ai.getArticleDate());
        TextView read = (TextView) v.findViewById(R.id.list_item_news_read);
        read.setText(ai.getReadNum() + " 已阅");
        //列表项添加监听器
        v.setOnClickListener(new OnClickListener() {
            @Override
            public void onClick(View arg0) {
                Intent intent = new Intent(context, ArticleActivity.class);
                intent.putExtra("articleId", ai.getAid());
                Log.i("Msg", "查看新闻 id= " + ai.getAid());
```

```
                context.startActivity(intent);
            }
        });
        return v;
    }
}
```

2. MiddleFragment 代码与布局文件

应用中心模块包括校园 App 中的所有独立功能,该项目中未对这些功能予以实现,只是完成了布局显示的效果,如图 12.5 所示。MiddleFragment 采用的布局文件是 fragment_middle.xml,包括一个 GridView 控件。

📖 Proj12 项目,fragment_middle.xml 布局文件

```xml
<?xml version = "1.0" encoding = "utf - 8"?>
<RelativeLayout xmlns:android = "http://schemas.android.com/apk/res/android"
    android:layout_width = "match_parent"
    android:layout_height = "match_parent"
    android:background = "#DDD" >
    <GridView
        android:id = "@ + id/gridView1"
        android:layout_width = "match_parent"
        android:layout_height = "wrap_content"
        android:numColumns = "3" >
    </GridView>
</RelativeLayout>
```

MiddleFragment 主要实现的功能是初始化 GridView 控件,创建菜单选项。菜单标题由 strings.xml 文件中的数组 md_menuTitle 提供,菜单图标是位于 drawable-ldpi 中的 PNG 图片。

📖 Proj12 项目,strings.xml 中的菜单标题数组

```xml
<string - array name = "md_menuTitle">
    <item>我的应用</item>
    <item>校园通知</item>
    <item>我的课表</item>
    <item>课程中心</item>
    <item>离线课件</item>
    <item>作业提交</item>
    <item>记事簿</item>
    <item>历史测试</item>
    <item>近期考试</item>
    <item>设置中心</item>
</string - array>
```

GridView 菜单项布局样式由文件 grid_item_menu.xml 确定,该布局文件比较简单,此处不再介绍。

📖 **Proj12 项目, edu/freshen/proj12/ MiddleFragment.java**

```java
public class MiddleFragment extends Fragment {
    GridView gv;                        //表格控件
    String titleData[];                 //菜单标题
    //菜单图标
    int iconData[] = {R.drawable.m_app, R.drawable.m_bord, R.drawable.m_calendar,
                R.drawable.m_courselist, R.drawable.m_files, R.drawable.m_major,
                R.drawable.m_note, R.drawable.m_search, R.drawable.m_test,
                R.drawable.m_menuset};
    @Override
    public View onCreateView(LayoutInflater inflater, ViewGroup container,
            Bundle savedInstanceState) {
        View v = inflater.inflate(R.layout.fragment_middle, null);
        gv = (GridView) v.findViewById(R.id.gridView1);
        return v;
    }
    @Override
    public void onActivityCreated(Bundle savedInstanceState) {
        super.onActivityCreated(savedInstanceState);
        //初始化菜单表格
        titleData = getResources().getStringArray(R.array.md_menuTitle);
        List<Map<String,Object>> data = new ArrayList();
        for (int i = 0; i < titleData.length; i++) {
            Map<String,Object> item = new HashMap();
            item.put("title", titleData[i]);
            item.put("icon", iconData[i]);
            data.add(item);
        }
        SimpleAdapter sa = new SimpleAdapter(getActivity(),data,
                R.layout.grid_item_menu,new String[]{"title","icon"},
                new int[]{R.id.grid_item_menu_title,R.id.grid_item_menu_icon} );
        gv.setAdapter(sa);
        gv.setOnItemClickListener(new OnItemClickListener() {
            @Override
            public void onItemClick(AdapterView<?> arg0, View arg1, int arg2, long arg3) {
                Toast.makeText(getActivity(), titleData[arg2] + "正在建设中!",
                                            Toast.LENGTH_LONG).show();
            }
        });
    }
}
```

3. RightFragment 代码与布局文件

RightFragment 实现聊天记录功能,该界面主要有上面的 ListView 控件显示聊天记录,下面的 EditText 和 Button 模拟发送聊天内容。聊天功能并未实现,此处仅实现了聊天的显示界面。如需实现实时聊天功能,则需要有服务器端的支持,该功能的实现可以参考第

10章的相关知识。RightFragment 采用的布局文件代码如下。

📖 Proj12 项目,fragment_right 布局文件

```xml
<?xml version = "1.0" encoding = "utf-8"?>
<LinearLayout xmlns:android = "http://schemas.android.com/apk/res/android"
    android:layout_width = "match_parent"
    android:layout_height = "match_parent"
    android:background = "#DDD"
    android:orientation = "vertical" >
    <ListView
        android:id = "@+id/mmf_lv"
        android:layout_width = "match_parent"
        android:layout_height = "wrap_content"
        android:layout_weight = "1.0"
        android:divider = "#DDD"
        android:dividerHeight = "4dp"
        android:stackFromBottom = "true" >
    </ListView>
    <LinearLayout
        android:layout_width = "match_parent"
        android:layout_height = "wrap_content"
        android:orientation = "horizontal"
        android:paddingBottom = "4dp" >
        <EditText
            android:layout_weight = "1.0"
            android:layout_width = "wrap_content"
            android:layout_height = "36dp"
            android:singleLine = "true"
            android:id = "@+id/mmf_msg" />
        <Button
            android:layout_width = "wrap_content"
            android:layout_height = "36dp"
            android:text = "发送"
            android:id = "@+id/mmf_bt" />
    </LinearLayout>
</LinearLayout>
```

RightFragment 主要完成聊天内容发送功能和加载聊天记录的功能。带有气泡的聊天背景实现方法有很多种,此处借助 ListView 进行了简单实现。气泡背景图片是 9.PNG 格式的,可以使用 SDK 附带工具 draw9patch.bat 进行编辑。项目中聊天背景气泡如图 12.10 所示。

图 12.10 聊天气泡背景图

📖 Proj12 项目,edu/freshen/proj12/ RightFragment.java

```java
public class RightFragment extends Fragment {
    ListView lv;                    //呈现聊天记录的列表
    EditText et;                    //文本输入框
    Button bt;                      //发送按钮
```

```java
    public List<Map<String,Object>> data = new ArrayList();        //聊天数据
    @Override
    public void onCreate(Bundle savedInstanceState) {
        super.onCreate(savedInstanceState);
        //创建模拟聊天数据
        Map<String,Object> m1 = new HashMap();
        m1.put("a", R.drawable.f6_02);
        m1.put("msg", "您好!");
        data.add(m1);
        Map<String,Object> m2 = new HashMap();
        m2.put("b", R.drawable.f6_16);
        m2.put("msg", "聊天内容测试,该泡泡框应该比较长!");
        data.add(m2);
        Map<String,Object> m3 = new HashMap();
        m3.put("a", R.drawable.f6_02);
        m3.put("msg", "还行吧!");
        data.add(m3);
    }
    @Override
    public View onCreateView(LayoutInflater inflater, ViewGroup container,
            Bundle savedInstanceState) {
        View v = inflater.inflate(R.layout.fragment_right, null);
        lv = (ListView) v.findViewById(R.id.mmf_lv);
        prepareLv();                                        //向ListView中添加模拟数据
        et = (EditText) v.findViewById(R.id.mmf_msg);
        bt = (Button) v.findViewById(R.id.mmf_bt);
        //聊天文本,发送按钮
        bt.setOnClickListener(new OnClickListener() {
            @Override
            public void onClick(View v) {
                if(et.getText().toString().length()<1)return;
                Map<String,Object> m = new HashMap();
                m.put("b", R.drawable.f6_02);
                m.put("msg", et.getText().toString());
                data.add(m);
                et.setText("");
                prepareLv();
            }
        });
        return v;
    }
    //使用自定义适配器ChatAdapter,准备填充聊天内容
    private void prepareLv() {
        ChatAdapter ca = new ChatAdapter(getActivity(), data, R.layout.list_item_rframe);
        lv.setAdapter(ca);
    }
}
```

聊天记录由自定义适配器 ChatAdapter 实现,列表项采用的布局文件是 list_item_

rframe,布局代码如下。聊天记录的背景气泡是 LinearLayout 控件(list_item_rframe_msgbox)的背景图片,根据聊天内容的来源选择不同的背景图片——chat01 或 chat02。

📖 Proj12 项目,list_item_rframe 列表项布局文件

```
<?xml version = "1.0" encoding = "utf-8"?>
<LinearLayout xmlns:android = "http://schemas.android.com/apk/res/android"
    android:layout_width = "match_parent"
    android:layout_height = "match_parent"
    android:orientation = "horizontal"
    android:background = "#DDD"
    android:gravity = "bottom"
    android:padding = "6dp">
    <ImageView
        android:layout_width = "36dp"
        android:layout_height = "36dp"
        android:id = "@+id/list_item_rframe_lface" />
    <LinearLayout
        android:id = "@+id/list_item_rframe_msgbox"
        android:layout_width = "wrap_content"
        android:layout_height = "wrap_content"
        android:layout_weight = "1.0"
        android:background = "@drawable/chat01"
        android:padding = "8dp" >
        <TextView
            android:id = "@+id/list_item_rframe_msg"
            android:layout_width = "wrap_content"
            android:layout_height = "wrap_content"
            android:text = "ddddd" />
    </LinearLayout>
    <ImageView
        android:id = "@+id/list_item_rframe_rface"
        android:layout_width = "36dp"
        android:layout_height = "36dp" />
</LinearLayout>
```

ChatAdapter 在填充聊天内容时需要根据聊天内容由谁发出动态地给 LinearLayout 控件选择背景图片,这样就可以使用不同的聊天气泡了。

📖 Proj12 项目,edu/freshen/proj12/ RightFragment.java

```
public class ChatAdapter extends BaseAdapter {
    public static List<Map<String,Object>> data;
    Context context;
    int layout;
    public ChatAdapter(Context context,List<Map<String,Object>> data, int layout) {
        super();
        this.context = context;
        this.data = data;
        this.layout = layout;
```

```
    }
    @Override
    public int getCount() {  return data.size();  }
    @Override
    public Object getItem(int arg0) {  return null;  }
    @Override
    public long getItemId(int index) {  return index;  }
    @Override
    public View getView(int index, View arg1, ViewGroup arg2) {
        LayoutInflater inflater = LayoutInflater.from(context);
        View v = inflater.inflate(layout, null);
        //解析控件并填充数据
        if(data.get(index).get("a")!=null){
            //处理其他人发言
            ImageView f = (ImageView) v.findViewById(R.id.list_item_rframe_lface);
            f.setImageResource(R.drawable.f6_16);
            LinearLayout msgb = (LinearLayout) v.findViewById(R.id.list_item_rframe_msgbox);
            msgb.setBackgroundResource(R.drawable.chat02);
            TextView msg = (TextView) v.findViewById(R.id.list_item_rframe_msg);
            msg.setText(data.get(index).get("msg").toString());
        }else{
            //处理回复发言
            ImageView f = (ImageView) v.findViewById(R.id.list_item_rframe_rface);
            f.setImageResource(R.drawable.f6_02);
            LinearLayout msgb = (LinearLayout) v.findViewById(R.id.list_item_rframe_msgbox);
            msgb.setBackgroundResource(R.drawable.chat01);
            TextView msg = (TextView) v.findViewById(R.id.list_item_rframe_msg);
            msg.setText(data.get(index).get("msg").toString());
        }
        return v;
    }
}
```

12.3.4　WebView 加载 HTML5 页面

在新闻列表中单击某一个新闻标题后，会跳转到 ArticleActivity，显示新闻页面。ArticleActivity 采用的布局文件是 activity_article.xml，该布局文件包含一个 WebView 控件，用于显示 HTML 页面，布局文件代码不再给出。

ArticleActivity 根据传入的参数获取新闻编号，访问服务器的资源。WebView 控件使用 loadUrl 方法，加载服务器返回页面，就可以显示服务器端的 detail.jsp 页面。

📖 Proj12 项目，edu/freshen/proj12/ ArticleActivity.java

```
public class ArticleActivity extends Activity {
    WebView wv;
    @Override
```

```
    protected void onCreate(Bundle savedInstanceState) {
        super.onCreate(savedInstanceState);
        setContentView(R.layout.activity_article);
        wv = (WebView) findViewById(R.id.webView1);
        Intent intent = getIntent();
        int aid = intent.getIntExtra("articleId", 1);
        Log.i("Msg", "接收新闻 id = " + aid);
        String queryUrl = UrlUtil.webRoot + "articleAct_findArticleById?aid = " + aid;
        wv.loadUrl(queryUrl);
        ActionBar ab = getActionBar();
        ab.setDisplayHomeAsUpEnabled(true);
    }
    @Override
    public boolean onOptionsItemSelected(MenuItem item) {
        if(item.getItemId() == android.R.id.home){
            this.finish();
        }
        return super.onOptionsItemSelected(item);
    }
}
```

12.4 习 题

对本章的校园 App 项目加以扩展。

（1）给校园 App 项目添加用户注册与登录功能，并可以记住用户等状态。

（2）实现阅读新闻时更新新闻阅读量的功能。例如，当前新闻阅读量为 5，当打开该新闻时，阅读量更新为 6。

（3）实现广告图片轮播功能。

（4）实现新闻列表分页功能，当新闻列表项滑动到底部时，动态加载下一页内容。

参考文献

[1] Cay S. Horstmann,Gary Cornell. Java 核心技术(卷 1,卷 2).9 版.北京:机械工业出版社,2013.
[2] Shawn Van Every. Android 多媒体开发高级教程.北京:清华大学出版社,2012.
[3] 李刚.疯狂 Android 讲义.北京:电子工业出版社,2015.
[4] 朱凤山. Android 移动应用程序开发教程.北京:清华大学出版社,2014.
[5] 王翠萍. Android 经典项目开发实战.北京:清华大学出版社,2015.
[6] 刘超.深入解析 Android 5.0 系统.北京:人民邮电出版社,2015.
[7] 林学森.深入理解 Android 内核设计思想.北京:人民邮电出版社,2014.
[8] 周胜韬. Android 安全技术揭秘与防范.北京:人民邮电出版社,2015.
[9] Joshua J.Drake,Pau Olive Fora. Android 安全攻防权威指南.北京:人民邮电出版社,2015.
[10] 王东华.精通 Android 网络开发.北京:人民邮电出版社,2016.
[11] Greg Nudelman. Android 应用 UI 设计模式.北京:人民邮电出版社,2013.
[12] 周建锋,朱凤山,等.网页设计与制作教程.北京:清华大学出版社,2015.
[13] 李东博. HTML5+CSS3 从入门到精通.北京:清华大学出版社,2013.
[14] Peter Lubbers. HTML5 程序设计.2 版.北京:人民邮电出版社,2012.
[15] 贾蓓,镇明敏,等. Java Web 整合开发实战.北京:清华大学出版社,2013.
[16] 李宁宁.基于 Android Studio 的应用程序开发教程.北京:电子工业出版社,2016.
[17] Theresa Neil.移动应用 UI 设计模式.2 版.北京:人民邮电出版社,2015.
[18] 唐汉明,翟振兴.深入浅出 MySQL 数据库开发优化与管理维护.2 版.北京:人民邮电出版社,2014.
[19] 朱元波. Android 传感器开发与智能设备案例实战.北京:人民邮电出版社,2016.
[20] 谢希仁.计算机网络.6 版.北京:电子工业出版社,2013.